Simply Rational

OXFORD SERIES IN EVOLUTION AND COGNITION

Simple Heuristics That Make Us Smart
Gerd Gigerenzer, Peter M. Todd, and ABC Research Group

Adaptive Thinking: Rationality in the Real World
Gerd Gigerenzer

Natural Selection and Social Theory: Selected Papers of Robert Trivers
Robert Trivers

In Gods We Trust: The Evolutionary Landscape of Religion
Scott Atran

The Origin and Evolution of Cultures
Robert Boyd and Peter J. Richerson

The Innate Mind: Structure and Contents
Peter Carruthers, Stephen Laurence, and Stephen Stich

The Innate Mind, Volume 2: Culture and Cognition
Peter Carruthers, Stephen Laurence, and Stephen Stich

The Innate Mind, Volume 3: Foundations and the Future
Peter Carruthers, Stephen Laurence, and Stephen Stich

Why Humans Cooperate: A Cultural and Evolutionary Explanation
Joseph Henrich and Natalie Henrich

Rationality for Mortals: How People Cope with Uncertainty
Gerd Gigerenzer

Simple Heuristics in a Social World
Ralph Hertwig, Ulrich Hoffrage, and ABC Research Group

The Shape of Thought: How Mental Adaptations Evolve
H. Clark Barrett

Simply Rational: Decision Making in the Real World
Gerd Gigerenzer

Simply Rational

Decision Making in the Real World

Gerd Gigerenzer

OXFORD
UNIVERSITY PRESS

OXFORD
UNIVERSITY PRESS

Oxford University Press is a department of the University of Oxford. It furthers the University's objective of excellence in research, scholarship, and education by publishing worldwide.

Oxford New York
Auckland Cape Town Dar es Salaam Hong Kong Karachi
Kuala Lumpur Madrid Melbourne Mexico City Nairobi
New Delhi Shanghai Taipei Toronto

With offices in
Argentina Austria Brazil Chile Czech Republic France Greece
Guatemala Hungary Italy Japan Poland Portugal Singapore
South Korea Switzerland Thailand Turkey Ukraine Vietnam

Oxford is a registered trademark of Oxford University Press in the UK and certain other countries.

Published in the United States of America by
Oxford University Press
198 Madison Avenue, New York, NY 10016

Library of Congress Cataloging-in-Publication Data
This title's Catalog-in-Publication Data is on file with the Library of Congress

ISBN 978–0–19–939007–6

CONTENTS

INTRODUCTION

Why don't most doctors know how to interpret patients' test results? Do complex problems need complex solutions, or can simple heuristics outperform "big data"? What is intuition, and do people believe that women or men are more intuitive? In this book, you will find answers to these and other practical questions; but, equally important, you will also discover a theoretical perspective on how to deal with risk and uncertainty.

This book is my third volume of collected papers—a sequel to *Adaptive Thinking* (2000) and *Rationality for Mortals* (2008). These papers originally appeared in medical, psychological, and economic journals, and are now easily accessible in one book.

The volume begins with a short personal account of how I ended up teaching risk literacy to medical doctors and federal judges, followed by three columns I wrote for the *British Medical Journal*. Part of a regular column I write on the art of risk communication, these short pieces are teasers for a longer review article on helping doctors and patients make sense of health statistics. They illustrate a general principle of my research program:

- Behavior is a function of mind *and* environment.

This stands in stark contrast to the traditional internal view that people come equipped with cognitive errors wired into their brains and are intrinsically hopeless when it comes to understanding risks. In my view, behavior should not be explained by internal states such as preferences, risk attitudes, or traits. For instance, failure to think statistically should not be attributed to cognitive limitations, but to the match between mind and environment—here, the external representation of statistical information. With this change in perspective, one can discover external representations that foster insight and make apparently inevitable cognitive errors "evitable."

In *Adaptive Thinking* (Gigerenzer, 2000), I reported first demonstrations that *natural frequencies*, unlike conditional probabilities, help people reason the Bayesian way. At that time, the research community took it for granted that people were "not Bayesian at all" (Kahneman & Tversky, 1972, p. 450). In *Rationality for Mortals* (Gigerenzer, 2008), I reported that even fourth graders can solve Bayesian problems when numbers are reported in natural frequencies. Similarly, in the first two sections of this volume, my colleagues and I show how natural frequencies and other transparent representations can help doctors better understand their patients' test results. These results confirm that the cause of reasoning failures is not simply inside the human mind. On the basis of this adaptive perspective, laypeople and professionals are shown how to understand health statistics. Such skills are particularly important in dealing with the big business of health care, where misleading statistics are often used to manipulate and persuade people, and where even well-meaning authorities see no other alternative than to "nudge" the public into reasonable behavior. But there is an alternative. This book provides a distinctively positive message:

- Risk literacy can be learned. Psychologists can teach doctors, judges, and other experts to become risk literate, using tools such as natural frequencies.

While classical decision theory still presents the laws of statistics as the only tool for rational decisions, it is important to understand that both statistics *and* heuristics are needed. The first six chapters of this volume deal with stable situations where risks can be calculated and where statistical thinking is sufficient. The chapters that follow deal with uncertain situations where not all risks can be calculated, and where heuristics can help to make better decisions. In his book *Risk, Uncertainty, and Profit* (1921), economist Frank Knight distinguished between "risk" and "uncertainty." In situations of risk, all alternatives are known, their consequences can be foreseen, and probabilities can be reliably measured. Playing roulette in the casino is an example; there, one can calculate the expected loss in the long run. Diagnostic tests based on epidemiological studies, as discussed in the first chapters of this book, provide another example of situations in which risk is calculable. In these situations, probability theory—such as Bayes' rule—and statistical thinking are sufficient. Not so in situations of uncertainty: Where to invest my money? Whom to marry? Whom to trust? By "uncertainty" I mean situations in which not all alternatives, consequences, or probability distributions are known, or can be reliably estimated. Here, the best course of action cannot be calculated, and believing otherwise can lead to disaster. As the financial crisis beginning in 2007 made crystal clear, the probability models used by rating agencies and the value-at-risk models used by large banks failed to predict and prevent it. Instead, by providing an illusion of certainty, they were part of the problem rather than its solution. Investment banks play in the real world of uncertainty, not in a casino.

In situations under uncertainty, humans and other animals rely on heuristics. My research group has dubbed the set of heuristics an individual or species has at its disposal *the adaptive toolbox*, and the study of the environmental conditions that favor a given heuristic *ecological rationality*.

Rationality for Mortals contains a chapter (with John Hutchinson) on the rules of thumb that animals use to find a mate, a nest site, or prey. For the current volume, I selected two articles on heuristics. The first, an overview article, distinguishes four classes of heuristics that individuals and institutions rely on, and the second, a specialized article, reviews the state of the art and the ongoing debates about the recognition heuristic. A key and surprising finding is that the "accuracy–effort trade-off" does not generally hold in situations of uncertainty—that is, that simple rules can lead to more accurate judgments than more effortful statistical models, including optimization models. The study of the ecological rationality of heuristics explains when and why these "less-is-more" effects emerge. This research has two implications:

1. Probability theory is not the only tool for rationality. In situations of uncertainty, as opposed to risk, simple heuristics can lead to more accurate judgments, in addition to being faster and more frugal.
2. Under uncertainty, optimal solutions do not exist (except in hindsight) and, by definition, cannot be calculated. Thus, it is illusory to model the mind as a general optimizer, Bayesian or otherwise. Rather, the goal is to achieve satisficing solutions, such as meeting an aspiration level or coming out ahead of a competitor.

Although I find these two points self-evident, they are still highly contested. Most of the time, the limits of optimization are ignored, probably because mathematically skilled researchers want to use differentials, which can only be done for sufficiently well-defined problems. As a consequence, researchers find themselves forced to create "small worlds" in which no surprises can happen and everything, including the consequences of one's actions and their probabilities, is certain. Examples include the trolley problems in moral psychology, the ultimatum game in experimental economics, and choices between monetary gambles in neuroeconomics.

It is not clear what the behavioral and neural data obtained in small worlds tell us about how the mind copes with uncertainty. Consider chess, where the optimal sequence of moves cannot be calculated by mind or machine. Instead of studying the heuristic strategies that players use, one could limit the chessboard to a four-by-four "small world," with only four figures per player, so that an optimal solution can be found. What this would tell us about winning the real game is up in the air.

Heuristics can deal with situations of uncertainty—that is, when optimal solutions cannot be computed. They are used consciously and unconsciously. In the latter case, people speak of intuition. For many professional fields, expert intuition is the key to success. For instance, after analyzing the facts, managers of large international companies make every second decision intuitively (Gigerenzer, 2014). In Western thought, intuition once was seen as the most certain form of knowledge, in which angels and spiritual beings excelled, but intuition today is linked to our bowels rather than our brains. For that reason, the very same managers would never admit to their gut feelings in public and instead hire a consulting firm to provide reasons after the fact. Similarly, since the 1970s, psychologists have scorned intuition as a fickle and unreliable guide to one's life that violates principles of logic or statistics. One prominent claim is that the intuitions of baseball coaches, players, and fans about the existence of a "hot hand" are wrong. However, the results of a study on the hot hand in volleyball, which I include here, indicate that the hot hand in fact exists among half of professional players and is used in an adaptive way.

For the final section, two papers on the nature of theories were chosen. One of them analyzes behavioral economics, a field that claims to add psychological realism to economics, but mostly just adds free parameters to classical expected utility maximization. Note that traditional ("neoclassical") economists explicitly state that expected utility models are *as-if* models—that is, their goal is to predict aggregate behavior, not to model psychological process. For instance, the Ptolemaic model of astronomy with the Earth in the center and planets orbiting around it in circles and epicycles is an as-if model, given that planets don't actually move in epicycles. By contrast, the Copernican model is based on the actual movements of the heavenly bodies. In science, progress usually consists of moving from as-if models to process models. However, much of today's neuroeconomics program reifies the as-if models of neoclassical economics and looks for their correlates in the brain. This is like sending a rocket into space in search of the planetary epicycles. Accordingly, the behavioral and neuroeconomics revolution itself remains largely "as-if." What it needs, to become truly revolutionary, are researchers who dare to study the correlates of psychological processes, such as heuristic search-and-stopping rules, in the brain. This would entail leaving the safe haven of optimization and facing the real world in which probabilities and values are not always known.

Beginning in the 17th century, the "probabilistic revolution" gave humankind the skills of statistical thinking, and eventually changed science and everyday life, from statistical mechanics to baseball statistics (Gigerenzer et al., 1989). Today, a second revolution is needed that would eventually provide humankind with the skills for dealing with uncertainty, not only with risk. What I envision is a "heuristic revolution" (Gigerenzer, 2014) that will enable us to understand the ecological rationality of both statistics and heuristics, and bring a dose of sanity to the study of rationality.

Simply Rational

Chapter 1

How I Got Started Teaching Physicians and Judges Risk Literacy

Like many unfortunate first-year psychology students, my first encounter with statistics began in a crowded lecture theater (in my case at the University of Munich), watching a flaky statistician scribble line after line of derivations on the blackboard. I dutifully copied the equations into my notebook, only to see him eventually reach for a sponge and wipe them away. We students had no clue what the equations were good for. We asked him to explain, but he knew only the math. Finally, after months of mutual suffering, he was fired. In the second semester, a psychology professor took over the job and began by hiring tutors. Thinking that teaching would be the best way to learn, I applied. At the interview, he showed me a chi-square equation and asked what it was. I had no idea, mistook the crudely drawn chi for an x, and responded that the equation decomposes an x on the left side into its additive components on the right side. That was good enough; I was hired. For many semesters, I was guilty of teaching other students what I naively believed to be statistical method, rejecting null hypothesis at the 5% significance level—a ritual I later called "mindless statistics" (Gigerenzer, 2004).

All that changed thanks to a regulation at the University of Munich requiring doctoral candidates to undergo additional oral examinations in two external fields. Virtually all of my peers chose psychiatry and education. I instead chose statistics in the mathematics department (and, admittedly, education). Those three years in the company of statisticians were an eye-opener. I had believed that statistics is statistics is statistics, but now I learned the difference between statistical *thinking* and *rituals*. My eyes were opened even wider at the Center for Interdisciplinary Research in Bielefeld, as a member of a one-year research group studying the "probabilistic revolution." From morning to

Originally published as Gigerenzer, G. (2014). How I got started teaching physicians and judges risk literacy. *Applied Cognitive Psychology, 4*, 612–614. References to chapters in the current volume were added.

midnight, I discussed the topic with historians of science, evolutionary biologists, physicists, economists, and statisticians (there was not much else to do in Bielefeld). You can find the exciting story about how probability changed science and everyday life in *The Empire of Chance* (Gigerenzer et al., 1989) and in *Cognition as Intuitive Statistics* (Gigerenzer & Murray, 1987). Once more I learned that there is more to statistics than the math.

How to Make Cognitive Illusions Disappear

One insight from the Probabilistic Revolution project proved highly relevant for my research. Probability, I'd discovered, has multiple meanings: relative frequency (counting in order to measure probability), propensity (designing a die or slot machine to determine the probability), and degrees of belief (measured by one's willingness to bet). And when I picked up a copy of *Judgment Under Uncertainty* (Kahneman, Slovic, & Tversky, 1982), published during my year in Bielefeld, I read the chapters from a different perspective than most of my peers.

One fascinating chapter was on overconfidence (Lichtenstein, Fischhoff, & Phillips, 1982). In this research, people were asked questions (e.g., "Which city has more inhabitants: Islamabad or Hyderabad?") and subsequently asked to rate their confidence that their answer was correct. Surprisingly, the average confidence was higher than the average proportion of correct answers—a phenomenon called overconfidence, attributed to people's self-delusion or other motivational deficits. My training, however, made me aware that the question asked was about a degree of belief (confidence) in a single event (the answer being correct or not), not about frequency. Thus, my colleagues and I conducted experiments in which we instead asked participants to estimate how many correct answers they gave. The result was another eye-opener. Overconfidence magically disappeared; estimated frequencies were as accurate as they could be: This result showed that what was called "overconfidence" is not a matter of personality or motivation but is based on cognitive inference mechanisms, which we spelled out in the probabilistic mental models theory (Gigerenzer et al., 1991). Extending these findings to other so-called cognitive illusions (Gigerenzer, Fiedler, & Olsson, 2012), we sparked a fruitful controversy about the nature of cognitive illusions.

Now that I understood the key to making cognitive illusions disappear, the next step was to leave the lab in order to share this knowledge and show experts, not only laypeople, how to understand risks.

How to Make Physicians Risk Literate

A most fertile discovery was that natural frequencies facilitate Bayesian reasoning. For years, influential psychologists claimed that people are lousy

Bayesians because they ignore base rates, and labeled this the base rate fallacy (Kahneman, 2011). In experiments, individuals were as a rule presented with conditional probabilities, such as hit rates and false alarm rates, and generally floundered when asked to estimate the Bayesian posterior probability. In our experiments, however, Ulrich Hoffrage and I showed for the first time that natural frequencies help students to make Bayesian inferences (Gigerenzer & Hoffrage, 1995, 1999; Kleiter, 1994). The problem was not simply, as had been suspected, in the human mind, but also in the representation of the information.

Inspired by a fascinating paper by David Eddy (1982), later consultant to the Clinton administration on health care reform, we began to systematically study physicians' ability to understand test results. If you test positive on a medical test, the physician should know how to estimate the probability that you actually have the disease (the "positive predictive value"). Our first study showed, however, that most physicians were in the dark (Hoffrage & Gigerenzer, 1998). For instance, their estimates of the probability that a patient has colorectal cancer given a positive FOB screening test ranged from 1% and 99%! Only one out of 24 physicians estimated correctly. But when the information was presented in natural frequencies, most physicians now understood what the positive predictive value was (see Chapters 3 and 5).

In order to get this and other related techniques out into the medical community, my collaborators and I published both in top medical journals (e.g., Gigerenzer & Edwards, 2003; Wegwarth et al., 2011) as well as in the low-impact-factor journals that doctors actually read. Even more effective in spreading the message were *Calculated Risk* (UK edition: *Reckoning with Risk* [Gigerenzer, 2002]) and *Risk Savvy* (Gigerenzer, 2014). These two trade books reached many health care providers, who would never have been reached if I had published exclusively in major scientific journals.

Once interest was raised in the medical field, the next step was to improve doctors' risk literacy. Over the last decade, I have taught about 1,000 physicians in their continuing medical education (CME) in risk literacy. These sessions covered natural frequencies versus conditional probabilities, absolute versus relative risks, frequency versus single-event statements, and mortality rates versus five-year survival rates. In each of these pairs, the first representation fosters insight, whereas the second supports innumeracy (Chapters 2–5). I should note that experienced physicians were one of the most appreciative audiences I have ever had.

I felt honored when natural frequencies began to be recommended by major medical organizations, including the Cochrane Collaboration, the International Patient Decision Aid Standards Collaboration, and the Medicine and Healthcare Products Regulatory Agency (the UK equivalent of the U.S. Food and Drug Administration). It is a rare event that a concept from cognitive psychology makes its way into medicine.

A British investment banker, David Harding, once read *Reckoning with Risk*, bought copies for all of his 200 employees and in 2007, over dinner, donated a generous sum to fund the *Harding Center of Risk Literacy* at the

Max Planck Institute for Human Development in Berlin. At the opening ceremony, Harding jokingly said that he had earned part of his wealth thanks to public statistical illiteracy and now wanted to give a portion back to make the public literate. In the years since, the Harding Center has been influential in motivating the editors of health brochures and medical journals to stop using misleading statistics, which often confuse rather than inform doctors and patients (Chapter 5; Gigerenzer & Muir Gray, 2011). For instance, in 2009 we helped the German Cancer Care write a new generation of pamphlets in which misleading relative risks and five-year survival rates were axed and replaced by transparent absolute numbers. After other national cancer associations followed suit, Germany became one of the few countries in which cancer screening pamphlets are free of misleading statistics. The Harding Center has also been involved in making statistical literacy and risk communication part of medical departments' core curricula.

Teaching Federal Judges

Doctors are not the only ones who receive inadequate training in understanding statistical evidence. Lawyers and judges get a mostly probability-free education. For that reason, at the O. J. Simpson trial, Alan Dershowitz, a renowned Harvard law professor who advised the defense team, could fool judges and jury about the probability that a man who had battered his wife (as Simpson had) actually murdered her (Gigerenzer 2002, Chap. 8; Good, 1995; Koehler, 1997). My first association with a law school was at the University of Virginia, where I taught a course on "How to Understand Probabilities." The students were extremely smart and articulate, but shockingly innumerate—and worse, did not think that it mattered. Yet that attitude changed the moment they realized that understanding how to present statistics would give them a competitive edge—over the prosecution if they were defense lawyers, or vice versa.

Here is an example. You (or your client) have been accused of committing a murder and are standing before the court. There is only one piece of evidence against you, but a damning one: Your DNA matches the traces found on the victim. The court calls an expert witness to explain what that match means. The expert testifies:

> "The probability that this match has occurred by chance is 1 in 100,000."

You can already see yourself behind bars. Yet there are always two ways to communicate statistical information: one transparent, the other misleading. The above statement phrases the evidence in terms of a single-event probability (that this match occurred by chance), which is potentially misleading. In contrast, a frequency statement is transparent:

> "Out of every 100,000 people, 1 will show a match."

If you live in a city with 2 million adult inhabitants, one can expect 20 whose DNA would match the sample on the victim. On its own, this fact seems very unlikely to land you behind bars. Thus, if you ever stand trial, make sure that the evidence is communicated in frequencies, not in single-event probabilities. The British Court of Appeal recommended that DNA evidence be presented in a frequency format, and other courts have followed (Gigerenzer, 2002).

In 2004 and 2005, I taught risk communication to about 50 U.S. federal judges in a continuing education program organized by the George Mason University School of Law. Today, however, both law and medical students still aren't taught to understand evidence adequately. Psychologists can offer effective tools for risk communication and teach experts how to use them (Chapters 2 to 5).

Risk Literacy

Basic research should be complemented by applied research, whenever possible. Research on risk literacy is a case in point. Heuristic decisions under uncertainty—when risks are unknown—is a second topic I am engaged with (Chapters 7 and 8; Gigerenzer, Hertwig, & Pachur, 2011). Applying this research to a variety of areas in the real world, from medicine to romance, has been fun and extended my horizon considerably. And it has helped to see the political consequences of what we are doing. For too long a time, some psychologists have argued that cognitive illusions are basically wired into our minds, meaning that there is little promise in trying to educate people (Kahneman, 2011). This stance has recently fueled a new wave of paternalism, according to which we need to be "nudged" into behaving sensibly by the few sane experts on earth (Thaler & Sunstein, 2008). Some kind of nudging can be useful, but not as a general philosophy of the 21st century. The debate between the nudgers and myself is covered in a *Nature* article entitled "Risk School" (Bond, 2009).

What my research shows, however, is that we can make people risk savvy—with the proper tools. With natural frequencies, even fourth graders can solve Bayesian problems, just as doctors and patients can learn to better understand health statistics (Gigerenzer, 2014). In teaching people these skills, psychologists can make a difference. Risk literacy needs to be taught not only to ongoing doctors and judges, but beginning at elementary school. Risk-savvy citizens are indispensable pillars of a modern democracy.

Part I

The Art of Risk Communication

Chapter 2

Why Do Single-Event Probabilities Confuse Patients?

The news reader announces a 30% chance of rain tomorrow. Thirty percent of what? Most people in Berlin think that it will rain tomorrow 30% of the time: for seven or eight hours (Gigerenzer, Hertwig, van den Broek, Fasolo, & Katsikopoulos, 2005). Others believe that it will rain tomorrow in 30% of the region, so probably not where they live. In New York the majority believes that it will rain on 30% of the days for which the prediction was made. That is, most likely it won't rain tomorrow.

A chance of rain tomorrow is a single-event probability. It refers to a unique event, such as rain tomorrow, and by definition does not specify a reference class. But people think in terms of classes: time, region, or days. These are not the only ones. As a woman in New York explained, "I know what 30% means: three meteorologists think it will rain, and seven not."

It is often said that people cannot think in terms of probabilities. But the real problem here is professionals' risk communication. Using a frequency statement instead of a single-event probability, meteorologists could state clearly that "it will rain on 30% of the days for which we make this prediction." New technologies have enabled meteorologists to add numerical precision to mere verbal statements ("it will be likely to rain tomorrow"), but little attention has been paid to the art of risk communication.

In health care the situation is similar. A psychiatrist used to prescribe fluoxetine to patients with depression. He always explained potential side effects, including loss of sexual interest and impotence: "If you take the medication, you have a 30–50% chance of developing a sexual problem" (Gigerenzer, 2002). When he finally realized that he was using a single-event probability, he switched to a frequency statement, which automatically specifies a reference class: "Of 10 patients to whom I prescribe the drug, three

Originally published as Gigerenzer, G., & Galesic, M. (2012). Why do single event probabilities confuse patients? *British Medical Journal, 344,* e245. This chapter has been slightly updated.

to five report a sexual problem." Now patients were less anxious about taking it. How had they initially understood the "30–50% chance?" Many had believed that something would go awry in 30–50% of their sexual encounters. The psychiatrist thought of his patients as the reference class, but the patients thought about their own sex life. If you always look at the sunny side of life, "three to five patients out of 10" doesn't make you nervous, because you think those three to five are the others. But even the brightest optimists are in trouble if the same numbers refer to their own sexual encounters. As a consequence, willingness to take the drug is reduced.

The ambiguity of single-event probabilities seems largely to have gone unnoticed. We could find only a single study in the medical risk literature (Slovic, Monahan, & MacGregor, 2000). If the problem is largely one of risk communication, then the usual culprit, numeracy, should not matter much. We asked 117 young and 73 elderly adults in Berlin what is meant by a "30–50% chance of developing a sexual problem," such as impotence or loss of sexual interest, after taking a popular drug for depression. Consistent with the ambiguity of the statement, people thought of different reference classes (Table 2.1), with interpretations varying more widely among the older group. Although misunderstanding is typically attributed to innumeracy, the respondents' level of numeracy made next to no difference. The problem is in the art of communication, not simply in people's minds.

Using probabilities without specifying a reference class is widespread in communication of risk in health care. For instance, the Mayo Clinic

Table 2.1: Interpretations of "30–50% Chance of Developing a Sexual Problem" after Taking a Drug

Interpretation	% of Respondents Aged 18–35 Years (n = 117)		% of Respondents Aged 60–77 Years (n = 73)	
	Low Numeracy* (n = 43)	High Numeracy* (n = 74)	Low Numeracy* (n = 52)	High Numeracy* (n = 21)
A: 30–50% of patients taking the drug will have sexual problems	65	78	33	38
B: Patients taking the drug will have a problem in 30–50% of their sexual encounters	9	8	33	33
C: Patients taking the drug will find sexual intercourse to be 30–50% less enjoyable than usual	12	6	21	10
D: Something else	14	8	13	19

* Numeracy defined as high or low according to median split across both groups on a numeracy rating consisting of the 12 items from Lipkus et al. (2001) and Schwartz et al. (1997).

announced: "The Food and Drug Administration (FDA) says that an extensive analysis of clinical trials showed that antidepressants may cause or worsen suicidal thinking or behavior in children and adolescents. The analysis showed that children taking antidepressants had about a 4% chance of developing suicidal thoughts or behavior, compared with only a 2% chance in children taking a sugar pill (placebo)" (Mayo Clinic, 2013).

What does it mean for a child to have a 4% chance of suicidal thoughts or behavior? It remains unclear. Some parents might think that this occurs to 4% of children who take antidepressants, while others might believe that their child will have suicidal thoughts 4% of the time or that 4% of the pills are flawed and cause suicidal thoughts.

The Centers for Disease Control and Prevention (2011) publicized that "condoms are 85–98% effective at preventing pregnancy." No reference class was specified on that page. A woman contemplating the use of condoms might think that:

a) she will get pregnant after 2–15% of times she has sex,
b) 2–15% of women relying on condoms get pregnant,
c) 2–15% of condoms are defective, or
d) 2–15% of men don't know how to use a condom safely.

Other websites make it clear that the effectiveness of birth control methods refers to "the number of women out of 100 who will have an unplanned pregnancy in the first year of using a method" (Healthwise Staff, 2010).

The official website of the U.S. Prostate Cancer Institute reported that "men have a 40% to 90% chance of experiencing retrograde ejaculation after prostatectomy" (Prostate.net, 2010). An ordinary man might think this estimate refers to the proportion of his sexual acts or to the proportion of men with prostatectomy where retrograde ejaculation occurs at least once—or to something else altogether.

In sum, single-event probabilities confuse patients because they do not specify a reference class. Good communication of risk requires a clear statement of what a probability refers to. Although necessary, this step alone is not sufficient, given that some patients misinterpret risks even when a reference class is given (Hanoch, Miron-Shatz, & Himmelstein, 2010). With the advance of personalized medicine and genetic counseling, doctors and patients will be overwhelmed by probabilities for individual patients. Frequency statements can help reduce potential confusion because they always refer to a class and are easily understood.

Chapter 3

HIV Screening

Helping Clinicians Make Sense of Test Results to Patients

In April 2013 the U.S. Preventive Services Task Force recommended that clinicians screen for HIV infection in people aged 15–65 years, revising its earlier position to screen only people at increased risk and pregnant women (Moyer, 2013). The proposal elicited discussion about the benefits and harms of antiretroviral treatment, the ethics of testing without people's explicit consent, and much else, but it neglected one crucial issue: risk literacy among clinicians.

When my colleagues and I tested 20 professional HIV counselors, 10 wrongly asserted that false-positive test results never occurred, and eight confused the test's sensitivity (the proportion of people with HIV who actually test positive for it) with its positive predictive value (the proportion of people who test positive who actually have HIV), with only two understanding what a positive test result meant (Gigerenzer, 2002). In a replication of this study in progress we see little improvement.

Does innumeracy among clinicians matter? No systematic studies of effects on patients exist—just anecdotal reports of people with false-positive test results engaging in unprotected sex with other HIV-positive people, believing that it would not matter anymore, and of people who committed suicide or who endured harmful effects of unnecessary antiretroviral treatment (Gigerenzer, 2002). A U.S. woman, newly married and pregnant, was told by her doctor to undergo HIV screening and tested positive on Western blotting. The doctors told her that the false-positive rate was five in 100,000, gave her handouts from the Internet about living with HIV, and sent her off to tell her husband and family the news. After a bad evening, she considered her low-risk lifestyle and went with her husband to a different clinic for a pinprick test; both partners have tested negative ever since (Gigerenzer, 2014).

Originally published as Gigerenzer, G. (2013). HIV screening: Helping clinicians make sense of test results to patients. *British Medical Journal, 347,* f5151.

How can we help clinicians understand the risk of false positives? Consider a low prevalence group in which the frequency of (undiagnosed) HIV infection is about one in 10,000, as in female U.S. blood donors (Centers for Disease Control and Prevention, 2001). If the test (such as enzyme immunoassay together with Western blotting) has a sensitivity of 99.95% and a specificity of 99.99%, what is the positive predictive value or P(HIV | pos)? To calculate this, medical students are taught to insert the prevalence, the sensitivity, and the false-positive rate into Bayes rule:

$$P(\text{HIV} \mid \text{pos}) = \frac{P(\text{HIV}) \times P(\text{pos} \mid \text{HIV})}{P(\text{HIV}) \times P(\text{pos} \mid \text{HIV}) + P(\text{no HIV}) \times P(\text{pos} \mid \text{no HIV})}.$$

In our case this gives P(HIV | pos) = 0.0001 × 0.9995/[0.0001 × 0.9995 + 0.9999 × 0.0001] or about 50%.

But this formula is not intuitive, which explains why even those who like to point out others' "probability blindness" are sometimes confused themselves, as exemplified by an MIT researcher who wrote that the sensitivity of the HIV test was 87% and that the false-positive rate was "the complement of 87 percent, or 13 percent" (Piattelli-Palmarini, 1991). A method that has been shown to improve insight is called natural frequencies (Aki, Oxman, Herrin, Vist, Terrenato, Sperati et al., 2011; Gigerenzer & Hoffrage, 1995, 1999). These can be represented as a tree (Figure 3.1). The top of the tree specifies a number of people, say 10,000. In the center of the tree these are split into one person who is expected to be infected (representing the prevalence) and those who are not. At the bottom these are split again into those who are expected to test positive or negative. Now it is easier to see that among those who test positive one is infected with HIV and one is not (panel A of figure). Thus the positive predictive value is 50%.

Prevalence, false-positive rates, and sensitivity can vary widely in HIV testing, depending on the risk group and the test used. The positive predictive value largely depends on the prevalence of HIV in the screening population and the false-positive rate of the test. The figure ignores sensitivity because it is of little use when the incidence of infection is so low. Panel B shows what happens when the prevalence of HIV among those screened increases to one in 1,000 (with the same false-positive rate).

Here, the positive predictive value increases to 10/11—that is, of every 11 people who test positive, we expect one false positive. Even if the prevalence is not exactly known, natural frequencies can help us to acquire a feeling for their order of magnitude.

Panels C and D show the same analysis for a test with an extremely low false alarm rate, one in 250,000 (Kleinman, Busch, Hall, Thomson, Glynn, Gallahan et al., 1998). Here we need to start with a larger group. In general, the minimum group size at the top of the tree can be determined from the prevalence or the false-positive rate, whichever is smaller, and its denominator is then the number on top of the tree. If the prevalence is one in 10,000

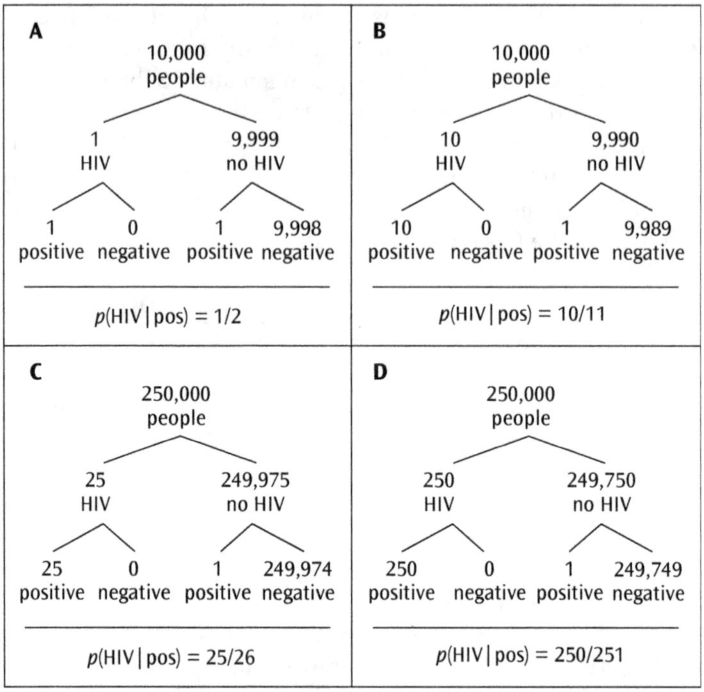

Figure 3.1: Natural frequencies help to understand the positive predictive value. In panel A the prevalence of HIV is one case in 10,000 people screened, and the false alarm rate is one in 10,000. The tree shows that a total of two positive test results is expected: one true positive and one false positive (the positive predictive value is 50%). Panel B shows the same tree for a prevalence of one in 1,000 people, resulting in a positive predictive value of 10/11 or 91%. Panels C and D show the same analysis for a false alarm rate of one in 250,000.

(panel C), 26 people are expected to test positive, one of them falsely. If the prevalence is one in 1,000 (panel D), the positive predictive value increases to 250/251.

Natural frequencies foster insight and should become part of the training of every medical student and HIV counselor.

Chapter 4

Breast Cancer Screening Pamphlets Mislead Women

Why should I have a mammogram? That question is regularly asked in pamphlets for screening. The answer is also regularly misleading. Women are told what they should do, but without being given the facts necessary to make informed decisions. This form of paternalism has a long tradition. In a campaign poster in the 1980s, the American Cancer Society declared: "If you haven't had a mammogram, you need more than your breasts examined."

Due to paternalism and pink ribbon culture, almost all women have a false impression of the benefit of mammography screening. For instance, 98% (!) of women in France, Germany, and The Netherlands overestimated its benefit by a factor of 10, 100, or more, or did not know (Gigerenzer, Mata, & Frank, 2009). Most surprisingly, those who frequently consulted their physicians and health pamphlets were slightly worse informed. Russian women gave the most realistic estimates among those in nine European countries studied—not because they have more information at their disposal but because there are fewer misleading pink ribbon pamphlets in Russia.

Misinformation needs to stop. All pamphlets should show a "fact box" that explains benefits and harms in a transparent way (Schwartz & Woloshin, 2009). Figure 4.1 shows one based on the most recent *Cochrane* review for women age 50 to 69 (Gøtzsche & Jørgensen, 2013):

Originally published as Gigerenzer, G. (2014). Breast cancer screening pamphlets mislead women. *British Medical Journal, 348*, g2636. This chapter has been slightly updated.

Breast Cancer Early Detection HARDING CENTER FOR RISK LITERACY

by mammography screening.
Numbers for women aged 50 years or older who participated in screening for 10 years on average.

	1,000 women without screening	1,000 women with screening
Benefits		
How many women died from breast cancer?	5	4*
How many women died from all types of cancer?	21	21
Harms		
How many women without cancer experienced false alarms, biopsies, or psychological distress?	–	100
How many women with non-progressive cancer had unnecessary treatments, such as complete or partial breast removal?	–	5

* This means that about 4 out of 1,000 women (50+ years of age) with screening died from breast cancer within 10 years – one less than without screening

***Source*:** Gøtzsche, PC, Nielsen, M(2011). *Cochrane database of systematic reviews* (1): CD001877.
Where no data for women above 50 years of age are available, numbers refer to wowen above 40 years of age.

www.harding-center.de

Figure 4.1 Fact box for breast cancer screening. Numbers are rounded.

In sum, the absolute reduction in mortality from breast cancer is about 1 in 1,000 women, but the reduction in total cancer mortality (including breast cancer) is 0. The difference between breast cancer and total cancer deaths is important because it is not always easy to determine the type of cancer from which a person died, and total cancer mortality is thus a more reliable measure.

A look at a sample of pamphlets reveals patterns in how the benefits of screening are actually communicated (for the sake of brevity, I do not deal with the harms). Four strategies are frequently used.

1. Zero-Number Policy: Tell Women What to Do without Stating Benefits

Even today, women are simply told to go to mammographic screening and are given no correct estimates of the benefit. In the United States, the Food and Drug Administration's Office of Women's Health leaflet (in pink) says on its first page that "Mammograms can help save lives." Similarly, the American Cancer Society's 2014 pamphlet *Breast*

> *Cancer: Early Detection* tells women on its first page, "Most doctors feel that early detection tests for breast cancer save thousands of lives each year, and that many more lives could be saved if even more women and their health care providers took advantage of these tests," and the National Cancer Institute's 2014 fact sheet says, "Screening mammography can help reduce the number of deaths from breast cancer among women ages 40 to 70, especially for those over age 50."*

In each case, no information is given about how large the benefit is. In the first two cases, the reduction in breast cancer mortality is misleadingly presented as "saving lives," even though there is no reduction in total cancer mortality (including breast cancer): no life is actually saved. Note the American Cancer Society's formulation that most U.S. doctors "feel" that lives are saved, which may be technically true. This zero-number policy appears to be widespread in the United States, unlike in Canada and the rest of the Western world.

2. Report Relative Risks Only

> The second strategy is to report the reduction in breast cancer mortality, but as a relative rather than absolute risk reduction. That is, the reduction from 5 in 1,000 to 4 in 1,000 is expressed as a 20% reduction, sometimes generously rounded up to over 30%. This makes the benefit look larger than the 0.1% absolute reduction. The Welsh NHS leaflet *Breast Screening Explained* says "Breast screening of women aged 50–70 has been shown to reduce the risk of dying from breast cancer by around 35%" (p. 2). And one by the New Zealand Breast Cancer Foundation (2011) claims that "Screening mammograms . . . reduce the chance of dying from breast cancer by approximately 33%" (p. 1).

None of these pamphlets tells women that there is no difference in total cancer mortality.

3. Report 5-Year-Survival Rates

> The third strategy is to use another misleading statistic, 5-year survival rates. It is well known that these rates say nothing about mortality reduction. In fact, increases in survival rates are not even correlated with decreases in mortality rates, r = 0.0. (Welch, Schwartz, & Woloshin, 2000). Lead-time bias (diagnosis of breast cancer through screening at an early stage that does nothing but advance the date of diagnosis) and overdiagnosis (diagnosis of a type of breast cancer that would never cause symptoms or death during a woman's lifetime) inflate

*An even more strongly worded message in pink from the FDA ("Mammography saves lives") can be found at http://www.fda.gov/ForConsumers/ByAudience/ForWomen/WomensHealthTopics/ucm117967.htm (retrieved November 5, 2014).

5-year survival rates without reducing mortality (Welch, Schwartz, & Woloshin, 2000; see also Chapter 5). Nevertheless, high survival rates continue to be used to impress women. For example, the Avon Foundation's breast health resource guide says, "There is a 97% 5-year survival rate when breast cancer is caught early before it spreads to other parts of the body" (p. 3).

4. Report Absolute Risk Reduction But Use Unrealistically High Numbers

Several pamphlets have stopped reporting misleading relative risks and 5-year survival rates. They report understandable absolute risks but inflate these. The leaflet produced by BreastScreen Australia states: "For every 1000 women who are screened every two years from age 50 to age 74 through BreastScreen (over 25 years): Around 8 (between 6 and 10) deaths from breast cancer will be prevented" (p. 13). And the NHS leaflet for England (2013) tells women, "Screening saves about 1 life from breast cancer for every 200 women who are screened" (p. 11).

One way to artificially inflate the absolute risk reduction (for about 10 years, as reported in the fact box) is to assume that the benefit will increase linearly if you consider 25 years (as BreastScreen does). But there is no evidence for this assumption. The only study that has actually investigated risk over 25 years found no reduction of breast cancer deaths at all (Miller et al., 2014).

A Right to Be Informed?

In Germany, the Harding Center for Risk Literacy (of which I am director) successfully exposed health organizations for misinforming the public about mammography screening. As a consequence, since about 2010, all deceptive relative risks and 5-year survival rates have been removed from German information literature, and harms are now reported in absolute numbers. Thus far, however, no German organization has dared to publish a fact box. In Austria, the Tyrolean Society for General Medicine did so in 2014 and was immediately attacked by representatives of the local gynecology departments. The leaflet of the Canadian Task Force *Should I be screened with mammography for breast cancer?* is another good example of how to inform women honestly.

I call on all women and women's organizations to tear up the pink ribbons and campaign for honest information. Only by correcting the current misinformation rate of 98% in various countries will women be in a position to make informed decisions.

Part II

Health Statistics

Chapter 5

Helping Doctors and Patients Make Sense of Health Statistics

Introduction

In a 2007 campaign advertisement, former New York City mayor Rudy Giuliani said, "I had prostate cancer, 5, 6 years ago. My chance of surviving prostate cancer—and thank God, I was cured of it—in the United States? Eighty-two percent. My chance of surviving prostate cancer in England? Only 44 percent under socialized medicine" (Dobbs, 2007). For Giuliani, these health statistics meant that he was lucky to be living in New York and not in York, since his chances of surviving prostate cancer appeared to be twice as high. This was big news. As we will explain, it was also a big mistake. High-profile politicians are not the only ones who do not understand health statistics or misuse them.

In this monograph, we—a team of psychologists and physicians—describe a societal problem that we call *collective statistical illiteracy*. In *World Brain* (1938/1994), H. G. Wells predicted that for an educated citizenship in a modern democracy, statistical thinking would be as indispensable as reading and writing. At the beginning of the 21st century, nearly everyone living in an industrial society has been taught reading and writing but not statistical thinking—how to understand information about risks and uncertainties in our technological world. The qualifier *collective* signals that lack of understanding is not limited to patients with little education; many physicians do not understand health statistics either. Journalists and politicians further contribute to the problem. One might ask why collective statistical illiteracy is not a top priority of ethics committees, medical curricula, and psychological research. One reason is that its very nature generally ensures that it goes undetected. Many of our readers might not have sensed that

Originally published as Gigerenzer, G., Gaissmaier, W., Kurz-Milcke, E., Schwartz, L. M., & Woloshin, S. W. (2007). Helping doctors and patients make sense of health statistics. *Psychological Science in the Public Interest, 8*, 53–96. This chapter has been slightly updated.

anything was wrong with Giuliani's conclusion, had we not highlighted it. Humans are facing a concealed societal problem.

In this monograph, we define statistical illiteracy in health care and analyze its prevalence, the damage it does to health and emotion, its potential causes, and its prevention. We argue that its causes are not simply inside the minds of patients and physicians—such as the lack of a math gene or a tendency to make hard-wired cognitive biases. Rather, we show that statistical literacy is largely a function of the outside world and that it can be fostered by education and, even more simply, by representing numbers in ways that are transparent for the human mind. To give the reader a sense of the problem, we begin with three examples.

I. Statistical Illiteracy in Patients, Physicians, and Politicians

The three cases that follow illustrate the three main points in this monograph: Statistical illiteracy (a) is common to patients, physicians, and politicians; (b) is created by nontransparent framing of information that may be unintentional (i.e., a result of lack of understanding) or intentional (i.e., an effort to manipulate or persuade people); and (c) can have serious consequences for health.

The Contraceptive Pill Scare

In October 1995, the U.K. Committee on Safety of Medicines issued a warning that third-generation oral contraceptive pills increased the risk of potentially life-threatening blood clots in the legs or lungs twofold—that is, by 100%. This information was passed on in "Dear Doctor" letters to 190,000 general practitioners, pharmacists, and directors of public health and was presented in an emergency announcement to the media. The news caused great anxiety, and distressed women stopped taking the pill, which led to unwanted pregnancies and abortions (Furedi, 1999).

How big is 100%? The studies on which the warning was based had shown that of every 7,000 women who took the earlier, second-generation oral contraceptive pills, about 1 had a thrombosis; this number increased to 2 among women who took third-generation pills. That is, the *absolute risk* increase was only 1 in 7,000, whereas the *relative* increase was indeed 100%. Absolute risks are typically small numbers while the corresponding relative changes tend to look big—particularly when the base rate is low. Had the committee and the media reported the absolute risks, few women would have panicked and stopped taking the pill.

The pill scare led to an estimated 13,000 additional abortions (!) in the following year in England and Wales. Figure 5.1 shows that, before the alert, abortion rates had been on the decline since 1990, but afterward, this trend was reversed (Furedi, 1999). Women's confidence in oral contraceptives was undermined, and pill sales fell sharply. For every additional abortion, there was also one extra birth, and the increase in both was particularly

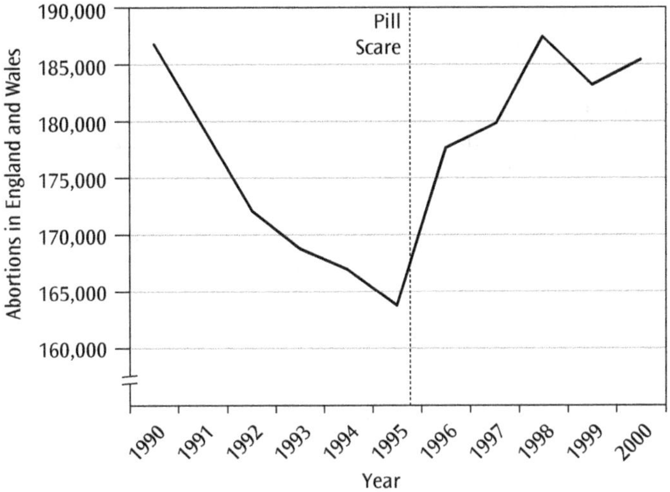

Figure 5.1: Relative risks scare women into unnecessary pregnancies and abortions. Reversal of downward trend in number of abortions in England and Wales following the 1995 pill scare.

pronounced in teenagers, with some 800 additional conceptions among girls under 16. The resulting cost increase to the National Health Service for abortion provision has been estimated at about £4–6 million ($6–8 million at that time). Ironically, abortions and pregnancies are associated with an increased risk of thrombosis that exceeds that of the third-generation pill. The pill scare hurt women, hurt the National Health Service, and even hurt the pharmaceutical industry. Among the few to profit were the journalists who got the story on the front page.

The 1995 pill scare was not the first one. Similar scares had occurred in 1970 and 1977, and after each one, the abortion rate rose (Murphy, 1993). And most likely, the 1995 scare will not be the last. Few citizens know the simple distinction between a relative increase ("100% higher") and an absolute increase ("1 in 7,000"). Medical journals, information brochures, and the media continue to inform the public in terms of relative changes, if only because big numbers make better headlines and generate more attention. But big numbers can also raise unnecessary anxieties and unrealistic hopes. When the next scare arrives, teenagers and adults will be as unprepared as ever to understand health statistics, creating another wave of abortions.

Few Gynecologists Understand Positive Mammograms

Since a large proportion of women participate in mammography screening, a key health statistic each gynecologist needs to know is the chances that a woman who tests positive actually has breast cancer. Mammography generates many false alarms. To avoid unnecessary anxiety or panic, women have a right to be informed what a positive test result means. Think of a woman who just received a positive screening mammogram and asks her doctor: Do I have breast cancer for certain, or what are the chances? Ninety-nine percent, 90%, 50%, or perhaps less? One would assume that every physician knows the answer. Is that so?

One of us (GG) trained about 1,000 gynecologists in risk communication as part of their continuing education in 2006 and 2007. At the beginning of one continuing education session in 2007, 160 gynecologists were provided with the relevant health statistics needed for calculating the chances that a woman with a positive test actually has the disease:

> Assume you conduct breast cancer screening using mammography in a certain region. You know the following information about the women in this region:
>
> - The probability that a woman has breast cancer is 1% (prevalence).
> - If a woman has breast cancer, the probability that she tests positive is 90% (sensitivity).
> - If a woman does not have breast cancer, the probability that she nevertheless tests positive is 9% (false-positive rate).
>
> A woman tests positive. She wants to know from you whether that means that she has breast cancer for sure, or what the chances are. What is the best answer?
>
> A. The probability that she has breast cancer is about 81%.
> B. Out of 10 women with a positive mammogram, about 9 have breast cancer.
> C. Out of 10 women with a positive mammogram, about 1 has breast cancer.
> D. The probability that she has breast cancer is about 1%.

Gynecologists could derive the answer from the health statistics provided, or they could simply recall what they should have known anyhow. In either case, the best answer is C—that is, that only about 1 out of every 10 women who test positive in screening actually has breast cancer. The other 9 are falsely alarmed (Kerlikowske, Grady, Barclay, Sickles, & Ernster, 1996a, 1996b). Note that the incorrect answers were spaced about an order of magnitude away from the best answer, in order to make it easier for the doctors. Figure 5.2 (left side) shows the 160 gynecologists' answers prior to training. Disconcertingly, the majority of them grossly overestimated the probability of cancer, answering "90%" or "81%." Another troubling result was the high variability in physicians' estimates, ranging between a 1% and 90% chance of cancer. The number of physicians who found the best answer, as documented in medical studies, was slightly less than chance (21%).

Do these physicians lack a gene for understanding health statistics? No. Once again, health statistics are commonly framed in a way that tends to cloud physicians' minds. The information is presented in terms of *conditional probabilities*—which include the sensitivity and the false-positive rate (or 1– specificity). Just as absolute risks foster greater insight than relative risks do, there is a transparent representation that can achieve the same in

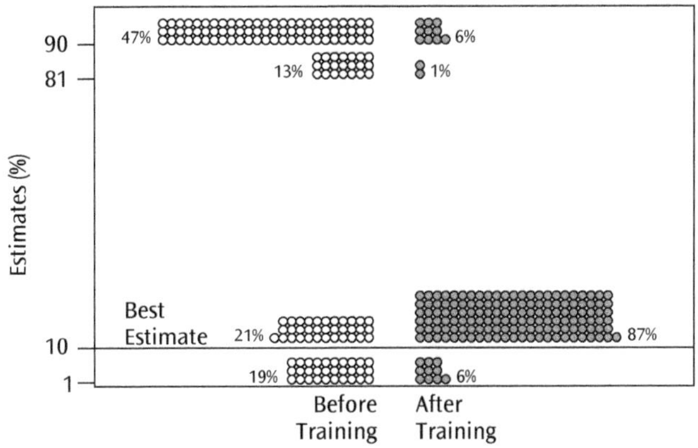

Figure 5.2: Few physicians understand test results: how to change that. Estimates by 160 gynecologists of the probability that a woman has breast cancer given a positive mammogram, before and after receiving training in how to translate conditional probabilities into natural frequencies.

comparison to conditional probabilities: what we call *natural frequencies*. Here is the same information from the above problem translated into natural frequencies:

> Assume you conduct breast cancer screening using mammography in a certain region. You know the following information about the women in this region:
>
> - Ten out of every 1,000 women have breast cancer.
> - Of these 10 women with breast cancer, 9 test positive.
> - Of the 990 women without cancer, about 89 nevertheless test positive.

After learning during the training session how to translate conditional probabilities into natural frequencies, the gynecologists' confusion disappeared; 87% of them now understood that 1 in 10 is the best answer (Fig. 5.2, right). How can this simple change in representation turn their innumeracy into insight? The reason is that natural frequencies facilitate computation, as explained in Figure 5.3. Natural frequencies represent the way humans encoded information before mathematical probabilities were invented in the mid-17th century and are easy to "digest" by our brains. Unlike relative frequencies and conditional probabilities, they are simple counts that are not normalized with respect to base rates (Gigerenzer & Hoffrage, 1995, 1999). That is, the four natural frequencies in Figure 5.3 (right side: 9, 1, 89, and 901) add up to the total number of 1,000 women, whereas the four conditional probabilities (left side) do not add up to 100%—instead each pair is normalized with respect to the base rates of cancer or no cancer, respectively.

This study illustrates a fundamental problem in health care: Many physicians do not know the probabilities that a person has a disease given a positive screening test—that is, the *positive predictive value*. Nor are they

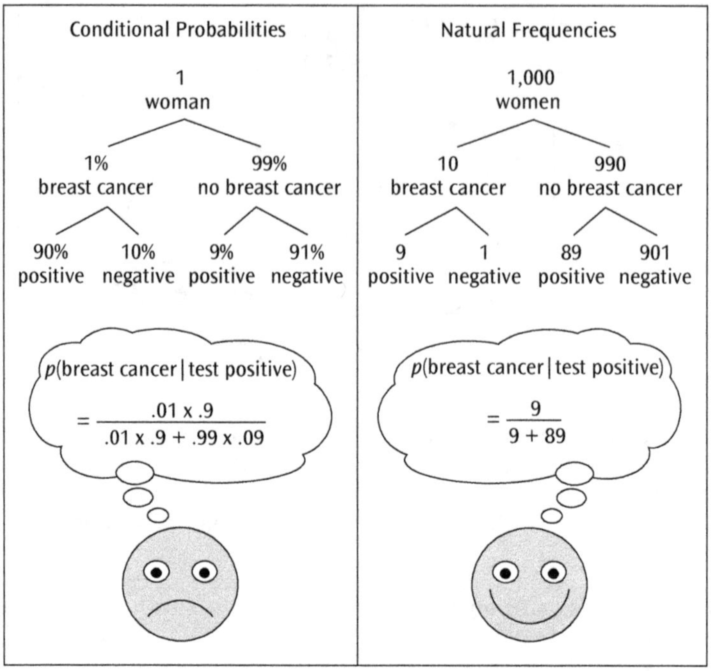

Figure 5.3: How do natural frequencies work? Shown are two ways of calculating the probability that a woman who tests positive in mammography screening actually has breast cancer (positive predictive value). The left side illustrates the calculation with conditional probabilities, and the right side with natural frequencies. The four probabilities at the bottom of the left tree are conditional probabilities, each normalized on base 100. The four frequencies at the bottom of the right tree are natural frequencies. The calculation using natural frequencies is simpler (smiling face) because natural frequencies are not normalized relative to base rates of breast cancer, whereas conditional probabilities (or relative frequencies) are, and need to be multiplied by the base rates. (The formula to calculate the positive predictive value is known as *Bayes rule.*)

able to estimate it from the relevant health statistics when those are framed in terms of conditional probabilities, even when this test is in their own area of specialty (Hoffrage & Gigerenzer, 1998). If you want to find out yourself if this is the case, ask your doctor. The result also shows that there is a fast and efficient cure. Yet doctors' and patients' collective innumeracy is a largely unknown problem in health care that continues to cause undue fear in the public. Months after receiving a false-positive mammogram, 1 in 2 women reported considerable anxiety about mammograms and breast cancer, and 1 in 4 reported that this anxiety affected their daily mood and functioning (Lerman et al., 1991). Everyone who participates in screening should be informed that the majority of suspicious results are false alarms. We face a large-scale ethical problem for which an efficient solution exists yet which ethics committees, focusing their attention instead on stem cells, abortion, and other issues that invite endless debates, have not yet noticed.

Higher Survival Does Not Mean Longer Life

Back to Rudy Giuliani. While running for president, Giuliani claimed that health care in the United States was superior to health care in Britain. Giuliani apparently used data from the year 2000, when 49 British men per 100,000 were diagnosed with prostate cancer, of which 28 died within 5 years—about 44% survived. Using a similar approach, he cited a corresponding 82% 5-year survival rate in the United States, suggesting that Americans with prostate cancer were twice as likely to survive as their British counterparts. Giuliani's numbers, however, are meaningless for making comparisons across groups of people that differ dramatically in how the diagnosis is made. In the United States, most prostate cancer is detected by screening for prostate-specific antigens (PSA), while in the United Kingdom, most is diagnosed by symptoms. The bottom line is that to learn which country is doing better, you need to compare mortality rates. To understand why, it is helpful to look at how "5-year survival" and mortality statistics are calculated. We'll start with survival.

Five-year survival is the most common survival statistic, but there is nothing special about 5 years. The statistic can be calculated for any time frame. Imagine a group of patients all diagnosed with cancer on the same day. The proportion of these patients who are still alive 5 years later is the 5-year survival rate. Here is the formula for the statistic:

$$\text{5-year survival rate} = \frac{\text{number of patients diagnosed with cancer still alive 5 years after diagnosis}}{\text{number of patients diagnosed with cancer}}$$

To calculate a mortality rate, imagine another group of people. The group is *not* defined by a cancer diagnosis. The proportion of people in the group who are dead after 1 year (the typical time frame for mortality statistics) is the "mortality rate." Here is the formula:

$$\text{Annual mortality rate} = \frac{\text{number of people who die from cancer over 1 year}}{\text{number of people in the group}}$$

The key difference to notice between these two kinds of statistics is the word *diagnosed*, which appears in the numerator and denominator of survival statistics but nowhere in the definition of mortality. Screening profoundly biases survival in two ways: (a) it affects the timing of diagnosis and (b) it affects the nature of diagnosis by including people with nonprogressive cancer. The first is called the *lead-time bias*, illustrated in Figure 5.4. Imagine a group of prostate cancer patients currently diagnosed on the basis of symptoms at age 67, all of whom die at age 70. Each survived only 3 years,

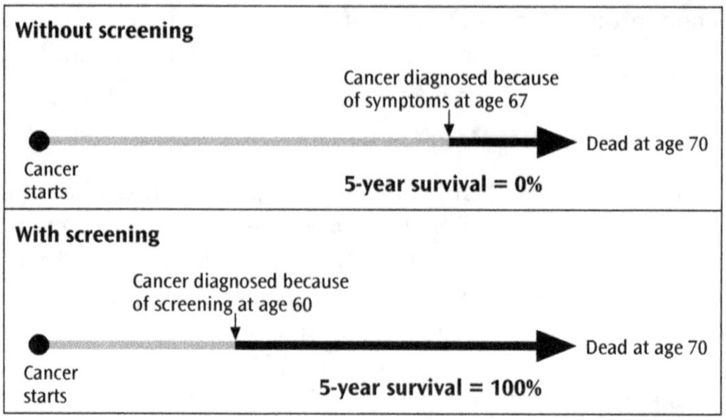

Figure 5.4: Lead-time bias. Even if the time of death is not changed by screening—and thus no life is saved or prolonged—advancing the time of diagnosis in this way can result in increased 5-year survival rates, causing such statistics to be misleading.

so the 5-year survival of this group is 0%. Now imagine that the same group is diagnosed with prostate cancer by PSA tests earlier, at age 60, but they all still die at age 70. All have now survived 10 years and thus their 5-year survival rate is 100%. Even though the survival rate has changed dramatically, nothing has changed about the time of death: Whether diagnosed at age 67 or at age 60, all patients die at age 70. This simple example demonstrates how survival rates can be increased by setting the time of diagnosis earlier, even if no life is prolonged or saved.

The second phenomenon that leads to spuriously high survival rates is the *overdiagnosis bias*, illustrated in Figure 5.5. Overdiagnosis is the detection

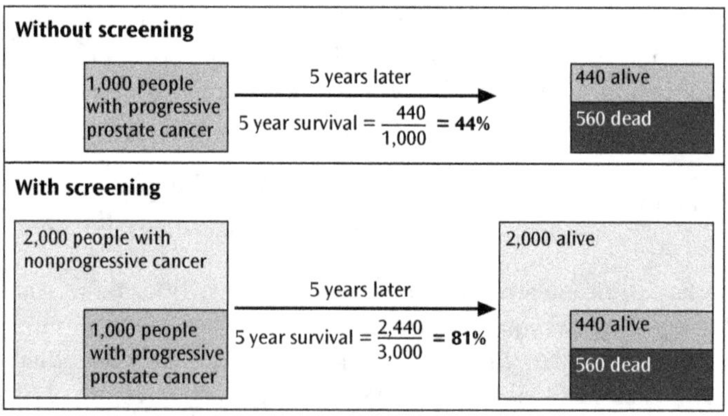

Figure 5.5: Overdiagnosis bias. Even if the number of people who die is not changed by screening—and thus no life is saved or prolonged—screening-detected nonprogressive cancers can inflate the 5-year survival rates, causing such statistics to be misleading.

of pseudodisease—screening-detected abnormalities that meet the pathologic definition of cancer but will never progress to cause symptoms in the patient's lifetime. These are also called nonprogressive cancers. Figure 5.5 (top) shows 1,000 men with progressive cancer who do not undergo screening. After 5 years, 440 are still alive, which results in a survival rate of 44%. Figure 5.5 (bottom) shows a population of men who participate in PSA screening and have cancer. The test detects both people with progressive and those with nonprogressive cancer. Imagine that screening detects 2,000 people with nonprogressive cancers—who by definition will not die of cancer in the following 5 years. These are now added to the 440 who survived progressive cancer, which inflates the survival rate to 81%. Note that even though the survival rate has changed dramatically, the number of people who die has not changed at all.

While the concept of nonprogressive cancer may seem implausible to many people, including clinicians, basic scientists have begun to uncover biological mechanisms that halt the progression of cancer (Folkman & Kalluri, 2004; Mooi & Peeper, 2006; Serrano, 2007). These mechanisms apply to many cancers—including one of the most dreaded, lung cancer. Amazingly, with computed tomography (CT) screening, almost as many nonsmokers were found to have lung cancer as smokers (Sone et al., 2001). Given that smokers are 15 times as likely to die from lung cancer, the computed tomography (CT) scans had to be finding abnormalities in nonsmokers that were technically cancer (based on their microscopic appearance) but that did not behave in the way lung cancer is expected to behave—as a progressive disease that ultimately kills (see also Welch, Woloshin, et al., 2007).

Due to overdiagnosis and lead-time bias, changes in 5-year survival rates have no reliable relationship to changes in mortality. For example, consider the 20 most common solid tumors in the United States over the last 50 years. Changes in 5-year survival were completely uncorrelated with changes in mortality (correlation coefficient = 0.0). That means that knowing about changes in survival tells you nothing about changes in mortality (Welch, Schwartz, & Woloshin, 2000)! In the context of screening, survival is always a biased metric. In the United States, screening for prostate cancer using the PSA test began in the late 1980s and spread rapidly, despite the lack of evidence that it saves lives. As a result, the number of new prostate cancer diagnoses soared. In Britain, PSA testing was introduced later and is still not routinely used. Consequently, new prostate cancer diagnoses (i.e., incidence) in Britain have risen only slightly. This largely explains why 5-year survival for prostate cancer is so much higher in the United States. The most recent figures (which differ from those cited by Giuliani) are 98% 5-year survival in the United States versus 71% in Britain.

But the real story is about mortality: Are American men half as likely to die from prostate cancer as British men are? The answer is no; the risk is about the same: about 26 prostate cancer deaths per 100,000 American men versus 27 per 100,000 in Britain (Shibata & Whittemore, 2001). If we use Giuliani's concern with prostate cancer for judging a health care system, the "socialist" English system appears to win since there are fewer diagnoses—that is, less overdiagnosis—but about the same mortality rate. Looking at the incidence and mortality data together suggests that many American men have been unnecessarily diagnosed (i.e., overdiagnosed) with prostate cancer during the PSA era and have undergone unnecessary surgery and radiation treatment, which often leads to impotence and/or incontinence.

Giuliani is not the only politician who has failed to appreciate the difference between survival rates and mortality rates. A report by the U.K. Office for National Statistics on cancer survival trends noted that 5-year survival for colon cancer was 60% in the United States compared to 35% in Britain. Experts dubbed this finding "disgraceful" and called for government spending on cancer treatment to be doubled. In response, then-Prime Minister Tony Blair set a target to increase survival rates by 20% over the next 10 years, saying, "We don't match other countries in its prevention, diagnosis and treatment" (Steimle, 1999, p. 1184). In fact, despite these large differences in 5-year survival, the mortality rate for colon cancer in Britain is about the same as the rate in the United States.

Conclusion

These three examples illustrate the theme of this monograph: the collective statistical illiteracy of patients, physicians, and politicians, as well as the considerable costs health systems pay as a consequence. The more widespread this illiteracy, the easier it is to manipulate the opinions of both doctors and patients, such as through campaigns promoting screening based on improved 5-year survival (see Part IV). We have also shown that there is a cure to this phenomenon that would be easy to implement: using transparent health statistics instead of the prevalent confusing ones, such as absolute risks instead of relative risks, natural frequencies instead of conditional probabilities, and mortality rates instead of 5-year survival rates when judging the value of screening (see Part VI). Framing information in a way that is most readily understood by the human mind is a first step toward educating doctors and patients in risk literacy.

II. What Is Statistical Literacy?

Statistical literacy in health does not require a degree in statistics. Rather, it means that citizens have basic competencies in understanding health statistics. For instance, statistical literacy implies that a person would recognize that comparing survival rates across countries where screening practices differ dramatically is nonsense and that the statistics cited by Giuliani do not mean that men in the United States are better off than in the United Kingdom.

It is desirable to define statistical literacy in concrete terms. We are aware that one could come up with a long textbook-like list, but a curriculum in statistics is precisely not our intention. What we are instead looking for are insights that can be taught in a short time and whose efficacy has been proved by psychological studies. To this end, we propose a list of insights that all patients and physicians should understand and questions that everyone should know to ask. We call this *minimal statistical literacy in health*.

Minimal Statistical Literacy in Health

Minimal statistical literacy applies to every medical decision, from whether a child's tonsils should be removed to whether an adult should take cholesterol-lowering medication. Minimal literacy focuses on the main concepts (like absolute risks) rather than the more advanced topics of variability (e.g., confidence intervals). Tables 5.1 and 5.2 serve as an illustration.

Learning to Live with Uncertainty *Understand that there is no certainty and no zero risk, but only risks that are more or less acceptable.*

For instance, the risk chart in Table 5.1 shows that women who never smoked have a much smaller risk of lung cancer than do smokers, but that risk still is not zero. Similarly, women with breast cancer genes BRCA-1 or BRCA-2, who face a high risk of breast cancer, do not necessarily develop breast cancer. And women who undergo radical bilateral mastectomy—despite lowering their breast cancer risk—can still develop it (Hartmann et al., 1999).

Questions to Ask about All Risks *Risk of what?* Understand the outcome to which the risk refers. For instance, the numbers in Table 5.1 refer to dying from disease, not getting the disease or developing a symptom.

Time frame? Understand the time the risk refers to. The frequencies of dying in Table 5.1 refer to a period of 10 years for all age groups. Time frames such as the "next 10 years" are easier to imagine than the widely used "lifetime" risks, are more informative because risks change over time, and are long enough to enable action being taken.

How big? Since there are no zero risks, size is what matters. Size should be expressed in absolute terms (e.g., 13 out of 1,000 women smokers age 50 die of heart disease within 10 years; see Table 5.1) or in comparative terms, relating the risk to a more familiar one. For example, for a 55-year-old American woman who is a smoker, the risk of dying from lung cancer in the next 10 years is about 10 times as high as dying from a car accident during the same time.

Does it apply to me? Check to see whether the risk information is based on studies of people like you—people of your age or sex, or people with health problems similar to yours. Table 5.1 shows that age matters for all causes of death, whereas whether one is a smoker or not is relevant for lung cancer but not colon cancer.

Table 5.1: Risk Chart for U.S. Women and Smoking. Find the line closest to your age and smoking status.[†] The numbers tell you **how many of 1,000 women will die in the next 10 years from . . .**

Age	Smoking	Vascular Disease		Cancer					Infection			Lung Disease	Accidents	All Causes Combined
		Heart Disease	Stroke	Lung Cancer	Breast Cancer	Colon Cancer	Ovarian Cancer	Cervical Cancer	Pneumonia	Flu	AIDS	COPD		
35	Never smoker	1			1						1		2	14
	Smoker	**1**	**1**	**1**	**1**						**1**		**2**	**14**
40	Never smoker	1			2	1		**Fewer than 1 death**			1		2	19
	Smoker	**4**	**2**	**4**	**2**						**1**	**1**	**2**	**27**
45	Never smoker	2	1	1	3	1	1				1		2	25
	Smoker	**9**	**3**	**7**	**3**	**1**	**1**		**1**		**1**	**2**	**2**	**45**
50	Never smoker	4	1	1	4	1	1						2	37
	Smoker	**13**	**5**	**14**	**4**	**1**	**1**		**1**			**4**	**2**	**69**
55	Never smoker	8	2	2	6	2	2	1	1			1	2	55
	Smoker	**20**	**6**	**26**	**5**	**2**	**2**	**1**	**1**			**9**	**2**	**110**
60	Never smoker	14	4	3	7	3	3	1	1			2	2	84
	Smoker	**31**	**8**	**41**	**6**	**3**	**3**	**1**	**2**			**18**	**2**	**167**
65	Never smoker	25	7	5	8	5	4	1	2			3	3	131
	Smoker	**45**	**15**	**55**	**7**	**5**	**3**	**1**	**4**			**31**	**3**	**241**
70	Never smoker	46	14	7	9	7	4	1	4			5	4	207
	Smoker	**66**	**25**	**61**	**8**	**6**	**4**	**1**	**7**			**44**	**4**	**335**
75	Never smoker	86	30	7	10	10	5	1	8			6	7	335
	Smoker	**99**	**34**	**58**	**10**	**9**	**4**		**14**			**61**	**7**	**463**

Note: Grey shading means fewer than 1 death per 1,000 women (from Woloshin, Schwartz, & Welch, 2008).

[†] A never smoker has smoked less than 100 cigarettes in her life and a current smoker has smoked at least 100 cigarettes or more in her life and smokes (any amount) now.

Table 5.2: Four Possible Test Outcomes

Test Result	Down Syndrome	
	Yes	No
Positive	82% Sensitivity	8% False-positive rate
Negative	18% False-negative rate	92% Specificity

Note: Testing for a disease (here: Down syndrome by measuring fetal nuchal translucency thickness) can have four possible outcomes: a positive result given disease, a positive result given no disease, a negative result given disease, and a negative result given no disease. The rates with which these four results occur are called sensitivity (or true positive rate), false-positive rate, false-negative rate, and specificity (true negative rate). The two shaded areas indicate the two possible errors, false positives and false negatives (data adopted from Snijders, Noble, Sebire, Souka, & Nicolaides, 1998).

Screening Tests *Understand that screening tests may have benefits and harms.* Benefits include the possibility of finding disease earlier, when treatment may be less invasive and/or more effective. Harms include costs, inconvenience, and false alarms—and, in our view, the most important harm of overdiagnosis. Overdiagnosis can be defined as the detection of pseudodisease or abnormalities that would never progress to cause symptoms in the patient's lifetime. For instance, it has been estimated that about 25% of breast cancers detected by mammography are overdiagnoses (Schwartz & Woloshin, 2007). The best evidence for overdiagnosis in lung cancer comes from studies of CT scans, which detected almost 10 times the amount of lung cancer than X-rays and, as mentioned before, diagnosed almost as many nonsmokers as smokers as having lung cancer (Sone et al., 2001).

Overdiagnosis leads to harm through overtreatment. The treatment of nonprogressive cancers results in unnecessary surgery and other invasive treatments—treatments that can only harm patients since they are being treated for a "disease" that would never have harmed them if left untreated.

Understand that screening tests can make two errors: false positives and false negatives. A false positive (false alarm) occurs when a test is positive (for example, a test for Down syndrome) in people who do not have the disease (no Down syndrome present). The false-positive rate is the proportion of positive tests among clients without the condition (Table 5.2). A false negative (miss) occurs when a test is negative in someone who does have the disease. The false-negative rate (miss rate) is the proportion of negative tests among clients with the condition.

Understand how to translate specificities, sensitivities, and other conditional probabilities into natural frequencies. Specificities and sensitivities continue to confuse physicians and patients alike. The specificity is the proportion of negative tests among clients without the condition; the sensitivity is the proportion of positive tests among clients with the condition (Table 5.2). Figure 5.3 illustrates how these can be translated into natural frequencies in order to facilitate deriving the positive predictive value.

Understand that the goal of screening is not simply the early detection of disease; it is mortality reduction or improvement of quality of life. Screening is testing for hidden disease in people without symptoms. It is only useful if early detection results in earlier treatment that is more effective or safer than later treatment. For instance, many smokers, current and past, wonder whether to get a CT scan to screen for lung cancer. While CT scans can clearly find more early-stage cancers, there is no evidence for reduced mortality rates. That is why no professional group currently recommends the test (in fact the American College of Chest Physicians now recommends against routine CT screening).

Treatment *Understand that treatments typically have benefits and harms.* Benefits include risk reduction—the lower probability of experiencing a feared outcome, such as getting or dying from disease. Treatment harms include bothersome or potentially even life-threatening side effects that result from medications or surgery. The value of treatment is determined by comparing the benefits (i.e., how much risk there is to reduce) and the harms.

Understand the size of the benefit and harm. Always ask for absolute risks (not relative risks) of outcomes with and without treatment.

Questions about the Science behind the Numbers *Quality of evidence?* A basic distinction is between evidence from a properly randomized controlled trial (Grade I evidence), well-designed cohort or case–control studies without randomization (Grade II), and opinions from respected authorities based on clinical experience (Grade III).

What conflicts of interest exist? Conflicts of interest can be inferred from the source that funded the study or from the goals of the institution that advertised the health statistics (see Part V).

III. How Widespread Is Statistical Illiteracy?

In health care, statistical illiteracy is typically presented as a problem faced by patients, sometimes by the media, and almost never by physicians. In this section, we analyze the collective statistical illiteracy of all three groups.

Do Patients Understand Health Statistics?

A citizen in a modern technological society faces a bewildering array of medical decisions. Should a pregnant woman undergo prenatal screening for chromosomal anomalies at age 35? Should parents send their teenage daughters for cervical cancer vaccination using Gardasil, despite reports that the vaccine could lead to paralysis? Whom should one trust? If citizens want to make informed decisions, they need more than trust: They need to understand health statistics. The evidence in this section documents, however, that most citizens (a) are not aware of basic health information, (b) do not understand the numbers if they encounter the information, and (c) tend to cherish the illusion of certainty about diagnostic results and treatments or follow the heuristic "trust your doctor"—both of which make risk literacy appear of little relevance. What follows is not an exhaustive overview but an analysis of the main issues. We begin with an elementary skill, called *basic numeracy*.

Basic Numeracy To analyze the prevalence of low numeracy and gauge the extent to which it impairs communication about health risks, Schwartz, Woloshin, Black, and Welch (1997) developed a simple three-question scale. The first question tests the respondent's ability to convert a percentage to a concrete number of people (out of 1,000), the second tests the ability to translate in the other direction, and the third tests basic familiarity with chance outcomes (Table 5.3). The test was applied to a random sample of female veterans in New England, 96% of whom were high school graduates, and whose average age was 68. Forty-six percent were unable to convert 1% to 10 in 1,000, 80% were unable to convert 1 in 1,000 to 0.1%, and 46% were unable to correctly estimate how many times a coin would likely come up heads in 1,000 flips, with the most common incorrect answers being 25, 50, and 250. The women's scores on this test strongly correlated with their ability to accurately interpret the benefit of mammography after being presented with standard risk reduction information: Only 6% of women answering just one basic numeracy question correctly could accurately interpret the data, compared to 40% of those answering all three questions correctly. Thus, basic numeracy seems to be a necessary precondition for minimal statistical literacy.

Table 5.3: The Basic Numeracy Assessment Scale

Task	Question
Convert a percent to a proportion	1. A person taking Drug A has a 1% chance of having an allergic reaction. If 1,000 people take Drug A, how many would you expect to have an allergic reaction? ____ person(s) out of 1,000
Convert a proportion to a percent	2. A person taking Drug B has a 1 in 1,000 chance of an allergic reaction. What percent of people taking Drug B will have an allergic reaction? ____%
Basic probability	3. Imagine that I flip a coin 1,000 times. What is your best guess about how many times the coin would come up heads in 1,000 flips? ____ times out of 1,000

Table 5.4: Percentage of U.S. Adults Aged 35 to 70 Giving Correct Answers to Basic Numeracy Questions (See Table 5.3), Overall and by Education Level

Question	Overall (n = 450)	Educational Attainment			
		High School Diploma or Less (n = 131)	Some College (n = 151)	College Degree (n = 103)	Postgraduate Degree (n = 62)
Convert 1% to 10 in 1,000	70	60	68	79	82
Convert 1 in 1,000 to 0.1%	25	23	21	30	27
How many heads in 1,000 coin flips?	76	62	76	87	86

Note: Schwartz and Woloshin (2000). Based on a nationally representative sample of U.S. citizens.

Table 5.4 shows the prevalence of low numeracy skills among U.S. adults—overall and stratified by educational attainment. The skills of the general adult public with high school education correspond roughly to those of the female veterans, whereas the skills of people with higher education are better on average. Note again the great difficulty large parts of the public, like the female veterans, have with translating small frequencies into percentages. Only 25% of the population could correctly convert 1 in 1,000 to 0.1%. Even among the highest education groups, at most 30% could solve this translation task. Lipkus, Samsa, and Rimer (2001) even found that only 21% of well-educated adults could answer this question correctly.

Medical Data Interpretation Test To test beyond basic numeracy, Schwartz, Woloshin, and Welch (2005) developed the medical data interpretation test (which includes some of the minimum statistical literacy introduced above). Its goal is to test the ability to make comparisons, such as between treatments—a fundamental requirement for informed decision making. Table 5.5 shows the answers of 178 participants with a broad range of educational attainment and backgrounds (recruited from advertisements in local newspapers, an outpatient clinic, and a hospital open house; the individual multiple-choice questions can be found in Schwartz et al. [2005]). Item nonresponses ("left blank") were low, suggesting that respondents understood the questions. Item difficulty varied widely, from 20% to 87% correct answers. The item that proved most difficult for the participants was number 5 in the section "knowledge basis for comparisons." The multiple-choice question was: "Which piece of information would be the best evidence that Gritagrel [a new drug against strokes] helped people?" Seventy percent of participants chose the answer "Fewer people died from strokes in the

Table 5.5: Proportion of Correct, Incorrect, and Missing Answers to the 18 Items on the Medical Data Interpretation Test for 178 Participants

	Answered Correctly (%)	Answered Incorrectly (%)	Left Blank (%)
Knowledge Basis for Comparisons			
Know that a denominator is needed to calculate risk	75	24	1
Know that denominators are needed to compare risks in 2 groups	45	54	1
Know that the base rate is needed in addition to relative risk to determine the magnitude of benefit	63	36	1
Know that a comparison group is needed to decide whether benefit exists	81	18	1
Know that lowering all-cause mortality provides better evidence of benefit than lowering a single cause of death	20	79	1
Comparison Tasks			
Select "1 in 296" as a larger risk than "1 in 407"	85	14	1
Inferred Items[a]			
Rate the riskiness of a 9 in 1,000 chance of death as the same as a 991 in 1,000 chance of surviving	61	37	2
Select a larger risk estimate for deaths from all causes than deaths from a specific disease	30	69	1
Select a larger risk estimate for a 20-year risk than for a 10-year risk	39	60	1
Calculations Related to Comparisons			
Calculate risk in intervention group by applying relative risk reduction to a baseline risk	87	11	2
Calculate 2 absolute risk reductions from relative risk reductions and baseline risks and select the larger	80	19	1
Calculate relative risk reduction from 2 absolute risks	52	46	2
Calculate absolute risk reduction from 2 absolute risks	77	19	4
Calculate the number of events by applying absolute risk to number in group	72	22	6
Context for Comparisons			
Know that age and sex of individuals in the source data are needed	47	51	2
Know that age of individuals in the source data is needed	60	39	1
Know that risk of other diseases is needed for context	62	35	3
Know that, for male smokers, the risk of lung cancer death is greater than prostate cancer death	60	37	3

[a]These items were based on a total of 5 separate questions.

Gritagrel group than in the placebo group" and only 20% correctly chose "Fewer people died for any reason in the Gritagrel group than in the placebo group." The distinction is important. Few medications have been shown to reduce the chance of death overall, and such a finding would reassuringly mean that (at least in this study) Gritagrel had no life-threatening side effects that substituted death from stroke with death from another cause. The medical data interpretation test appears to have reasonable reliability and validity (Schwartz et al., 2005).

There is no single study that tests all aspects of minimal statistical literacy, and in what follows we review studies that address selected issues.

The Illusion of Certainty The first item in minimal statistical literacy is learning to live with uncertainty. To appreciate the importance of health statistics, patients need to understand that there is no certainty in the first place. As Benjamin Franklin (1789/1987) once said: "In this world, there is nothing certain but death and taxes." The term *illusion of certainty* refers to an emotional need for certainty when none exists. This feeling can be attached to test results that are taken to be absolutely certain and to treatments that appear to guarantee a cure.

Even very good tests make errors. For instance, a 36-year-old American construction worker tested negative on ELISA tests 35 times before it was established that he was infected with HIV (Reimer et al., 1997). A series of what appears to be 35 misses in a row is an extreme case. Yet in one-time applications of tests, both false positives and misses are typical. In a nationwide survey in 2006, 1,000 German citizens over 18 were asked: "Which of the following tests are absolutely certain?" (Fig. 5.6). While only 4%

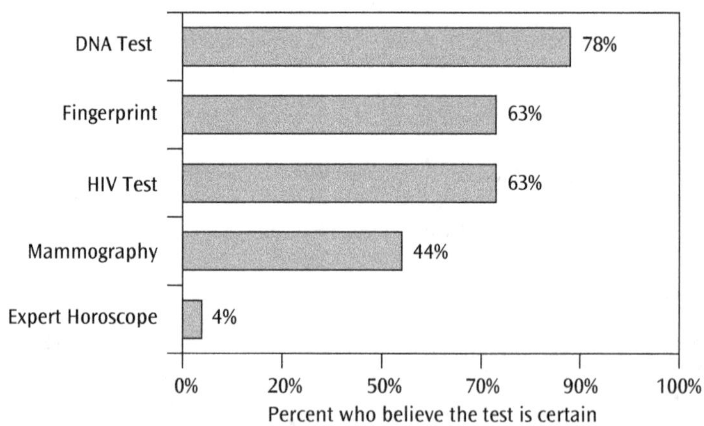

Figure 5.6: The illusion of certainty. Shown are results from face-to-face interviews conducted in 2006, in which a representative sample of 1,016 German citizens was asked: "Which of the following tests are absolutely certain?" (Gigerenzer, 2014).

believed an expert horoscope to give absolutely accurate results, a majority of citizens believed that HIV tests, fingerprints, and DNA tests were absolutely certain, even though none of these are (Gigerenzer, 2002, 2014). In contrast to these tests, which tend to make relatively few errors, the much less reliable result of a mammography (positive or negative mammogram) was rated as "absolutely certain" by 46% of the women and by 42% of the men. Yet its miss rate is about 10%, and the false-positive rate is almost as high. A university education is only a slight safeguard against the illusion of certainty: One out of three women with a university degree also believed that mammograms are absolutely certain.

When women participate in a 10-year program of annual mammography, the chances of a false alarm multiply: Every other woman without cancer can expect one or more false-positive test results (Elmore et al., 1998). Schwartz, Woloshin, Sox, Fischhoff, and Welch (2000) asked a stratified sample of 479 American women without breast cancer to estimate the chance of a false-positive result during a 10-year program. The median answer was 20% (an underestimate, but in the right ballpark), with 99% of the women believing that false positives occur. The fact that so many German women say that a singular test result is absolutely certain, whereas almost all the American women respond that false positives can occur in a series of 10 tests, may be related to the different perception of the singular as opposed to the repeated test. At the same time, given that the German women were asked for certainty of result and most mammography results are negative, their response may largely reflect the belief that if the test result is negative, one can be sure of not having cancer. In fact, many women say that they participate in screening to be sure that they do not have cancer. Similarly, genetic testing is often perceived as infallible: In a survey in the Netherlands, one third of respondents failed to understand that a prenatal test such as amniocentesis is not absolutely certain, as well as that if a person has a genetic predisposition for a disease, this person will not necessarily get the disease (Henneman, Timmermans, & van der Wal, 2004, pp. 11–12).

The illusion of certainty may also result from confusion between early detection and prevention. Proscreening campaigns in various countries have used the term "cancer prevention," wrongly suggesting that early detection could prevent the risk of getting cancer. In a cross-cultural study, over 4,000 randomly sampled women aged 15 and above were asked whether it is correct that "regular mammography every 2 years in women who are well prevents the risk of contracting breast cancer" or that mammography "reduces the risk" or "does not have any influence on the risk" (the correct answer). Noteworthy proportions of women in Switzerland (10%), the United Kingdom (17%), the United States (26%), and Italy (33%) shared the illusion of certainty that screening would prevent cancer (Domenighetti et al., 2003).

Screening is intended to detect existing cancers at an early stage. So it does not reduce the risk of getting breast cancer; it increases the number of positive diagnoses. Nevertheless, 57%, 65%, 69%, and 81% of the same random sample of women in the United States, Switzerland, the United Kingdom, and Italy, respectively, believed that screening reduces or prevents the risk of getting breast cancer (Domenighetti et al., 2003). An equally astounding 75% of a representative sample of German women who participated in mammography screening wrongly believed that screening reduces the risk of developing breast cancer (*Apotheken Umschau*, 2006).

Understanding Basic Risks Patients at Auckland Hospital, New Zealand, were asked: "What do you feel is the likelihood of you having a heart attack over the next 12 months?" This likelihood depends on individual risk factors, such as age, sex, a previous cardiac event, a family history of coronary heart disease, diabetes, smoking, and other known factors. Yet patients' risk estimates showed no correlation with any of these factors (Broadbent et al., 2006). The authors reported that there was also no optimistic bias, in which individuals tend to systematically underestimate threats to their health; perceived risks were simply unrelated to the actual risk. In a study in Switzerland, people were shown to lack even minimum medical knowledge of the risk factors for stroke, heart attack, chronic obstructive pulmonary disease, and HIV/AIDS. No participant was able to answer all questions correctly—on average, they got only one third right. The number correct was only moderately higher for people with personal illness experience (Bachmann et al., 2007).

Why do patients in these studies know so little about their risk factors? One possibility is that clinicians may be ineffective in communicating risks and do not notice how inaccurate their patients' perceptions of future risks are. Other studies indicate that patients may still have a good qualitative sense of their risk, whereas their quantitative judgments are strongly influenced by the framing of the questions asked (Woloshin, Schwartz, Black, & Welch, 1999).

Another potential reason why patients lack understanding of basic risks is that they rarely ask questions. Audiotapes of 160 adult patients' visits to doctors in North Carolina revealed that in only one out of four visits did the patient and doctor actually discuss risks or benefits (Kalet, Roberts, & Fletcher, 1994). Only few (about one in six) of these discussions were initiated by the patient, and in the majority of the discussions, the physician stated the risk with certainty (e.g., "You will have a heart attack if you don't lose weight"). Moreover, of the 42 patients who said that they actually had discussed risks with their doctors, only 3 could recall immediately after the discussion what was said. Yet almost all (90%) felt that they had their questions answered, had understood all that was said, and had enough information. Similarly, Beisecker and Beisecker (1990) reported that only few patients actively engage in information-seeking behavior in their consultations with physicians, and Sleath, Roter, Chewning, and Svarstad (1999) concluded that patients often do not ask questions about medications. In a review of 20 interventions directed at increasing patient participation, 11 assessed patient asking behavior. Congruent with the results reported above, question-asking behavior was generally low, and it was not easy to increase it: Out of the 11 interventions, only 5 resulted in significant increases in question asking (Harrington, Noble, & Newman, 2004). In contrast, patients who more actively engage during their encounters with physicians are more likely to understand treatment rationales and recommendations, are more satisfied with their health care, and even have better clinical outcomes (e.g., Roter & Hall, 1993; Street, 2001). In sum, the few studies available suggest that many patients are reluctant to ask questions, which is at odds with the goal of shared decision making.

Understanding That Screening Tests May Have Benefits and Harms Sir Muir Gray, knighted by the British Queen for his contribution to health care issues, is known for saying that "All screening programmes do harm; some do good as well, and, of these, some do more good than harm at reasonable cost" (Gray, Patnick, & Blanks, 2008, p. 480). What does the public know about the benefits? Consider mammography screening, where the absolute risk reduction of dying from breast cancer is in the order of 1 in 1,000 women. Let us take any estimate between 0 and 5 in 1,000 as correct. Only 6% of the women in random samples in four countries had the correct information. In contrast, 60%, 44%, 37%, and 37% of the women in the United States, Italy, the United Kingdom, and Switzerland, respectively, believed that out of 1,000 women the absolute risk reduction is 80 women or more (Domenighetti et al., 2003). A similar overestimation of benefits has been reported for PSA screening (Gigerenzer, Mata, & Frank, 2009). Whereas in these studies no information about relative risk reduction was given, Gigerenzer (2014) posed the following problem to a representative sample of 1,000 German citizens: "Early detection with mammography reduces the risk of dying from breast cancer by 25%. Assume that 1,000 women aged 40 and older participate regularly in screening. How many fewer would die of breast cancer?" Figure 5.7 shows the large variability in the understanding of this health statistic and the small proportion of citizens who understand that it means around 1 in 1,000. The most frequent estimate was 500 out of 1,000—that is, an overestimation by orders of magnitudes.

What does the public know about the harms? Schwartz et al. (2000) asked a stratified sample of 479 American women and found them to be quite knowledgeable about false positives, tending to view them as an acceptable consequence of screening. Yet very few had ever heard of other potential harms. Ninety-two percent believed that mammography could not harm a woman without breast cancer. Only 7% agreed that some breast cancers grow so slowly that these would never affect a women's health, and only 6% had ever heard of ductal carcinoma in situ, even after the researchers explained what that means: a breast abnormality that can be picked up by mammograms but that does not always become invasive. Nevertheless, almost everyone with ductal carcinoma in situ is treated by surgery. This problem—the detection of "pseudodisease"—is arguably the most important harm of screening, as it results in unnecessary surgery and radiation (Welch, 2004).

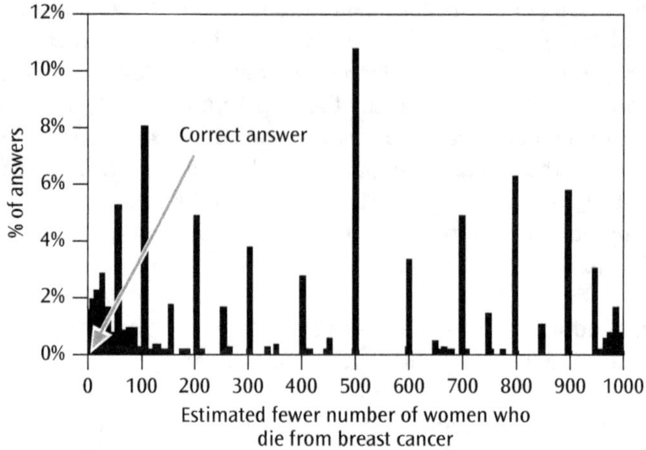

Figure 5.7: What does a 25% relative risk reduction mean? A representative sample of 1,000 German citizens was asked: "Early detection with mammography reduces the risk of dying from breast cancer by 25%. Assume that 1,000 women aged 40 and older participate regularly in screening. How many fewer would die of breast cancer?" The best estimate is about 1 in 1,000, but most people grossly overestimated.

This unbalanced view of screening may have important consequences for new screening tests. A random sample of 500 Americans was asked whether they would rather receive $1,000 in cash or a free total-body CT scan. Seventy-three percent said they would prefer the CT scan (Schwartz, Woloshin, Fowler, & Welch, 2004). Yet total-body CT scans are not endorsed by any professional medical organization and are even discouraged by several because screening tests like this can result in important harm.

Understanding Test Results Patients in a clinic in Colorado and in a clinic in Oklahoma were asked about standard tests for diseases such as strep throat infection, HIV, and acute myocardial infarction (Hamm & Smith, 1998). Each patient judged (a) the probability that a person has the disease before being tested (base rate), (b) the probability that a person tests positive if the disease is present (sensitivity), (c) the probability that a person tests negative if the disease is absent (specificity), and (d) the probability that a person has the disease if test results are positive (positive predictive value). Most patients estimated the four probabilities to be essentially the same—independent of whether the base rate was high or low or the test accurate or not. This result held independently of whether the patients had been tested or treated for the disease or had accompanied a family member or friend who had been tested or treated for it at a doctor's office. The fact that even experienced patients did not understand health statistics suggests that their doctors either never explained the risks or failed to communicate them properly. Studies with university students show that they too have difficulties drawing conclusions from sensitivities and specificities (Cosmides & Tooby, 1996; Gigerenzer & Hoffrage, 1995).

Understanding Treatment Outcomes More treatment is not always better. From the 1890s until about 1975, in the footsteps of surgeon William Halsted, the standard treatment for breast cancer was mastectomy, which involves complete removal of the breast, surrounding tissues, and lymph nodes. Systematic studies, however, indicated that lumpectomy, a less invasive procedure, is as effective as mastectomy but with less harm to the patient (National Institutes of Health Consensus Conference, 1991). Despite this "good news," many physicians and women nevertheless stick with mastectomy. Even after being reminded of the equivalent beneficial effects, half of the surgeons surveyed said they would choose mastectomy over breast-conserving surgery for themselves (Collins, Kerrigan, & Anglade, 1999). This may have been an informed decision on their part (perhaps because of their desire to reduce their chance of recurrence) but also could have been based on the illusion that more invasive treatment is more effective.

A prominent example is the former First Lady Barbara Bush, who underwent a mastectomy in 1987 despite her physician's recommendation for a lumpectomy. Many American women copied her decision, which led to a significant drop in breast-conserving surgery that had been on the increase beforehand (Wong & King, 2008). Interviews with these women indicate that most believe mastectomy to provide certainty that the cancer cannot recur, and feel personally responsible to do everything possible to ensure this. Family members who share the belief that more aggressive treatment is always better tend to support or even demand it. A 53-year-old communications director with a graduate degree, for instance, reported the reaction of her three daughters to her diagnosis: "'Mom, just have them both off. Just please, we want you around, just please have it taken care of.' By that, they meant mastectomy" (Wong & King, 2008, p. 586).

Understanding the Difference between Relative and Absolute Risk Reduction Is perceived treatment efficacy influenced by framing information in terms of relative and absolute risk reduction? In a telephone survey in New Zealand, respondents were given information on three different screening tests for unspecified cancers (Sarfati, Howden-Chapman, Woodward, & Salmond, 1998). In fact, the benefits were identical, except that they were expressed either as a *relative risk reduction*, as an *absolute risk reduction*, or as the *number of people needed to be treated* (screened) to prevent one death from cancer (which is 1/absolute risk reduction):

- Relative risk reduction: If you have this test every 2 years, it will reduce your chance of dying from this cancer by around one third over the next 10 years

- Absolute risk reduction: If you have this test every 2 years, it will reduce your chance of dying from this cancer from around 3 in 1,000 to around 2 in 1,000 over the next 10 years
- Number needed to treat: If around 1,000 people have this test every 2 years, 1 person will be saved from dying from this cancer every 10 years

When the benefit of the test was presented in the form of relative risk reduction, 80% of 306 people said they would likely accept the test. When the same information was presented in the form of absolute risk reduction and number needed to treat, only 53% and 43% responded identically. Medical students also fall prey to this influence (Naylor, Chen, & Strauss, 1992), as do patients (Malenka, Baron, Johansen, Wahrenberger, & Ross, 1993), and ordinary people are found to make more "rational" decisions about medication when given absolute risks (Hembroff, Holmes-Rovner, & Wills, 2004). In contrast, Sheridan, Pignone, and Lewis (2003) reported that relative risk reduction would lead to more correct answers by patients, but this is apparently a consequence of improper phrasing of the absolute risks, which was "treatment A reduces the chance that you will develop disease Y by 10 per 1,000 persons" (p. 886). This awkward statement is a hybrid between a single-event probability (it is about "you") and a frequency statement yet is not an absolute risk reduction (Gigerenzer, 2003).

A review of experimental studies showed that many patients do not understand the difference between relative and absolute risk reduction and that they evaluate a treatment alternative more favorably if benefits are expressed in terms of relative risk reduction (Covey, 2007).

In summary, the available studies indicate that very few patients have skills that correspond to minimum statistical literacy in health (cf. Reyna & Brainerd, 2007). Many seek certainty in tests or treatments, benefits of screening are wildly overestimated and harms comparatively unknown, early detection is confused with prevention, and basic health statistics such as the differences between sensitivity and specificity and between absolute and relative risks are not understood. This lack of basic health literacy prevents patients from giving informed consent.

Do Journalists Help the Public to Understand Health Statistics?

The press has a powerful influence on public perceptions of health and health care; much of what people—including many physicians—know and believe about medicine comes from the print and broadcast media. Yet journalism schools tend to teach everything except understanding numbers. Journalists generally receive no training in how to interpret or present medical research (Kees, 2002). A survey of health reporters at daily newspapers in five midwestern states (70% response rate) found that over 80% had no training in covering health news or interpreting health statistics (Voss, 2002). Not surprisingly, few (15%) found it easy to interpret statistical data, and under a third found it easy to put health news in context. This finding is similar to that of a survey by the Freedom Forum, in which nearly half of the science writers agreed that "reporters have no idea how to interpret scientific results" (Hartz & Chappell, 1997).

The American Association for the Advancement of Science (AAAS) asked more than 1,000 reporters and public information officers what science news stories are most interesting to reporters, their supervisors, or news consumers (AAAS, 2006). The top science topic in the U.S. media is medicine and health, followed by stem cells and cloning, and psychology and neuroscience. In Europe, where national and local newspapers devote many more pages to covering science, topic number one is also medicine and health, followed by environment and climate change. Thus, a minimum statistical literacy in health would do journalists and their readers an excellent service.

Problems with the quality of press coverage, particularly in the reporting of health statistics about medical research, have been documented (Moynihan et al., 2000; Ransohoff & Harris, 1997; Rowe, Frewer, & Sjoberg, 2000; Schwartz, Woloshin, & Welch, 1999a). The most fundamental of these include failing to report any numbers, framing numbers in a nontransparent way to attract readers' attention, and failing to report important cautions about study limitations.

No Numbers As shown in Table 5.6, one disturbing problem with how the media report on new medications is the failure to provide quantitative data on how well the medications work. In the United States, Norway, and Canada, benefits were quantified in only 7%, 21%, and 20% of news stories about newly approved prescription medications, respectively. In place of data, many such news stories present anecdotes, often in the form of patients describing miraculous responses to a new drug. The situation is similar when it comes to the harms of medications: Typically less than half of stories name a specific side effect and even fewer actually quantify it.

Nontransparent Numbers Table 5.6 also demonstrates that when the benefits of a medication are quantified, they are commonly reported using only a relative risk reduction format without providing a base rate. Reporting relative risk reductions without clearly specifying the base rates is bad practice because it leads readers to overestimate the magnitude of the benefit. Consider one medication that lowers risk of disease from 20% to 10% and another that lowers it from 0.0002% to 0.0001%. Both yield a 50% relative risk reduction, yet they differ dramatically in clinical importance.

Sometimes there is another level of confusion: It is not clear whether a "percent lower" expression (e.g., "Drug X lowers the risk of heart attack by 10%") refers to a relative or an absolute risk reduction. To avoid this confusion, some writers express absolute risk reductions as "percentage points" (e.g., "Drug X reduced the risk of heart attack by 10 percentage points"). This approach may be too subtle for many readers. The frequency format

Table 5.6: Percentage of Media Reports Presenting Benefits and Harms of Medications and Other Interventions

Media	Medications/ Setting	Benefit		Harm Mentioned
		Quantitative Information Provided	Relative Risk Reduction Only*	
Newly Approved Medications				
U.S. newspaper[a] ($n = 15$)	Ropinirole (Requip)	7	0	29
Major Norwegian newspapers[b] ($n = 357$)	18 newly released medications	21	89	39
Canadian newspaper[c] ($n = 193$)	Atorvastatin, Celexicob Donepezil, Oseltamivir, Raloxifene	20	39	32
Other Medications and Interventions				
U.S. newspaper/ television[d] ($n = 200$)	Pravastatin, Alendronate, Aspirin	60	83	47
Australian newspaper[e] ($n = 50$)	All medical interventions	40	n/a	44
Major international newspapers and U.S. national radio/ TV[f] ($n = 187$)	Research results from 5 major scientific meetings	60	35	29

*Percentage among the subset where benefit was quantified; [a]Woloshin & Schwartz, 2006a; [b]Høye, 2002; [c]Cassels et al., 2003; [d]Moynihan et al., 2000; [e]Smith, Wilson, & Henry, 2005; [f]Woloshin & Schwartz, 2006b.

may make this distinction clearer (e.g., "For every 100 people who take drug X, 10 fewer will have a heart attack over 10 years"). But the most important way to clarify risk reductions is to present the fundamental information about the absolute risks in each group (e.g., "Drug X lowered the risk of heart attack by 10 in 100: from 20 in 100 to 10 in 100 over 10 years").

Harms are mentioned in only about one third of reports on newly approved medications, and they are rarely if ever quantified. While benefits are often presented in a nontransparent format, harms are often stated in a way that minimizes their salience. This is most dramatic in direct-to-consumer advertisements, which often display the relative risk reduction from the medication in prominent, large letters (without the base rate), but present harms in long lists in very fine print. TV ads typically give consumers more time to absorb information about benefits (typically qualitative claims about the drug, like "It worked for me") than about side effects, resulting in better recall of purported benefits (Kaphingst, DeJong, Rudd, & Daltroy, 2004; Kaphingst, Rudd, DeJong, & Daltroy, 2005). A second technique is to report benefits in relative risks (big numbers) and harms in absolute risks (small numbers). This asymmetry magnifies benefits and minimizes harm. A simple solution (again) is to present both benefits and harms in the same format—in absolute risks.

No Cautions All studies have limitations. If the press is to help the public understand the inherent uncertainties in medical research, they should state the major limitations and important caveats. Unfortunately, this happens only rarely. In a content analysis of the high-profile media coverage of research presented at five scientific meetings (Woloshin & Schwartz, 2006b), few stories included cautions about studies with inherent limitations. For example, only 10% of stories about uncontrolled studies noted that it was impossible to know if the outcome really related to the exposure.

These problems are a result not only of journalists' lack of proper training but also of press releases themselves, including those from medical schools. Press releases are the most direct way that medical journals communicate with the media, and ideally they provide journalists with an opportunity to get their facts right. Unfortunately, however, press releases suffer from many of the same problems noted above with media coverage of medical news (Woloshin & Schwartz, 2002). They often fail to quantify the main effect (35% of releases), present relative risks without base rates (45% of those reporting on differences between study groups), and make no note of study limitations (77%). Although medical journals work hard to ensure that articles represent study findings fairly and acknowledge important limitations, their hard work is hence partially undone by the time research findings reach the news media. Better press releases could change this, helping journalists write better stories.

A few newspapers have begun to promote correct and transparent reporting in place of confusion and sensationalism. And there are a number of efforts to teach journalists how to understand what the numbers mean. In Germany, for example, one of us (GG) has trained some 100 German science writers, and in the United States there are MIT's Medical Evidence Boot Camp and the Medicine in the Media program sponsored by the National Institutes of Health and the Dartmouth Institute for Health Policy and Clinical Practice's Center for Medicine and the Media (where two of us, LS and SW, teach journalists from around the world).

Do Physicians Understand Health Statistics?

It is commonly assumed that only patients have problems with health statistics, not their physicians. Most psychological, legal, and medical articles on patient–doctor communication assume that the problem lies in the patient's mind. Doctors may be said to pay insufficient attention to their patients' feelings or not listen carefully to their complaints, consult with them only 5 minutes on average, or withhold information—but rarely is it considered that many doctors might be statistically illiterate (e.g., Berwick, Fineberg, & Weinstein, 1981; Rao, 2008).

Why do doctors need minimum statistical literacy? One important skill that doctors should have is to be able to critically assess the findings of a study in the relevant literature, as is expected from every psychologist or economist. If unable to do so, doctors are more dependent on hearsay or leaflets provided by the pharmaceutical industry to update their knowledge. In entering this largely unknown territory, we begin with a test of basic numeracy.

Basic Numeracy Schwartz and Woloshin (2000) tested physicians at Dartmouth Hitchcock Medical Center on basic numeracy. Compared to the general public (Table 5.4), physicians were better in basic numeracy (Table 5.7). Nevertheless, only 72% of the physicians could answer all three questions correctly. Just as for laypeople, the most difficult operation for the physicians was to convert 1 in 1,000 into a percentage: One out of four physicians got it wrong. Similar results have been obtained by Estrada, Barnes, Collins, and Byrd (1999), who reported that only 60% of medical staff got all three questions correct.

The Illusion of Certainty Physicians need to inform patients that even the best tests are not perfect and that every test result therefore needs to be interpreted with care or the test needs to be repeated. Some test results are more threatening than others and need to be handled particularly carefully. One terrifying example is a positive HIV test result. At a conference on AIDS held in 1987, former Senator Lawton Chiles of Florida reported that of 22 blood donors in Florida who had been notified that they had tested positive with the ELISA test, 7 committed suicide. A medical text that documented this tragedy years later informed the reader that "even if the results of both AIDS tests, the ELISA and WB [Western blot], are positive, the chances are only 50-50 that the individual is infected" (Stine, 1999, p. 367). This holds for people with low-risk behavior, such as blood donors. Indeed, consider a test (consisting of one or two ELISA tests and a Western blot test, performed on a single blood sample) with an extremely high sensitivity of about 99.9% and specificity of

Table 5.7: Percentage of Physicians Answering Basic Numeracy Questions Correctly

Question	Physicians at Grand Rounds (n = 85)
Convert 1% to 10 in 1,000	91
Convert 1 in 1,000 to 0.1%	75
How many heads in 1,000 coin flips?	100

From Schwartz and Woloshin (2000).

about 99.99% (numbers vary, because various criteria have been used that maximize specificity at the expense of sensitivity, or vice versa). Nonetheless, due to a very low base rate on the order of 1 in 10,000 among heterosexual men with low-risk behavior, the chance of infection can be as low as 50% when a man tests positive in screening. This striking result becomes clearer after these percentages are translated into natural frequencies: Out of every 10,000 men, it is expected that one will be infected and will test positive with high probability; out of the other, noninfected men, it is expected that one will also test positive (the complement to the specificity of 99.99%). Thus, two test positive, and one of these is infected (Fig. 5.8). AIDS counselors need to properly inform everyone who takes the test (see Chapter 3).

To investigate the quality of counseling of heterosexual men with low-risk behavior, an undercover client visited 20 public health centers in Germany to take 20 HIV tests (Gigerenzer, Hoffrage, & Ebert, 1998). The client was explicit about the fact that he belonged to a no-risk group, like the majority of people who take HIV tests. In the mandatory pretest counseling session, the client asked: "Could I possibly test positive if I do not have

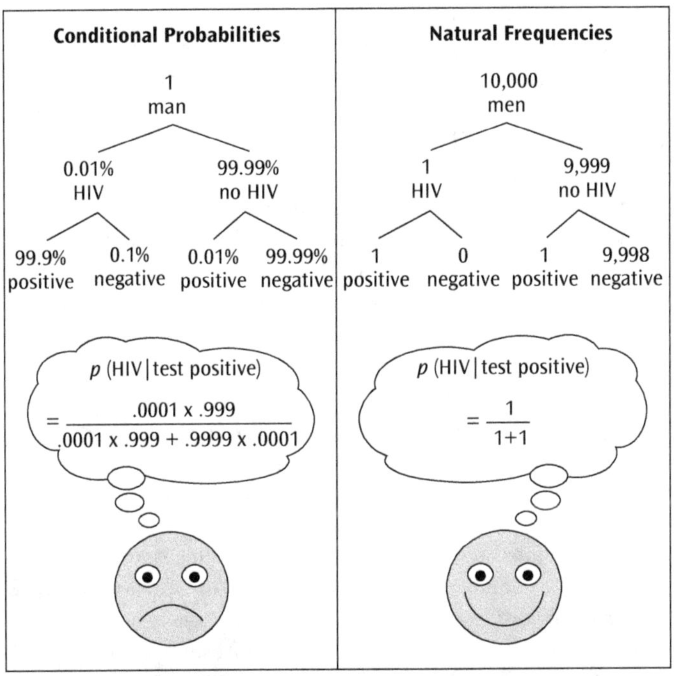

Figure 5.8: What does a positive HIV test mean? Shown here are two ways of calculating the chances that a heterosexual man with low-risk behavior who gets a positive HIV test result (positive ELISA test and positive Western blot test) is actually infected with HIV. The information on the left is presented in terms of conditional probabilities. The information on the right is presented in terms of natural frequencies, which simplify the computations and foster insight.

the virus? And if so, how often does this happen? Could I test negative even if I have the virus?" Table 5.8 shows the answers of 20 professional counselors, mostly physicians, to the first question. The first 13 counselors exhibited the illusion of certainty—although counselor 10 had a more differentiated view. Counselors 14 to 16 also initially claimed that no false-positive test results ever happened, but when the client asked again whether this was absolutely true, they changed their mind (in contrast to the others, who insisted on their standpoint). Only three counselors (17–19) immediately told the client that false positives can occur since the specificity is not perfect although very high. Counselor 20 provided no concrete information but insisted on blind trust. Note that if no false positives occur, a positive test would imply an HIV infection with certainty. After we sent copies of our article reporting this state of affairs to hundreds of counseling centers, some have begun to train their counselors how to understand HIV test statistics.

PSA Counseling In 2004, *Stiftung Warentest*, the German equivalent of the U.S. *Consumer Reports*, went beyond testing computer screens and cell phones and began to test the quality of doctors. In the first study, a 60-year-old man (a physician) paid undercover visits to 20 urologists in Berlin, drawn randomly from a total of 135 urologists, and asked for advice on PSA screening. Medical society guidelines call for thorough and systematic counseling before the first PSA test: For instance, counseling should explain that the PSA test can miss cancers or cause false alarms. It should also inform the patient that even in the event of a true positive, not every cancer needs to be treated (i.e., that overdiagnosis exists); there is instead a danger of overtreatment, whereby the treatment does not help the patient but may lead to harms such as incontinence and impotence. The patient should also know that there is no proof that early detection of prostate cancer

Table 5.8: Answers by 20 AIDS Counselors to the Client's Question: "If One Is Not Infected with HIV, Is It Possible to Have a Positive Test Result?"

1. "No, certainly not"	11. "False positives never happen"
2. "Absolutely impossible"	12. "With absolute certainty, no"
3. "With absolute certainty, no"	13. "With absolute certainty, no"
4. "No, absolutely not"	14. "Definitely not" . . . "extremely rare"
5. "Never"	15. "Absolutely not" . . . "99.7% specificity"
6. "Absolutely impossible"	16. "Absolutely not" . . . "99.9% specificity"
7. "Absolutely impossible"	17. "More than 99% specificity"
8. "With absolute certainty, no"	18. "More than 99.9% specificity"
9. "The test is absolutely certain"	19. "99.9% specificity"
10. "No, only in France, not here"	20. "Don't worry, trust me"

prolongs life (*Stiftung Warentest*, 2004). Only 2 of the 20 urologists knew the relevant information and were able to answer the patient's questions (and were graded A), and 4 others knew some of the information (grade C). The majority, 14 urologists (half of these graded D and F), could not answer most of the patient's questions, wrongly argued that it was scientifically proved that PSA screening prolongs life, and were not aware of any disadvantages. As one explained to the client, "There is nothing to ponder; at your age you must take the test" (*Stiftung Warentest*, 2004, p. 86).

Physicians Are Confused by Sensitivities and Specificities Hoffrage and Gigerenzer (1998) tested 48 physicians with an average professional experience of 14 years, including radiologists, internists, surgeons, urologists, and gynecologists. The sample had physicians from teaching hospitals slightly overrepresented and included heads of medical departments. They were given four problems; one of these was screening for colorectal cancer with the fecal occult blood test (FOBT). Half of the physicians were given the relevant information in conditional probabilities (a sensitivity of 50%, a false-positive rate of 3%, and a prevalence of 0.3%), which is the form in which medical studies tend to report health statistics. The physicians were then asked to estimate the probability of colorectal cancer given a positive test result. Each point in Figure 5.9 (left) represents one physician. Note that their estimates ranged between a 1% and a 99% chance of cancer! If patients knew this striking variability, they would be rightly concerned. Note that the physicians' answers were not random. The modal answer was 50% (the sensitivity), and four physicians deducted the false-positive rate from the sensitivity (arriving at 47%). When interviewed about how they arrived at their answers, several physicians claimed to be innumerate and in their embarrassment felt compelled to hide this fact from patients by avoiding any mention of numbers.

Yet when the information was provided in natural frequencies rather than conditional probabilities, those who believed themselves to be innumerate could reason just as well as the others. The information was presented as follows: 30 out of every 10,000 people have colorectal cancer. Of these 30, 15 will have a positive FOBT result. Of the remaining people without cancer, 300 will nonetheless test positive. As Figure 5.9 (right) shows, most physicians estimated the positive predictive value precisely, and the rest were close. Similar results were found for the three other problems (Fig. 5.10). Thus, the problem is not so much in physicians' minds but in an inadequate external representation of information, which is commonly used in medicine.

Only 18% of physicians and medical staff could infer the positive predictive value from probability information in a study by Casscells, Schoenberger, and Grayboys (1978). Eddy (1982) reported that 95 out of 100 physicians overestimated the probability of cancer after a positive screening mammogram by an order of magnitude. Similarly, Bramwell, West, and Salmon (2006) found only 1 out of 21 obstetricians being able to estimate the

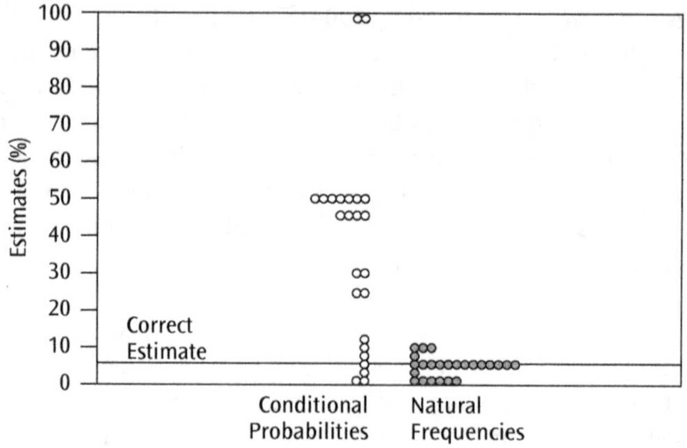

Figure 5.9: How to reduce the variability in physicians' judgments. Shown are individual estimates by physicians that a person has colorectal cancer given a positive fecal occult blood test when information was given in conditional probabilities (left) versus natural frequencies (right). Variability decreased dramatically and the correct answer was given more often when numerical information was in natural frequencies (Hoffrage & Gigerenzer, 1998).

probability of an unborn actually having Down syndrome given a positive test, with those giving incorrect responses being fairly confident in their estimates. When the same information was given in natural frequencies, 13 out of 20 obstetricians arrived at the correct answer. In one Australian study, 13 of 50 physicians claimed they could describe the positive predictive value,

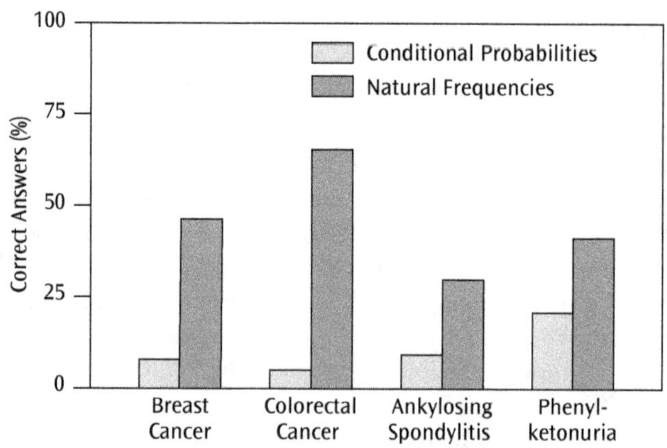

Figure 5.10: The percentage of physicians' correct estimates of positive predictive values for a range of tests/diseases when information was given in conditional probabilities versus natural frequencies. Natural frequencies fostered diagnostic insight in across all four diagnostic tasks (Hoffrage & Gigerenzer, 1998).

but when directly interviewed, only 1 could do so (Young, Glasziou, & Ward, 2002). Similar effects were reported for members of the U.S. National Academy of Neuropsychology (Labarge, McCaffrey, & Brown, 2003). Ghosh and Ghosh (2005) reviewed further studies that showed that few physicians were able to estimate the positive predictive value from the relevant health statistics.

Studies of legal professionals who evaluated criminal court files involving rape and murder showed similar results. When judges and professors of law had to estimate the probability that the defendant was the source of a DNA trace found on a victim, given the sensitivity and false-positive rate of DNA fingerprinting and base rate information, only 13% could reason correctly. When the DNA statistics were presented in natural frequencies, 68% of the professionals were successful (Hoffrage, Lindsey, Hertwig, & Gigerenzer, 2000; Koehler, 1996a; Lindsey, Hertwig, & Gigerenzer, 2003).

Relative Risk Reductions Can Cause Exaggerated Perceptions of Treatment Effects In one of the earliest studies published on this topic, Naylor et al. (1992) found that physicians rated the effectiveness of a treatment higher when the benefits were described in terms of a relative risk reduction ("A medical intervention results in a 34% relative decrease in the incidence of fatal and nonfatal myocardial infarction") rather than as an absolute risk reduction ("A medical intervention results in a 1.4% decrease in the incidence of fatal and nonfatal myocardial infarction—2.5% vs. 3.9%"; p. 920) or a number needed to treat ("77 persons must be treated for an average of just over 5 years to prevent 1 fatal or nonfatal myocardial infarction"; p. 920). Yet one cannot blame this misunderstanding on the physicians alone, since the authors of the study themselves incorrectly specified the absolute risk reduction as "a 1.4% decrease" (p. 920) instead of a decrease by 1.4 percentage points (see above). More recently, Mühlhauser, Kasper, and Meyer (2006) presented results from three diabetes prevention studies to participants in European diabetes conferences (160 nurse educators, 112 physicians, 27 other professionals). When results were presented as relative risk reduction, 87% of the health professionals evaluated the effect of the preventive intervention as important or very important. However, when the same results were presented by giving the corresponding fasting plasma glucose values, only 39% of the health professionals evaluated the effect similarly.

After interviewing one of us (GG) on the confusion caused by relative risks, an editor of a medical journal who also heads a teaching hospital in Switzerland asked all 15 gynecologists in his department what the widely known 25% risk reduction by mammography really means. How many fewer women die of breast cancer? One physician thought that 25% means 2.5 out of 1,000, another, 25 out of 1,000; the total range of the answers was between 1 and 750 in 1,000 women (Schüssler, 2005). A group of 150 gynecologists who took a course in risk communication by GG as part of their continuing education were also asked what the 25% risk figure meant. Using an interactive voting system, the physicians could choose between four alternatives:

> Mammography screening reduces mortality from breast cancer by about 25%. Assume that 1,000 women age 40 and over participate in mammography screening. How many fewer women are likely to die of breast cancer?
>
> - 1 [66%]
> - 25 [16%]
> - 100 [3%]
> - 250 [15%]

The numbers in the brackets show the percentage of gynecologists who gave the respective answer. Two thirds understood that the best answer was 1 in 1,000. Yet 16% believed that the figure meant 25 in 1,000, and 15% responded that 250 fewer women in 1,000 die of breast cancer. The overestimation of the benefit was most pronounced among physicians in their 50s and 60s, with 21% and 27%, respectively, estimating "250 out of 1,000." After the training session in risk communication, all physicians understood the correct estimate—except one, who still insisted that the answer had to be 250 out of 1,000.

Do physicians understand the number needed to treat, which is defined as the number of patients that must be treated in order to save the life of one patient? It is also called "number needed to harm," since treatments typically have side effects. Few studies have been conducted on this question (Covey, 2007). In a survey of 50 Australian physicians, only 8 could understand and explain number needed to treat to others (Young et al., 2002). Studies in the United States and Europe have consistently shown that physicians and medical students prefer relative risk reductions to number needed to treat (see Ghosh & Ghosh, 2005). British researchers submitted four identical proposals for funding a cardiac rehabilitation and a breast cancer screening program, except that the benefit was presented either in relative risk reduction, absolute risk reduction, the absolute values from which the absolute risk reduction is computed, or number needed to treat (Fahey, Griffiths, & Peters, 1995). Only 3 out of the 140 reviewers (members of the Anglia and Oxford health authorities) noticed that the four proposals were equivalent, and when the benefits were described in relative risk reductions, the authorities saw the program as having the greatest merit and were most willing to fund it.

In her meta-analysis on the effect of presenting information in terms of absolute risks versus relative risks, Covey (2007) analyzed 13 experiments that investigated physicians and 3 experiments that investigated other health professionals, which show how physicians and health professionals can be consistently manipulated by framing the treatment effect differently. The results reviewed in this section demonstrate that even professionals are likely to evaluate effects as more beneficial when they are presented as relative risk reduction.

Geography Is Destiny If medical practice were always founded on the best scientific evidence, then practices involving similar patients would not differ largely between hospitals and regions, with every patient receiving the most appropriate treatment known. Reality is different, however. Medical practice is often based not on scientific evidence but rather on local habits. The *Dartmouth Atlas of Health Care* (Center for the Evaluative Clinical Sciences Staff, 1996) documents the striking variability in the use of surgical treatments across all regions in the United States. For instance, the proportion of women in Maine who have undergone a hysterectomy ranges from less than 20% to more than 70% between regions. Similarly, 8% of the children in one community in Vermont had their tonsils removed, whereas this figure was as high as 70% in others. In Iowa, the proportion of men who have had prostate surgery varies between 15% and more than 60% (Center for the Evaluative Clinical Sciences Staff, 1996).

These numbers indicate that surgical treatments are often not based on evidence. Population differences that would necessitate disparities in treatments as large as those reported within the same state are unlikely. Instead, the tendency to follow local custom is the single most important explanation for regional differences in medical practice (Eddy, 1996). These local customs may be the result of the uncertainty about the outcome of many medical treatments. Unlike new medications, which the U.S. Food and Drug Administration (FDA) ensures are tested, surgical procedures and medical devices are not systematically subjected to evaluation (although even with FDA approval, use of medication is still extremely variable).

Collective statistical illiteracy may be one major reason why regional customs outweigh evidence. If evidence is neither understood nor communicated properly, few will be able to recognize that something might be wrong with what their local peers are usually doing. Improved statistical skills might provide doctors and patients with the momentum to reduce this unwanted geographical variation and to practice shared decision making based on the best scientific evidence, a huge and necessary step toward evidence-based medicine (Barry, Fowler, Mulley, Henderson, & Wennberg, 1995).

Specialty Is Destiny Similarly, if treatments are based on the scientific evidence, it should barely matter which specialist one happens to consult. However, aside from geography, the physician's specialization all too frequently determines treatment. The treatment of localized prostate cancer in the United States, for instance, generally depends on whom the patient visits. A study found that some 80% of urologists recommended radical surgery, whereas some 90% of radiation oncologists recommended radiation treatment (Center for the Evaluative Clinical Sciences Staff, 1996, p. 135). This pattern of variation suggests that doctors treat patients according to their specialty and that patients are not generally advised about their options in a way that encourages them to participate in decision making.

Collective Statistical Illiteracy

In the previous section, we showed that statistical illiteracy exists among patients, physicians, and journalists. The high degree of this form of innumeracy is often striking. We call this phenomenon collective illiteracy, and it is collective in two senses. First, it exists among all three groups simultaneously, and second, the groups influence each other. Doctors influence patients' understanding of health issues, and the media influence both. In this way, shared statistical illiteracy becomes a stable phenomenon whose existence is rarely noticed.

IV. Consequences of Statistical Illiteracy

Consumers are bombarded with messages promoting the latest new test, drug, or treatment. Many of these messages employ techniques that deliberately and insidiously exploit limited statistical literacy in order to convince the audience that they are at high risk of illness (and do not know it) and would be foolish or irresponsible not to buy the advertised service or product. We discuss two consequences of misleading advertising in this section: emotional manipulation and impediments to informed consent and shared decision making.

Susceptibility to Manipulation of Anxieties and Hopes

The advertisements in Figure 5.11 are an illustrative sample of those that try to raise anxieties or hopes. In the first example, one of the most prestigious cancer centers in the United States informs the reader that "as national mortality rates for prostate cancer fluctuated between 1960 and 1990, five-year survival rates for prostate cancer among M.D. Anderson patients continued to improve." The implication is that higher 5-year survival rates would mean that more lives are saved, as Giuliani implied. Yet as we have shown, there is no relationship between the survival rate and the mortality rate. The ad compares the survival rates at M.D. Anderson with the mortality rates in the United States. The statistically illiterate reader, who may not notice the difference and has never heard of lead-time bias and overdiagnosis bias, is led to conclude that the center has made considerable progress in treating patients.

In each of the advertisements, the message explicitly or implicitly overstates a risk, a benefit, or both. Such ads contribute to a climate of anxiety and concern, even when the event is as rare as brain cancer. Whereas readers with adequate statistical literacy would know which questions to ask (e.g., how large is the risk, how large is the benefit, what is the state of the evidence), readers without these skills are likely to accept the messages at face value and undergo testing or treatment that is not in their best interest. Some may think that it is better to play it safe, even when an illness is rare. But these additional tests trigger a cascade of unnecessary medical intervention, overdiagnosis, and overtreatment that may result in harm, which means there is nothing "safe" about this strategy. For the severely ill, these harms generally pale in comparison to the potential benefits. But for those experiencing mild symptoms (or who have mild forms of disease), the harms become much more relevant. And for the many labeled as having predisease, or for those who are "at risk" but destined to remain healthy, or for those who have pseudodisease, treatment can only cause harm. An epidemic of diagnoses can be as dangerous to our health as disease is (Welch, Schwartz, & Woloshin, 2007).

Informed Consent and Shared Decision Making Undermined

In April 2007, the American College of Physicians—the largest medical specialty society in the United States—issued new guidelines on screening

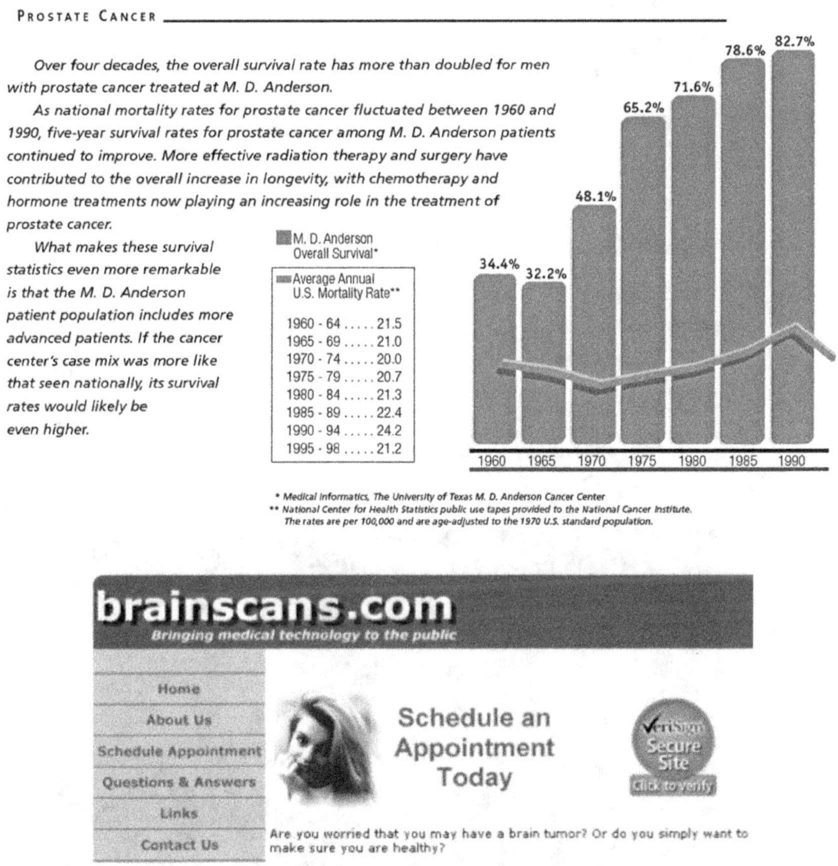

Figure 5.11: Tactics used in a selection of health messages to manipulate consumers' anxieties and hopes, and the consequences of such manipulation.

Figure 5.11: Continued

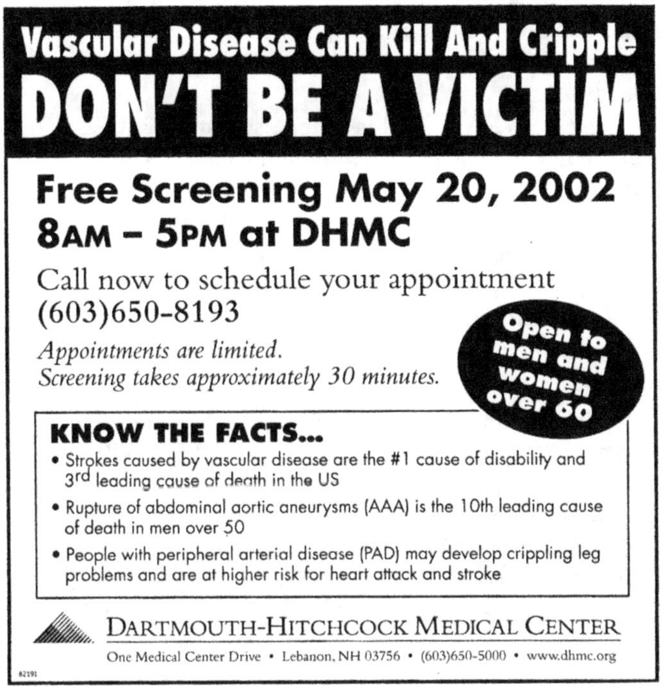

Figure 5.11: Continued

mammography for women aged 40 to 49. Rather than calling for universal screening, the guidelines recommend that women make an informed decision after learning about the benefits and harms of mammography (Schwartz & Woloshin, 2007). Yet many doctors do not understand the potential benefits and harms of mammography, including what a positive mammogram means. Collective statistical illiteracy makes informed consent science fiction.

The term *informed consent* refers to an ideal of how doctors and patients interact. Patients should be informed about the pros and cons of a treatment and its alternatives, and should decide on this basis whether they want to undergo treatment. To emphasize that the goal of informed consent is not simply obtaining patients' consent to doctors' decisions, the term *shared decision making* is often used instead (Moumjid, Gafni, Bremond, & Carrere, 2007). Yet studies indicate that clinicians rarely communicate the uncertainties about risks and benefits of treatments to patients (Braddock, Edwards, Hasenberg, Laidley, & Levinson, 1999). Shared decision making can be seen as a middle ground between "doctor knows best" paternalism and rampant consumerism. Although there is no unanimous definition, key aspects are the exchange of information between the physician and the patient and the involvement of both patient and physician in making the decision (Towle & Godolphin, 1999). Informed shared decision making thus requires that patients and doctors understand the benefits and harms of different treatment options. The classical view is that the technical knowledge about risks and benefits is held by the physician and is shared with the patients to enable them to decide according to their preferences (Charles, Gafni, & Whelan, 1997).

As we have reviewed in this article, statistical illiteracy not only is typical for patients but also exists among physicians. Thus, even with goodwill, some doctors would not be able to inform their patients adequately without two essential skills: understanding health statistics and communicating these in a transparent form. If both patients and physicians do not have minimal literacy in health statistics, an effective risk communication cannot take place and informed shared decision making is impossible.

This fundamental obstacle for the ideal of shared decision making has been rarely noticed, and is not a major topic at conferences on shared decision making and patient information. Their focus instead tends to be on patients as the problem, due to either their lack of knowledge or their emotional distress when forced to deal with uncertainty. Moreover, many physicians are concerned that their patients would no longer trust them if they disclosed their own uncertainty (Politi, Han, & Col, 2007). Similarly, the legal doctrine of informed consent deals with voluntary consent to biomedical research and medical treatment, the question of how much information suffices (an issue in malpractice trials), the patient's competence, and the right to refuse treatment. In contrast, doctor's statistical literacy has not yet been recognized as an issue, but is simply taken for granted. Physicians protect themselves against patients who might turn into plaintiffs by having them give their written consent. But informed consent involves more than just signing a form.

V. Causes of Statistical Illiteracy

Why does collective statistical illiteracy persist? And why is it not more of an issue at medical conferences, including those on informed consent and shared decision making? One obvious reason is the lack of training in statistical thinking in primary education and medical training, which we discuss in Section VI. In the present section we analyze factors specific to the patient–physician relationship and the health care environment.

Today, health statistics and randomized trials are an indispensable part of clinical practice. Yet medicine in fact has held a long-standing antagonism toward statistics. For centuries, treatment was based on "medical tact" in relation to the individual patient and on an ethic of personal trust rather than quantitative facts, which were dismissed as impersonal or irrelevant to the individual. The numerical method was alien to European therapeutic ethos, and equally so to 19th-century American medical practice, which presumed that disease was specific to the "natural" constitution of the individual (Warner, 1986). Some of the rare and mostly neglected early advocates for statistical thinking in medicine are described in Coleman (1987).

When averages became accepted much later, in 20th-century medicine, statistics redefined health as the "normal" rather than the "natural" state, with normality characterized by averages. Even in the 1940s and 1950s, Sir Austin Bradford Hill (1897–1991), who introduced the first large-scale clinical trials, spoke of medical opposition to statistics in his lectures at medical schools (Porter, 1995).

In 1937, an anonymous editorial in *The Lancet* stressed the importance of statistics for both laboratory and clinical medicine, and criticized physicians' "educational blind spot" (Fig. 5.12). In 1948, the British Medical Association (BMA) Curriculum Committee recommended the inclusion of statistics in medical education. They proposed 10 lectures with additional time for exercises, ranging from teaching core concepts such as chance and probability to interpreting correlations (Altman & Bland, 1991). Yet two decades passed before the General Medical Council (GMC), in 1967, echoed the BMA recommendation (Morris, 2002). Not until 1975 did statistics become a mandatory subject in medical schools within the University of London, and it took 10 more years in Austria, Hungary, and Italy (Altman & Bland, 1991, p. 230). By comparison, in psychology and other social sciences, statistics were already institutionalized as part of university curricula in the 1950s (Gigerenzer & Murray, 1987). Doctors working on higher degrees such as an MD were thereafter encouraged to do their own research. Yet the quality of this research has been criticized by statisticians as being the product of inexperienced researchers in a hurry or of "Mickey Mouse trials" published solely to decorate curricula vitae (Altman & Bland, 1991, p. 224). The problem is less the physicians themselves than the organization of medicine and the academic structure of biostatistics. Young biostatisticians are rewarded for theoretical work, less so for applications to medicine. The new emerging relation between patient, physician, and biostatistician is depicted in a cartoon from 1978 (Fig. 5.13).

The long and enduring opposition to health statistics can be traced back to the struggle between three 19th-century visions of the physician: artist, determinist, or statistician (Gigerenzer et al., 1989, chaps. 2 and 4). We argue that these professional ideals go hand in hand with patients' corresponding ideals, which even today fuel the mixture of feelings about health statistics: The artist embodies paternalism and requests blind trust from the patient, the determinist strives for perfect knowledge of causes and invites the illusion of certainty in patients, and the statistician relies on facts rather than medical charisma, paving the way for shared decision making. The first two ideals effectively deter interest in health statistics.

Paternalism and Trust

Physicians who think of themselves as artists place their trust in charisma, personal experience, and skill. They rely on personal intuition rather than impersonal numbers and exhibit characteristic faith in their own judgment. Risueño d'Amador (1836, pp. 634–635) argued before the Royal Academy

THE LANCET] MATHEMATICS

THE LANCET

LONDON: SATURDAY, JANUARY 2, 1937

MATHEMATICS AND MEDICINE

STATISTICS are curious things. They afford one of the few examples in which the use, or abuse, of mathematical methods tends to induce a strong emotional reaction in non-mathematical minds. This is because statisticians apply, to problems in which we are interested, a technique which we do not understand. It is exasperating, when we have studied a problem by methods that we have spent laborious years in mastering, to find our conclusions questioned, and perhaps refuted, by someone who could not have made the observations himself. It requires more equanimity than most of us possess to acknowledge that the fault is in ourselves. A colleague of a famous contemporary statesman has been quoted as complaining that he used figures as though they were adjectives. The medical profession as a whole would not find it easy to counter a similar charge. For most of us figures impinge on an educational blind spot. We have never been taught to recognise them for what they are. This is a misfortune, because simple statistical methods concern us far more closely than many of the things that we are forced to learn in the six long years of the medical curriculum. Many of our problems *are* statistical; and there is no other way of dealing with them. In preventive medicine this is so obvious that it has acquired general recognition. In laboratory work, though recognition has come more slowly, it is now widely realised that it is very unsafe to base conclusions on statistically inadequate data. In clinical medicine recognition is coming more slowly still; so slowly that many avoidable errors, and a sad waste of material, still hinder progress.

Figure 5.12: Excerpt from a *Lancet* 1937 editorial ("Anonymous," 1937) documenting the emerging realization that statistics is actually relevant for both laboratory and clinical medicine.

Figure 5.13: The new Holy Trinity in medicine (Rimm & Bortin, 1978). This cartoon was a reaction to the statistical revolution in medicine in the 1970s. The physician continues to play God for the patient, but no longer for himself. For him, God's voice is in the verdict of the biostatistician, "significant" (i.e., "$p < .05$") or "not significant." The biostatistician, finally, sees God in the mirror.

Figure 5.13: Continued

of Medicine in Paris that the use of statistics was antimedical, for it aimed "not to cure this or that disease, but to cure the most possible out of a certain number." Following the law of the majority would condemn individual patients to death. Therefore, the physician must rely on intuition, not on the mechanical collection and use of health statistics. In this view, the use of statistics was antiscientific—it presupposed a level of homogeneity among patients that might be appropriate for physics but was utterly unrealistic in medicine.

For the physician-as-artist, the patient resembles a unique sculpture that is molded and shaped and therefore essentially passive. The artist assumes responsibility for the patient's body, and the patient engages in a paternalistic relationship of trust and obedience. Paternalism is defined as a hierarchical relationship in which a figurehead (the father, *pater* in Latin) makes decisions on behalf of others (the children) for their own good. Today, paternalism remains widespread, but it would be wrong to simply attribute it to physicians with an antiquated sense of their profession. Involved are *two* players who react to each other's expectations in a game of trust. Discussions among physicians indicate that many are ambivalent about being regarded as omniscient and omnipotent godlike figures, and would instead prefer being able to admit when they are uncertain about the best treatment (Gigerenzer, 2002). Yet they also tend to believe that patients want a father figure and might switch to another doctor who is willing to play this role. As mentioned above, medical organizations—including the American College of Physicians, the U.S. Preventive Services Task Force, and the Academy of Family Physicians—explicitly recommend that every man should weigh the pros and cons of PSA screening because the benefits (mortality reduction) are unclear, while severe harms (incontinence and impotence) occur in one third to two thirds of surgeries following a positive test. Yet among patients who participated in PSA screening, 68% said that it was because their doctor told them to, and 16% reported that their wife or girlfriend influenced their decision (Federman, Goyal, Kamina, Peduzzi, & Concato, 1999). The paternalist heuristic "If you see a white coat, trust it" is decidedly not a decision strategy of the uneducated public only. Consider neoclassical economists, whose doctrine includes weighing all pros and cons of alternatives, emphasizing rational choice rather than trust. Yet two thirds of more than 100 American economists surveyed said that they had not weighed any pros and cons of PSA screening but only followed their doctor's recommendation. Another 7% said that their wives or relatives had exerted an influence on the decision (Berg, Biele, & Gigerenzer, 2013).

Paternalism is practiced in many forms. *Concealed paternalism* is an extreme form in which physicians do not even inform patients about tests or treatments performed on them. It is not infrequent in the United States, where doctors routinely do PSA screening tests on men without obtaining their consent. For instance, about one third of men without prostate cancer were unaware that their physician had ordered a PSA test (Federman et al., 1999). Concealed paternalism is in part a reaction to the unpredictabilities of the U.S. legal system that encourages physicians to practice defensive medicine—to protect themselves against potential lawsuits—rather than do what they consider best for the patient (there are no legal consequences for overdiagnosis, only for underdiagnosis). For instance, in 2003, Daniel Merenstein, a young family physician in Virginia, was sued because he did not automatically order a PSA test for a patient. Merenstein had followed the recommendations of the medical organizations and informed the man about the pros and cons; he then declined to take the test. The patient unfortunately developed a horrible, incurable form of prostate cancer. The plaintiff's attorney claimed that the PSA test was standard in the Commonwealth of Virginia and that Virginia physicians routinely do the test without informing their patients. The jury exonerated Merenstein, but his residency was found liable for $1 million. After this experience, he felt he had no choice but to overdiagnose and overtreat patients, even at the risk of causing unnecessary harm: "I order more tests now, am more nervous around patients; I am not the doctor I should be" (Gigerenzer, 2007, p. 161).

A glamorous version of paternalism is found in public health messages that replace physicians with celebrities as trustworthy authorities.

Once again, the goal is to persuade people to do the "right" thing rather than encourage them to make informed decisions. For example, celebrity endorsements of cancer screening typically consist of messages asserting that the celebrity's life was saved by screening or that the life of a loved one was lost due to failure to be screened. In the United States, these celebrity messages are widely heard and have increased the number of people undergoing screening (Larson, Woloshin, Schwartz, & Welch, 2005).

Paternalism and its counterpart, trust in authority, make patients' grasp of health statistics superfluous. Moreover, patients who desire a paternalistic relationship want care, not cure by numbers—so they would be unable to detect whether or not their physician understands health statistics. Paternalism is one potential cause of collective statistical illiteracy.

Determinism and the Illusion of Certainty

The second vision of the physician is that of a determinist who relies on experimentation to find the true causes of disease and eventually will be able to treat these with certainty. This view, like that of the physician-as-artist, has been hostile to health statistics. To understand why, it is important to realize that, before the early 20th century, experiment and statistics were opposed practices. For experimenters, collecting numbers was evaluated as unscientific. Science was about causes, not chances. The determinist believed that through careful experiments, science could teach the physician to control every detail, so that averages and medical intuition alike would be rendered otiose. In Paris, the famous physiologist Claude Bernard vehemently opposed the "medical tact" promoted by Risueño d'Amador as charlatanism, but also rejected statistics as proposed by P.C.A. Louis (1787–1872). Bernard argued that being content with an average means failing to deal with the variation that is of supreme importance when curing patients. There exists, he insisted, no average pulse, but only a resting, working, or eating pulse. Nor is there average urine, for urine during fasting is different from urine during digestion. How could a physician interested in curing each patient, and not just some proportion, remain content with averages? In Bernard's (1865/1957, pp. 137–138) own words:

> A great surgeon performs operations for [a kidney] stone by a single method; later he makes a statistical summary of deaths and recoveries, and he concludes from these statistics that the mortality law for this operation is two out of five. Well, I say that this ratio means literally nothing scientifically and gives us no certainty in performing the next operation; for we do not know whether the next case will be among the recoveries or the deaths. What really should be done, instead of gathering facts empirically, is to study them more accurately, each in its special determinism.

Determinism prevailed, although some medical researchers, such as Louis in Paris and Ignaz Semmelweis (1818–1865) in Vienna, collected numbers. Louis, known as the father of modern medical statistics, showed that bloodletting in pneumonia had no effect on outcome. Semmelweis discovered that the incidence of fatal puerperal fever could be drastically cut from about 20% to 1% by requiring physicians to wash their hands between examinations. Semmelweis's discovery of a general cause, cleanliness, was largely ignored at a time when each patient and thus each cause of death were believed to be unique. Outraged by the indifference or outright hostility of the medical profession, Semmelweis eventually had a mental breakdown and was confined to an institution where he died shortly after—ironically, by what appears to have been a wound infection. Louis and Semmelweis are today considered to be forerunners of "evidence-based medicine."

It is to the credit of Sir Ronald Fisher (1890–1962) that the opposition between the experimenters and the statisticians was finally dissolved in the 1920s. Fisher joined experimentation with statistics, and after they had become two sides of the same coin, experimentation radically changed, now being defined by randomization, repetition, and other statistical concepts (Gigerenzer et al., 1989). Based on Fisher's work, Sir Austin Bradford Hill (1897–1991) promoted the new union between experimentation and statistics as an indispensable part of medicine.

Although statistics suppressed determinism, its traces have not been entirely wiped out. Specifically, determinism has survived in the form of the *illusion of certainty* in patients' minds, fostered by information brochures and advertisements. An illusion of certainty is defined as the belief that some event is absolutely certain even when such certainty does not exist. It is a major emotional obstacle toward learning to live with uncertainty.

Figure 5.6 showed that large proportions of the general public have illusory certainty about the perfection of tests, including HIV testing and mammography. This illusion is not simply a product of the individual mind but, as we have seen, has its historical origins in deterministic medical science. Today, it is fueled by health messages that claim or suggest certainty. For instance, the philanthropic Burda Foundation has established a network against colorectal cancer; according to its website: "It has been proven that with early detection, almost 100% of colorectal cancer cases could be prevented or cured" (Felix Burda Stiftung, 2008). When we inquired about where to find this evidence, the head of the foundation's marketing and communication department responded that he could not recall the precise study, but that researchers—mostly in U.S. studies—found that 60% to 90% of colorectal cancers can be prevented. Since physicians always overlook something, he explained, it follows that colorectal cancer is theoretically 100% curable. An example of a more suggestive illusion of certainty is the brain scan advertisement in Figure 5.11, where the reader is asked: "Do you simply want to make sure you are healthy?"

A subtle way to induce the illusion of certainty is by analogies, such as combat metaphors that liken "war" on cancer to recent military triumph (Wong & King, 2008). In this militarized narrative, cancer is the enigmatic enemy, described as "lawless," "savage," and "relentless." This suggests that one can "slash," "burn," or "poison" the cancer cells with surgery, radiation therapy, and chemotherapy, respectively. Once the cancer is killed, the enemy is beaten, and the war is won. And the earlier the enemy is detected and the more slashing and burning that take place, the faster and more decisive the victory will be.

To summarize, determinism and its psychological counterpart, the illusion of certainty, make health statistics appear to be a wasted enterprise. The goal is certainty, rather than learning how to live with uncertainty. Like paternalism and trust, this ideal is incompatible with the quest for health statistics. Yet these factors are not the only ones. Conflicts of interest ensure that physicians and patients learn about only part of the relevant health statistics, which are framed in a way to serve particular purposes rather than to create an informed citizenship.

Conflicts of Interest

There are various players in public health with goals that can conflict with transparent risk communication—goals such as pushing a political agenda, attracting media attention, selling a new drug, increasing compliance with screening, or trying to impress physicians. Conflicts of interest lead to omission of relevant information and the use of nontransparent framing.

At issue is the distinction between content and form. All information can be communicated in several forms. The degree of transparency is empirically defined by the proportion of people in a population who can correctly understand it. Transparency is relative to expertise. For instance, when information necessary to estimate the chances that a baby has Down syndrome was presented in terms of conditional probabilities, obstetricians, midwives, and patients alike found it to be nontransparent. When the information was instead given in the form of natural frequencies, it proved to be much more transparent to the obstetricians than to the other groups (Bramwell et al., 2006). When we speak of transparent versus nontransparent forms in this article, we thus oversimplify what is a gradual matter and dependent on population. Transparent forms include absolute risks, natural frequencies, mortality rates, and, in general, statements about frequencies or depictions of frequencies in pictures. Nontransparent forms include relative risks, conditional probabilities such as sensitivities and specificities, survival rates, and statements about single events that do not specify the reference class. As the case of Giuliani illustrates, misunderstandings by nontransparent information go largely unnoticed since the issue has not yet been subject to public awareness.

Do Medical Journals Provide Transparent Information? Where do nontransparent statistics come from? One hypothesis is that they originate from innumerate physicians, patients, and journalists, who are both manufacturers and victims of statistical confusion. Yet surprisingly, nontransparent health statistics such as relative risks without the base rate often appear in leading medical journals, and it is often from these sources that the numbers spread to physicians, the media, and the public. Nuovo, Melnikow, and Chang (2002) analyzed 359 articles that reported randomized trials in the years 1989, 1992, 1995, and 1998 that were published in *Annals of Internal Medicine, British Medical Journal* (*BMJ*), *Journal of the American Medical Association* (*JAMA*), *The Lancet*, and *The New England Journal of Medicine*. Only 25 articles reported absolute risk reduction, and 14 of these 25 also included the number needed to treat, which is simply the inverse of the absolute risk reduction. That is, only about 7% of the articles reported the results in a transparent way. The same journals, along with the *Journal of the National Cancer Institute*, were analyzed again in 2003/2004 (Schwartz, Woloshin, Dvorin, & Welch, 2006). Sixty-eight percent of 222 articles failed to report the absolute risks for the first ratio measure (such as relative risks) in the abstract; about half of these did report the underlying absolute risks elsewhere in the article but the other half did not. An analysis of *BMJ, JAMA*, and *The Lancet* from 2004 to 2006 found that in about half of the articles, absolute risks or other transparent frequency data were not reported (Sedrakyan & Shih, 2007). These analyses indicate that one reason why physicians, patients, and journalists talk about relative risk reductions in isolation is because the original studies regularly provide the information in this nontransparent form. Fortunately, the major medical journals, through initiatives like CONSORT and the international peer review congresses, are paying increasing attention to these issues.

Yet readers can be misled more directly than just via nontransparent framing. In some cases, benefits and harms of treatments are reported in different currencies: benefits in big numbers (relative risk reduction), but harms in small numbers (absolute risk increases). We call this technique *mismatched framing*. For instance, the *Guide to Clinical Preventive Services* of the U.S. Preventive Services Task Force (2002) states the relative risk reduction (not the absolute risk reduction) when describing the benefits of screening—"sigmoidoscopy screening reduced the risk of death by 59% for cancers within reach of the sigmoidoscope" (p. 93); but when the harms associated with the procedure are described, these are reported in absolute risks—"Perforations are reported to occur in approximately 1 of 1,000–10,000 rigid sigmoidoscopic examinations" (p. 94). An analysis of three major medical journals, *BMJ, JAMA*, and *The Lancet* from 2004 to 2006 revealed that when both benefits and harms of therapeutic interventions were reported, 1 in 3 studies used mismatched framing and did not report the benefits in the same metric as the harms. In most cases, relative risks were reported for benefits, and absolute frequencies were reported for harms (Sedrakyan & Shih, 2007).

The prevalent use of relative risks (and odds ratios) is sometimes defended on the basis that these ratio measures are transportable to different populations with different baseline risks, or that they summarize two numbers in one. But these features are also their main weakness, since they conceal the underlying absolute risks. Relative risk estimates are meaningless for understanding the chances of experiencing either a benefit or a harm. Even when readers understand relative risks, they cannot judge the clinical significance of the effect unless the underlying absolute risks are reported. As mentioned before, a relative risk reduction of 50% is compatible with both a substantial mortality reduction from 200 to 100 in 10,000 patients and a much smaller reduction from 2 to 1 in 10,000 patients. If the absolute risks are reported, the relative risks can be derived from these, but not vice versa. Randomized trials provide some of the best information in medicine, but unless the results are reported adequately, assessing and comprehending the information is difficult.

Why do medical journals not make transparency a requirement for submissions? One answer is competing interests. One third of the trials published in the *BMJ* and between two thirds and three quarters published in the major North American journals were funded by the pharmaceutical industry (Egger, Bartlett, & Juni, 2001). Richard Smith (2005), former editor of the *BMJ* and former chief executive of the *BMJ Publishing Group*, explained the dependency between journals and the pharmaceutical industry:

> The most conspicuous example of medical journals' dependence on the pharmaceutical industry is the substantial income from advertising, but this is, I suggest, the least corrupting form of dependence. . . . For a drug company, a favourable trial is worth thousands of pages of advertising . . . . Publishers know that pharmaceutical companies will often purchase thousands of dollars' worth of reprints, and the profit margin on reprints is likely to be 70%. Editors, too, know that publishing such studies is highly profitable, and editors are increasingly responsible for the budgets of their journals and for producing a profit for the owners. . . . An editor may thus face a frighteningly stark conflict of interest: publish a trial that will bring US$100,000 of profit or meet the end-of-year budget by firing an editor.

It is in the very interest of pharmaceutical companies to present the results in a way that is most likely to impress the readers and, particularly, the doctors who receive the reprints. And relative risk reductions for the benefits of one's drug are an efficient means toward this end. "Journals have devolved into information laundering operations for the pharmaceutical industry," wrote Richard Horton (2004, p. 9), editor of *The Lancet*.

Are Patients Likely to Find Transparent Information in Medical Pamphlets and Websites? *Pamphlets.* Information on breast cancer screening should provide information about the potential benefits and harms, so that a woman can make an informed decision whether she wants to participate or not. If she participates, she also needs information about the positive predictive value. An investigation of 58 pamphlets informing women about breast cancer screening in Australia (Slaytor & Ward, 1998) found that a majority of pamphlets (35, or 60%) included information about the lifetime incidence rate, but only 1 pamphlet included the risk of actually dying of breast cancer (Table 5.9). Naturally, the incidence rates loom larger than the mortality

Table 5.9: Percentage of Informational Materials That Provide Specific Pieces of Information about Breast Cancer Screening to Patients in Various Countries

	Pamphlets (Australia)[a] $n = 58$	Pamphlets (Germany)[b] $n = 27$	Pamphlets (Austria)[c] $n = 7$	Websites (8 countries)[d] $n = 27$	Invitations (7 countries)[e] $n = 31$
Baseline Risk					
Lifetime risk of developing breast cancer	60	37	43	44	32
Lifetime risk of dying from breast cancer	2	4	0	15	n/a
Benefits from Screening					
Relative risk reduction of death from breast cancer	22	7	0	56	23
Absolute risk reduction of death from breast cancer	0	7	0	19	0
Number needed to screen to avoid one death from breast cancer	0	4	0	7	0
Harms					
Overdiagnosis and overtreatment (e.g., carcinoma in situ)	n/a	11	n/a	26	0
Harms from X-rays	n/a	44	100	15	n/a
Psychological distress related to false-positive results	n/a	11	n/a	37	n/a
Test Properties					
Proportion of women who are recalled (positive tests)	14	11	14	44	19
Proportion of breast cancers detected by mammography (sensitivity)	26	19	0	26	23
Proportion of women who test negative among those without breast cancer (specificity)	0	4	0	0	0
Proportion of women with breast cancer among those who test positive (positive predictive value)	0	15	0	15	0

Note: The table lists all mentions of the respective piece of information, independent of whether the piece of information was given correctly. It is based on different studies, and not all studies assessed all pieces of information (n/a).

[a]Slaytor & Ward (1998); [b]Kurzenhauser (2003); [c]Rásky & Groth (2004); [d]Jorgensen & Gøtzsche (2004); [e]Jorgensen & Gøtzsche (2006).

rates and thus contribute to raising anxiety; campaigns selectively reporting incidence rates have been criticized for this reason (Baines, 1992). Most important, the mortality rate, not the incidence rate, is relevant for screening, since the goal of screening is to reduce mortality, whereas it cannot reduce incidence. The information about benefits and harms that women would need to make an informed decision, in contrast, was scarce in these pamphlets (consistent with patients' lack of knowledge; see Part II). Only 22% of the Australian pamphlets reported the benefit in quantitative terms, always in relative risk reductions, and never in a transparent form, such as in absolute risk reductions. No information about potential harms was available. The most important information about the test quality, that about 9 out of 10 women who test positive do not have cancer, was never mentioned. An analysis of German brochures (Kurzenhäuser, 2003) revealed a similar picture, apart from the specific attention given to the dangers of X-rays. A few German pamphlets did, however, provide information about benefits and harms in a transparent way. In Austrian pamphlets, in contrast, there was a striking absence of relevant information (Rásky & Groth, 2004), except for constant assurances that the potential harms of X-rays are negligible and that mammography can save lives. Like in Australia, information about the positive predictive value was never provided. All 7 of the Austrian pamphlets mentioned that early detection increases the chance for complete recovery, but all were mute on the size of this increase. It is telling that when a German pamphlet (from the women's health network Nationales Netzwerk Frauen und Gesundheit; not included in Table 5.9) informed women about screening in a more comprehensive and transparent way, the Austrian Association of Physicians asked their members to remove it from their shelves because they feared it would lead to lower compliance (Noormofidi, 2006). This is the same association that, when *The Lancet* published a meta-analysis finding homeopathy to have no effect (Shang et al., 2005), responded that meta-analyses are an interesting instrument for theoretical science but of little relevance to clinical practice (Österreichische Ärztekammer, 2005).

Mismatched framing also occurs in pamphlets and leaflets. Yet as Table 5.9 shows, it can only occur in the few that actually provide information about both benefits and harms. For instance, one leaflet explained that hormone replacement therapy "has been proven to protect women against colorectal cancer (by up to more than 50%)" whereas the risk of breast cancer "may possibly increase by 0.6% (6 in 1,000)" (see Gigerenzer, 2002, p. 206). Looking up the absolute risk reduction, which was not reported, one finds that the 50% benefit corresponds to an absolute number that is less than 6 in 1,000. In a study, this leaflet was given to 80 women between age 41 and 69; 75% of these incorrectly understood the numbers to mean that hormone replacement therapy prevents more cases of cancer than it produces, whereas only 4% correctly understood that the opposite was the case (Hoffrage, 2003).

Invitations for Screening. In countries with publicly funded screening, eligible citizens are often made aware of these programs by letters of invitation.

Thus, by sheer numbers of citizens reached, such letters are—alongside physicians—potentially the most important source of information about screening. Invitation letters would be the ideal opportunity to provide the patients with balanced, transparent information about screening, so that they can make informed decisions. Yet there is a conflict of interest built into the system: Those who are responsible for the screening program are also responsible for designing the invitations, which puts their goal of increasing compliance at odds with increasing transparency. For example, German health authorities, addressing women between 50 and 69, said that it is important that as many women as possible participate and this is best reached by personal invitations (Bundesministerium für Gesundheit, 2002b). The official leaflet sent to all women in Germany in this age group contained much useful information, including that 5% will be recalled (i.e., test positive) and that 80% of these do not have cancer, but included no information about the size of the potential benefit (Kassenärztliche Bundesvereinigung, 2004). If women were told that it is indeed unclear whether the benefits of mammography screening outweigh its harms, some might decide against it; thus, transparent health statistics are likely to decrease compliance in this case.

Jørgensen and Gøtzsche (2006) investigated letters of invitation to breast cancer screening in seven countries with publicly funded screening: Australia, Canada, Denmark, New Zealand, Norway, Sweden, and the United Kingdom (Table 5.9). Most of the invitations (97%) stated the major benefit of screening, the reduction in breast cancer mortality. However, the very few (23%) that also mentioned the size of the benefit always did so in terms of relative risk reductions rather than absolute risk reductions. None of the invitations included information about potential harms or the positive predictive value. Instead, most invitations used persuasive wording and prespecified appointments. Thus, the invitation letters clearly aim at compliance rather than at informing the public.

If citizens look for additional information on the Internet, does this provide a more balanced perspective?

Websites. A study of 27 Scandinavian– and English–speaking websites demonstrated that all those of advocacy groups and governmental institutions (24 websites in total) recommended screening and favored information that shed positive light on it (Jørgensen & Gøtzsche, 2004). Only a few mentioned the major potential harms of screening: overdiagnosis and overtreatment. Three websites of consumer organizations had a more balanced perspective on breast cancer screening and included information on both the potential benefits and harms. In total, very few sites met the standards of informed consent, as specified by the General Medical Council's (1998) guidelines for patient information.

Mismatched framing was also used in the National Cancer Institute's Risk Disk, intended to help women make informed decisions about whether to use tamoxifen for the primary prevention of breast cancer (Schwartz, Woloshin, & Welch, 1999b). The benefit of tamoxifen is stated with the following relative risk reduction: "Women [taking tamoxifen] had about 49%

fewer diagnoses of invasive breast cancer." In contrast, the harm of more uterine cancer was presented as "the annual rate of uterine cancer in the tamoxifen arm was 30 per 10,000 compared to 8 per 10,000 in the placebo arm" (National Cancer Institute, 1998). And in fact, the Breast Cancer Prevention Study Fact Sheet (National Cancer Institute, 2005) presents the 49% statistic and no numbers for the increased risk of uterine cancer.

This problem is not limited to information about cancer. For example, advice on the World Wide Web about how to manage fever in children at home was similar: Complete and accurate information was rare, and some websites contained advice that should in fact be discouraged (Impiccatore, Pandolfini, Casella, & Bonati, 1997). Rigby, Försstrom, Roberts, and Wyatt, for the TEAC-Health Partners (2001) estimated that one quarter of the messages disseminated by Internet health information services are false. These results are alarming, given that many people use the Internet to acquire information about health issues—in the European Union, this number was already 44% in 2005 and is increasing steadily (see Andreassen et al., 2007).

How Accurate Are Leaflets Distributed to Doctors? For the busy physician with limited time to keep abreast of medical research, advertisement leaflets by the pharmaceutical industry are a major source of further education. These are directly sent to doctors or personally handed to them by well-dressed representatives. A leaflet typically summarizes the results of a published study for the physician in a convenient form. Do doctors get accurate summaries? Researchers from the German Institute for Quality and Efficiency in Health Care searched for the original studies and compared these with the summaries in 175 leaflets (Kaiser et al., 2004). The summaries could be verified in only 8% of the cases (!). In the remaining 92% of cases, key results of the original study were often systematically distorted or important details omitted. For instance, one pamphlet from Bayer stated that their potency drug Levitra (Vardenafil) works up to 5 hours—without mentioning that this statistic was based on studies with numbed hares. Should doctors have wanted to check the original studies, the cited sources were often either not provided or impossible to find. In general, leaflets exaggerated baseline risks and risk reduction, enlarged the period through which medication could safely be taken, or did not reveal severe side effects of medication pointed out in the original publications.

The spread of advertising for medical products reflects the increase in the commercialization of medicine—and profits from the statistically illiterate, who are unlikely to ask the tough questions. Even for advertisements placed in medical journals, selective reporting of results has been documented (Villanueva, Peiró, Librero, & Pereiró, 2003). In the United States, direct-to-consumer advertising constitutes the single largest effort to inform the public about prescription drugs—on which pharmaceutical companies spent more than $4 billion in 2010. These ads typically assert the benefit of the drug with personal statements (e.g., "It works for me") or with data on popularity of the drug ("Over a million people have begun to take this drug to manage their diabetes"). But the ads fail to provide the most fundamental information consumers need to make informed decisions: How well does the drug work, and what are the side effects? (Woloshin, Schwartz, Tremmel, & Welch, 2001). The education of patients and physicians alike is too important to be left to the pharmaceutical industry and pseudoeducational campaigns that promote sales.

Do Political Institutions Promote Informed Citizens? In 2001, the German government proposed mammography screening for all women between ages 50 and 69: "Mammography screening could reduce mortality from breast cancer by 30%, that means, every year about 3500 deaths could be prevented, ca. 10/day" (cited in Mühlhauser & Höldke, 2002, p. 299). Note the use of a relative risk reduction, suggesting a big benefit, instead of the absolute risk reduction, which is on the order of 1 in 1,000. Furthermore, the public was not informed that there is no evidence that the total mortality is reduced by screening—that is, that any lives are saved. The estimated 3,500 women are the decreased number of women who die of breast cancer within 10 to 15 years, whereas the total number of deaths remains the same in this period for women who participate in screening or not (Gøtzsche & Nielsen, 2006). The Berlin Chamber of Physicians (Ärztekammer Berlin, 2002, March 21) protested in a 2002 press release against a general screening program on the grounds that there is no scientific evidence that the potential benefits of screening are higher than its harms, and that the parliament's health committee overstated benefits and downplayed harms. Two days later, the German Minister of Health, Ulla Schmidt, responded in a press release that there is sufficient evidence in favor of screening because "there is an up to 35% reduction in breast cancer mortality" (Bundesministerium für Gesundheit, 2002a). Note once again the use of relative risk reduction. When one of the authors (GG) clarified what this number means in an interview in the German weekly *Die Zeit*, the advisor of the Secretary of Health, Professor Karl Lauterbach, defended the use of relative risk reduction by responding that "in justifying the programs, the Secretary of Health does not inform individual women, but the public. If an individual doctor advises patients, he should, as Mr. Gigerenzer, state the absolute risk and its reduction" (Lauterbach, 2002, p. 16). According to this logic, transparency is for individual women, not for the public. It is a pity that a democratic government confuses taxpayers about the benefits of a program that they ultimately finance. But political interests reign over transparency in health in other countries, too.

In 1997, the National Institutes of Health Consensus Development Conference on Breast Cancer Screening for Women Ages 40 to 49 was convened at the request of the director of the National Cancer Institute (NCI). The expert panel reviewed the medical studies and concluded with a 10-to-2 vote that there is insufficient evidence to recommend screening for this age group and that "a woman should have access to the best possible relevant information regarding both benefits and risks, presented in an understandable and usable form" (National Institutes of Health Consensus Development Panel, 1997, p. 1015). At the news conference, Richard Klausner, Director of the NCI, said he was "shocked" by this evidence, and that night a national television program began its news coverage with an apology to American women for the panel's report. Eventually, the Senate voted 98 to 0 for a nonbinding solution in favor of mammography for women in their 40s. The director of the NCI asked the advisory board to review the panel's report, a request that they first declined, but in March 1997, the board voted 17 to 1 that the NCI should recommend mammography screening every one or two years for women in this age group—against the conclusion of its own expert panel (Fletcher, 1997). The voting members of the NCI advisory board are appointed by the U.S. president, not by the medical experts in the field, and are under great pressure to recommend cancer screening.

In 2002, new studies became available that again indicated that the benefits of mammograms may not outweigh the risks, and Donald Berry, chairman of the department of biostatistics at M.D. Anderson Cancer Center explained this result to the Senate, but to no avail. The Bush administration restated the recommendation and Andrew von Eschenbach, the director of the NCI at that time, announced that women in their 40s should get mammograms (Stolberg, 2002).

The mesh between medicine and politics is visually captured in two stamps (Figure 5.14). The U.S. Postal Service has used commemorative stamps depicting matters of historical, social, and cultural importance to the nation. The mechanisms for choosing stamps were designed to insulate the Postal Service from special interest groups. But in 1996, a California surgeon and founder of a nonprofit advocacy organization for breast cancer research approached Representative Vic Fazio (D-Calif.) with the idea of issuing a fund-raising stamp (Woloshin & Schwartz, 1999). In August 1997, the Breast Cancer Research Stamp Act was signed into U.S. law, against the objections of the Postal Service. The denomination was 40 cents, of which 8 cents went to federal research on breast cancer. The nation's first-ever fund-raising stamp was issued in 1998 at a White House ceremony hosted by First Lady Hillary Rodham Clinton and Postmaster General William Henderson. The idea for a prostate cancer stamp emerged in Congress in reaction to the breast cancer stamp. The Postal Service once more opposed the bill calling for a new semipostal stamp, and eventually a regular stamp that promoted "annual checkups and tests" was released.

Evidence did not seem to matter. Just 2 years before the stamp's release, in 1996, the U.S. Preventive Services Task Force (1996) had concluded that "routine screening for prostate cancer with digital rectal examinations, serum tumor markers (e.g., prostate-specific antigen), or transrectal ultrasound is not recommended" (p. 119). Against the scientific evidence, the Postal Service became a vehicle for special interest groups.

Summary

In this section we argued that there is a network of causes for collective statistical illiteracy. Statistical thinking is a latecomer in medical practice and research, which had been dominated by two conflicting models

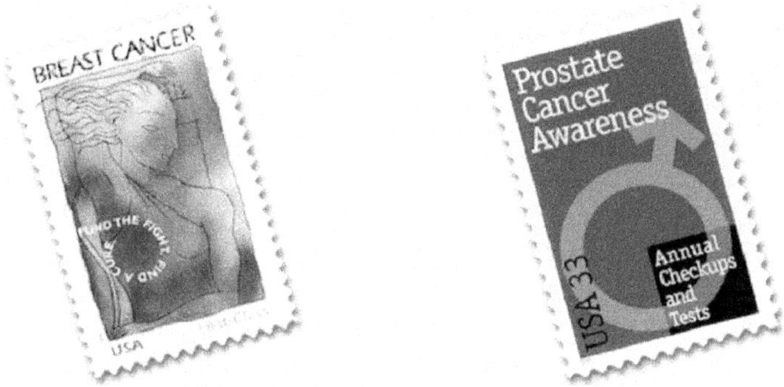

Figure 5.14: U.S. Postal Service stamps promoting breast and prostate cancer screening—an illustration of the intersection between medicine and politics.

of physicians: the godlike artist and the scientific determinist, both of whom rejected statistics. These ideals go hand in hand with unconditional trust and illusions of certainty in patients, for whom statistical information appears of little relevance. Now that these two visions of the patient–physician relationship are beginning to crumble in the age of information, organizations with other interests spend much subtle energy in preventing citizens from receiving the relevant information about potential benefits and harms of medical treatments in a transparent form. The sad part of this story is that, to a considerable degree, democratic governments and medical organizations that disseminate information pamphlets play their part in this game.

VI. Therapy

The network of factors we have described—competing interests, trust, paternalism, and illusion of certainty—provides a challenge for change. Yet if we can change one fundamental factor, some of the other obstacles might fall like a row of dominos. In our opinion, this would be education of the public in statistical thinking combined with training in transparent framing. An educated citizenship will know what questions to ask, what information is missing, and how to translate nontransparent statistics into transparent ones. But that necessitates rethinking how statistical thinking is taught.

Medical doctors tend to think of psychologists as therapists, useful for the emotionally disturbed but not for members of their own trade. Research and training in transparent risk communication, however, is a field in which cognitive psychologists can actually help doctors. In this last section, we define the task that psychological and medical researchers should address: the efficient training of pupils, medical students, and doctors in understanding risks and uncertainties. We also discuss sources of resistance.

Teach Statistical Literacy in School

Statistical thinking is the most useful part of mathematics for life after school. Today, however, almost all of the available time is spent on the mathematics of certainty—from algebra to geometry to trigonometry. If children learned to deal with an uncertain world in a playful way, much of collective statistical illiteracy would be history. But for the teacher, like for the doctor, statistical thinking is a late arrival: Elementary and high schools have been "probability free" even longer than medical schools have been. In 1992, when Michael Shaughnessy reviewed the situation in the United States, he reported that only 2% of college-bound high school students had taken a course in probability and statistics, whereas 90% of these students had taken a course in algebra (Shaughnessy, 1992). The Quantitative Literacy Project (Gnanadesikan, Scheaffer, & Swift, 1987) and the Middle Grades Mathematics Project (Philips, Lappan, Winter, & Fitzgerald, 1986) were among the pioneering programs to make some inroads into the teaching of probability and statistics in the middle grades.

National school systems differ profoundly in the time allotted to different areas within mathematics. Germany's educational system, for instance, traditionally paid very little attention to teaching data analysis and probability. In recent years this has changed, and competencies in data analysis and probability are now a mandatory part of national curricula from elementary school to grade 12. Yet that alone does not solve the problem. Many teachers are simply not prepared to teach statistics. Performance of German students in statistics and probability as measured by the 2003 Programme for International Student Assessment (PISA) continued to be relatively weak. PISA documented a relatively stronger performance for American 15-year-olds in the area of "uncertainty" as compared to "quantity" and "shape and space." However, this result has to be seen against the low overall performance of the U.S. students, putting their competence in dealing with "uncertainty" at a similar unsatisfactory level as that of the German students. The U.S. National Council of Teachers of Mathematics (NCTM) has announced its commitment to teaching data analysis and probability in grades prekindergarten to 12, as described in its *Principles and Standards for School Mathematics* (NCTM, 2000), and declared data analysis and probability its "Professional Development Focus of the Year," providing additional resources and continuing education. The NCTM prefaced its *Principles* with a simple truth: "Young children will not develop statistical reasoning if it is not included in the curriculum."

Today, the mathematics curriculum in many countries includes probability and statistics. Yet research on the effect of teaching has shown that while students can learn how to compute formal measures of averages and variability, they rarely understand what these statistics represent or their importance and connection to other concepts (Garfield & Ben-Zvi, 2007).

Few pupils learn to see a connection between statistics in school and what is going on in their world. Why do schools contribute so little to statistical literacy? We believe that there are four factors. Statistical thinking is taught

(a) too late in school,
(b) with representations that confuse young minds,
(c) with boring examples that kill motivation, and
(d) by teachers who are unversed in statistical thinking

Statistical Literacy Should Be Taught as Early as Reading and Writing An essential requirement for starting early is a discrete (not continuous) concept of probability. Children can easily understand natural numbers, whereas proportions and continuous quantities are more difficult (Butterworth, 1999; Gelman & Gallistel, 1978). Yet many mathematics educators insist that probability needs to be introduced as a continuous variable, along with continuous distributions. This theoretical vision is a major obstacle to a successful head start with statistical thinking. For instance, at a conference on teaching statistics in school, where we showed that children can easily understand statistics with discrete representations (such as the absolute number of cases, as in Figs. 5.3 and 5.8), a mathematics professor asked why the frequentistic, discrete concept of probability was being emphasized, as opposed to the subjective, continuous concept (according to which a continuous probability distribution describes a person's degree of belief in a proposition, such as that the next president of the United States will be Republican; see Savage, 1972). He seems to have been thinking about philosophical schools of probability, not about children.

In recent years, a consensus has emerged from the recommendations of professional associations (e.g., the NCTM and the German Gesellschaft für Didaktik der Mathematik) that instruction in statistics and probability should begin in primary school. This understanding is new and revolutionary, given that generations of students in the 20th century have learned statistics and probability only in their later secondary and tertiary education.

Start with Transparent Representations Teaching statistics early is not sufficient. It is also essential to represent probabilistic information in forms that the human mind can grasp. To this end, visual and hands-on material can enable a playful development of statistical thinking. For instance, tinker-cubes are Lego–like units that first graders can use to represent simple events, to combine to represent joint events, and to count to determine conditional frequencies (Kurz-Milcke, Gigerenzer, & Martignon, 2008; Kurz-Milcke & Martignon, 2007). At a later age, visualization software such as Fathom (Finzer & Erickson, 2006; www.keypress.com/x5656.xml) and TinkerPlots (Konold & Miller, 2005; www.keypress.com/x5715.xml; Biehler, Hofmann, Maxara, & Prömmel, 2006) are available for exploring and manipulating data sets (Garfield & Ben-Zvi, 2007). By starting with concrete representations of risks, children can build up confidence in understanding the basic concepts, and will less likely develop a math phobia when continuous concepts are introduced at a later point.

Consider a particularly challenging task: Bayesian inference, which is needed in medicine to derive the positive predictive value from a prior probability (e.g., the base rate of a disease) and from the sensitivity and the false-positive rate of a test (see Fig. 5.3). For decades, psychologists had concluded that even adults are doomed to fail—"man is apparently not a conservative Bayesian: he is not a Bayesian at all" (Kahneman & Tversky, 1972b, p. 450), and "our minds are not built (for whatever reason) to work by the rules of probability" (Gould, 1992, p. 469). Yet when the information is presented in natural frequencies rather than conditional probabilities, even fourth to sixth graders can reliably solve these tasks (Zhu & Gigerenzer, 2006). Computer-programmed tutorials showed that people can learn how to translate conditional probabilities into natural frequencies in less than 2 hours (Sedlmeier & Gigerenzer, 2001). Most important, learning was not only fast but also remained stable after weeks of subsequent tests, whereas students who were taught how to insert probabilities into Bayes rule (see Fig. 5.3, left side) forgot fairly quickly what they had learned (see also Ruscio, 2003). Statistical literacy is more than learning the laws of statistics; it is about representations that the human mind can understand and remember.

Teach Real-World Problem Solving, Not Applying Formulas to Toy Problems

People love baseball statistics, are interested in graphs about stock indices, have heard of probabilities of rain, worry about the chance of a major earthquake, and are concerned about cholesterol and blood pressure. How safe is the contraceptive pill? What is the error margin for polls and surveys? Is there a probability that extraterrestrial life exists? Personal relevance is what makes statistics so interesting.

To build up motivation, curricula should start with relevant everyday problems and teach statistics as a problem-solving method. However, in most curricula, statistics is taught as a formal mathematical discipline, where problems are purely decorative. One begins with a law of probability and then presents problems that can be safely answered by this law—which is why the use of randomizing devices such as coins, dice, and urns abound. Even when a textbook gives itself an applied feel, the content is more often than not only secondary. This approach leads to a continuous stream of more or less boring examples that do their best to kill young people's curiosity and motivation.

Is lack of motivation the reason students learn so little about statistics? The sparse evidence available suggests that the answer is no (Martignon & Wassner, 2005). Forty mathematics teachers who taught at German Gymnasien (grades 5–13) were asked to rate their students' interest, attentiveness, motivation, and comprehension when being taught probability and statistics compared to the rest of mathematics education. Many teachers reported that their students were more interested, attentive, and motivated when being taught probability and statistics than they were when being taught other types of mathematics (Fig. 5.15). Yet, strikingly, this did not lead to better comprehension. We believe that this dissociation can largely be overcome by beginning with real-world problems and transparent representations, and recently textbooks have incorporated these principles from psychological research (Gigerenzer, 2002). For instance, one secondary school textbook (Jahnke & Wuttke, 2005) introduces Bayes rule with the real story of a 26-year-old single mother who tested positive in a routine HIV test at a Virginia hospital, lost her job, moved into a halfway house with other HIV-positive residents, had unprotected sex with one of them, eventually developed bronchitis, and was asked by her new doctor to take the HIV test again. She did, and the result was negative, as was her original blood sample when it was retested. The poor woman had lived through a nightmare because her physicians did not understand that there are false alarms even when both the ELISA and the Western blot test are positive. After hearing this example, the students are given the relevant information in natural frequencies and can compute that the positive predictive value of the two tests combined was only about 50%, not 100% as the original physicians had assumed. Here, students are taken from a real and gripping problem to statistical thinking. As a next step, they can learn where to find the relevant information themselves and how to ask questions about the assumptions for applying statistical principles to the real

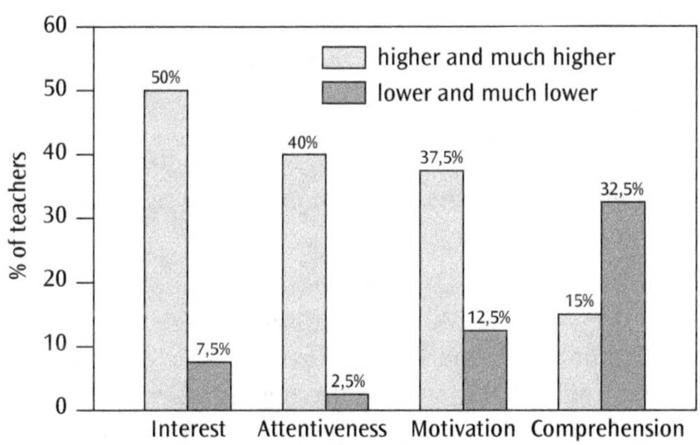

Figure 5.15: Mathematics teachers' judgments about students' attitudes to and comprehension of probability and statistics compared to the rest of mathematics. For instance, 50% of the teachers said that students' interest was higher for probability and statistics, 7.5% estimated it as lower, and the others as equal. Note the discrepancy between interest, attentiveness, and motivation on the one hand, and comprehension on the other (Martignon & Wassner, 2005).

world—questions that do not arise when only toy problems (e.g., involving cards or dice) are used.

Statistical literacy demands rethinking the teaching of statistics. Is mathematical statistics an end in itself or a useful tool for solving problems? In our opinion, to be helpful for patients, physicians, and journalists alike, it should be taught as a disciplined problem-solving technique. One great disappointment of motivated students is when they find out that school statistics has little to do with their own world.

Teach Teachers First Studies on preservice and in-service K–12 teachers suggest that both groups have troubles in understanding and teaching statistics (for an overview, see Garfield and Ben-Zvi [2007]). For instance, elementary school teachers have difficulties in finding out the median of data sets presented graphically (Bright & Friel, 1998). As long as teachers themselves do not understand, they are likely to resist. Similar to what occurs in medical training, resistance to statistics education is rarely articulated openly in print but is indirectly present through the lack of support for its actual attainment. Unexpressed concerns can be detrimental to an undertaking. For this reason, we are making an effort to explicate four major concerns of teachers, beginning in the early grades and continuing on through the middle ones:

- Concern #1: *There are simply more important things in the elementary math curriculum; in other words, something else would suffer from it.*
- Concern #2: *Statistics is about games of chance and touches upon content that is simply not appropriate for children in the elementary grades.*
- Concern #3: *We experience difficulties teaching probability and statistics to high school and even college students, let alone to students in elementary school.*
- Concern #4: *In spite of my education as a math teacher, I know very little about data analysis, probability, and teaching in this area of mathematics.*

The first concern is one of mathematics educators who do not seem to realize that statistical thinking is indispensable. It also reflects the hierarchy within the mathematics profession, with abstract mathematics at the top and empirical statistics at the bottom. In our view, the traditional emphasis on Latin as a foreign language in schools provides an apt comparison. After 4 years of Latin, although students showed improved skill in grammar-related activities, such as letter-exact reading and forming complex sentences, they did not learn a modern Romance language (e.g., Spanish) more easily than did a group lacking proficiency in Latin (Haag & Stern, 2003). Learning the mathematics of certainty cannot be assumed to simply transfer to readily learning statistics and probability, nor can it be assumed to be more important.

The second concern is very peculiar to statistics education. Historically, games of chance were an early topic of probability theory, but not the only one, the others being the analysis of mortality tables for insurance and the evaluation of the reliability of testimony in court (Daston, 1988). Yet the connection with games of chance can evoke moral protest. In the 1980s, Israeli psychologist Ruma Falk devised a hands-on probability game in which young children could develop their intuitions. Children had to choose one of two disks (like two roulette wheels) to spin before making a move on a game board. Each disk was divided into sectors of two colors, one color favorable and one unfavorable. The challenge was to identify and spin the disk with the higher probability of a favorable outcome.

> The game was sharply criticized by parents and educators as being "uneducational." They objected to the notion of a game in which one might make a correct choice (of the disc with a higher probability of success) and yet obtain an unfavorable outcome, while on the other hand, an incorrect decision may be rewarded. Obviously, they wished for a consistent, "just" system. Implied in their criticism was the expectation that good decisions would always be reinforced, while bad ones would never be.
>
> (Falk & Konold, 1992)

This concern involves a double misunderstanding. Statistics is not only about games of chance but about health and other everyday issues as well. And real life is not always fair in every instance, even if it hopefully is in the long run.

The third and fourth concerns need to be addressed in teacher training. A radical solution would be to take teaching of statistical thinking out of the hands of mathematics teachers and turn it into a problem-solving field. Such a new field could be called "statistical reasoning" and might help young people make better decisions about health, drugs, alcohol use, driving, biotechnology, and other relevant issues. This teaching revolution is related to Moore's (1997) "new pedagogy" designed to overcome the "professional fallacy" that introductory courses are a step in the training of formal statisticians.

How Can Primary and Secondary School Contribute to Statistical Literacy? We recommend that primary and secondary schools begin teaching statistical thinking as a problem-solving discipline in its own right, not as an appendage to math education. In this way, a majority of citizens could reach minimal or even higher levels of statistical literacy. With this basic knowledge, patients, physicians, and journalists would no longer be as easily confused by numbers, which could directly impact on some of the other causes mentioned in Part V. Statistical thinking as a problem-solving discipline puts the solution of individual and social problems first, using statistical tools as a means toward that end. The goals of this discipline include the following:

- To learn that societal problems can be solved by critical thinking instead of mere belief, trust in authority, or violence
- To develop empirical thinking by formulating competing hypotheses and collecting and analyzing data to test them
- To develop critical thinking skills in evaluating the applicability of various statistical models to real-world problems
- To learn to use transparent representations and computer-based visualization techniques

Teaching statistical thinking as problem solving can be directly connected to teaching health in schools. Steckelberg, Hülfenhaus, Kasper, Rost, and Mühlhauser (2007, 2009) developed a curriculum and a test of critical health literacy for grade 11 secondary school students, both as a 1-week project and over a longer period. The curriculum contains six modules, ranging from recognizing fallacies and misinterpretations of data representations to designing experiments to understanding systematic reviews to appraising patient information. The curriculum was well accepted by students, who perceived it as personally beneficial, and increased their competence in health literacy.

Teach Statistical Literacy in Medical Training

As described in the previous section, not until in the late 20th century did medical schools begin to teach statistics, and there are still medical organizations, physicians, and students who tend to see statistics as inherently mathematical and clinically irrelevant for the individual patient (Altman & Bland, 1991; Gigerenzer, 2002). This attitude is reinforced by curricula focusing on analysis of variance and multiple regression techniques; transparent risk communication is rarely recognized as an essential part of medical training and is not part of the general medical curriculum in Germany and the United States. To check whether there have been any changes, we contacted the Association of American Medical Colleges (AAMC), the national association that accredits U.S. medical schools, and asked if there "are any ongoing AAMC initiatives addressing numeracy (sometimes called 'statistical literacy') in medical school education?" The answer was "There are currently no AAMC initiatives in this area."

Statisticians have long criticized the fact that many introductory statistics texts in medicine are not written by experts on statistics and, furthermore, that this lack of expertise is even sold as a strength, as the renowned British statistician Michael J.R. Healy noticed:

> I do not know a single discipline other than statistics in which it is a positive recommendation for a new text book, worthy of being quoted

> on the dust cover, that it is not written by a specialist in the appropriate field. Would any medical reader read, would any medical publisher publish, my new introduction to brain surgery—so much simpler and more clearly written than those by professional brain surgeons, with their confusing mass of detail? I trust not.
>
> (Healy, 1979, p. 143)

As a result, some textbooks contain gross errors (see Altman & Bland, 1991; Eddy, 1982; Schönemann, 1969). Errors in textbooks and journals include confusion of conditional probabilities, as when equating the positive predictive value with the sensitivity, or the *p* value with the probability that the null hypothesis is correct. These errors, however, also have a long history in psychology (Gigerenzer, 2004).

Yet it is important to go beyond this common critique. A curriculum with standard statistical techniques does not guarantee understanding health statistics, as we demonstrate in Part III. In contrast, teaching medical students transparent representations does foster understanding (Hoffrage, Gigerenzer, Krauss, & Martignon, 2002; Kurzenhäuser & Hoffrage, 2002). We believe that statistical literacy is more important for clinical practice than specific statistical techniques are (Appleton, 1990). In the end, medical schools need to ensure that every graduate has minimal statistical literacy, if not a more advanced understanding.

Transparency

With the spread of democracies in the last century, transparency has become as highly valued as free speech and free press, for instance when fighting against corruption or for public access to disclosed information. The Vienna philosopher and political economist Otto Neurath (1882–1945) is one of the fathers of this social movement, who in the 1920s and '30s developed a strikingly beautiful symbolic way to represent economic facts to the largely uneducated Viennese public. This method allowed everyone to understand statistics in a "blink of an eye" by using pictorial representations called "isotypes" that conform to the psychology of vision (e.g., Neurath, 1946). Neurath's isotypes have not yet been adapted to health statistics, but various graphic representations are in use (Elmore & Gigerenzer, 2005; Galesic, Garcia-Retamero, & Gigerenzer, 2009; Paling, 2003; Kurz-Milcke et al., 2008; Lipkus, 2007; Schapira, Nattinger, & McHorney, 2001). Here we focus on transparent tables and numbers (see also Fagerlin, Ubel, Smith, & Zikmund-Fisher, 2007; Peters, Hibbard, Slovic, & Dieckmann, 2007).

Numbers, Not Only Words An important response to statistical illiteracy is to give the public more numbers. Patients have a right to learn how big benefits and harms of a treatment are. Qualitative risk terms are notoriously unclear. There are attempts to standardize verbal expressions, such as the

EU guideline for drug labels and package leaflets, where specific terms are defined for frequency intervals. However, people seem to overestimate the frequencies of side effects based on those labels (Steckelberg, Berger, Köpke, Heesen, & Mühlhauser, 2005). Moreover, terms such as "unlikely" are interpreted differently from context to context. For example, more severe side effects are estimated to occur less frequently than less severe side effects described by the same qualitative term (Fischer & Jungermann, 1996). Patients tend to overestimate risks when disclosed verbally, and are less likely to comply if information is given numerically (Young & Oppenheimer, 2006). For both written and verbal information, patients had a more accurate perception of risk when it was numerical as opposed to verbal (see the review by Trevena, Davey, Barratt, Butow, & Caldwell, 2006). Therefore, risk should always be specified numerically.

Contrary to popular belief, studies report that a majority of patients do prefer numerical information to care only (Hallowell, Statham, Murton, Green, & Richards, 1997; Wallsten, Budescu, Zwick, & Kemp, 1993). Some studies have addressed differences between patients who do and do not prefer to see numbers. For instance, men who prefer to communicate with their physicians in words only ("no numbers, please") more often also prefer early aggressive surgery for prostate cancer over watchful waiting (Mazur, Hickam, & Mazur, 1999).

Data Tables: Drug Facts Boxes While tables are routinely used to communicate data in scientific articles, there seems to be a hesitancy to use them in communicating with the general public. But tables are a practical way to look at and compare a series of numbers. To be efficient, such a table should be simple—that is, focus on the relevant information (see Fig. 4.1). We have developed a one-page summary of drug information at the heart of which is a study-findings table summarizing the benefit and side effect data from trials used in the Food and Drug Administration's (FDA's) drug approval process (Schwartz, Woloshin, & Welch, 2007). Compare the drug box on tamoxifen (Table 5.10) with the original advertisement (Fig. 5.16).

The table format provides a structure for readers to help them think about drug performance. By being given data outcomes side by side, readers are reminded that understanding an effect entails comparing what would happen with and without the drug. Similarly, presented with information about benefit and harm on the same page, readers are reminded that judging whether a drug is "worth it" means comparing good and harmful effects. Benefit needs to be judged in the context of harm, and vice versa. A small benefit may not be seen as sufficient if there are significant harms. Alternatively, significant harms may be tolerable in the context of substantial benefit. Another positive effect of presenting data symmetrically (i.e., providing absolute event rates for outcomes with and without the drug) is that information about benefit and harm is given equal weight: The numerical information is given in both percentages and frequencies. We have tested the drug box in two studies and both have demonstrated that people (even

Table 5.10: Drug Facts Box Summarizing Benefits and Side Effects of a Drug So That Comparison Is Made Easy

Prescription Drug Facts: NOLVADEX (Tamoxifen)

What is this drug for?	Reducing the chance of getting breast cancer
Who might consider taking it?	Women at high risk of getting breast cancer (1.7% or higher risk over 5 years). You can calculate your breast cancer risk at http://bcra.nci.nih.gov/btc
Who should *not* take it?	Women who are pregnant or breastfeeding
Recommended testing	Have a yearly checkup that includes a gynecological examination and blood tests
Other things to consider doing	No other medicines are approved to reduce breast cancer risk for women who have not had breast cancer

NOLVADEX Study Findings

13,000 women at high risk of getting breast cancer were given either NOLVADEX or a sugar pill for 5 years. Here's what happened:

What difference did NOLVADEX make?	Women given a sugar pill	Women given NOLVADEX (20 mg a day)
Did NOLVADEX help?		
Fewer women got invasive breast cancer (16 in 1,000 fewer due to drug)	3.3% 33 in 1,000	1.7% 17 in 1,000
No difference in dying from breast cancer	About 0.09% in both groups or 0.9 in 1,000	
Did NOLVADEX have side effects?		
Life-threatening side effects		
More women had a blood clot in their leg or lungs (additional 5 in 1,000 due to drug)	0.5% 5 in 1,000	1.0% 10 in 1,000
More women got invasive uterine cancer (additional 6 in 1,000 due to drug)	0.5% 5 in 1,000	1.1% 11 in 1,000
No difference in having a stroke	About 0.4% in both groups or 4 in 1,000	
Symptom side effects		
More women had hot flashes (additional 120 in 1,000 due to drug)	68% 680 in 1,000	80% 800 in 1,000
More women had vaginal discharge (additional 200 in 1,000 due to drug)	35% 350 in 1,000	55% 550 in 1,000
More women had cataracts needing surgery (additional 8 in 1,000 due to drug)	1.5% 15 in 1,000	2.3% 23 in 1,000
Bottom Line		
No difference in deaths from all causes combined	About 1.2% in both groups or 12 in 1,000	

How long has the drug been in use?

NOVALDEX *was first approved by the FDA in 1982.* Studies show that most serious side effects or recalls of new drugs happen during their first 5 years of approval.

From Schwartz, Woloshin, & Welch (2007).

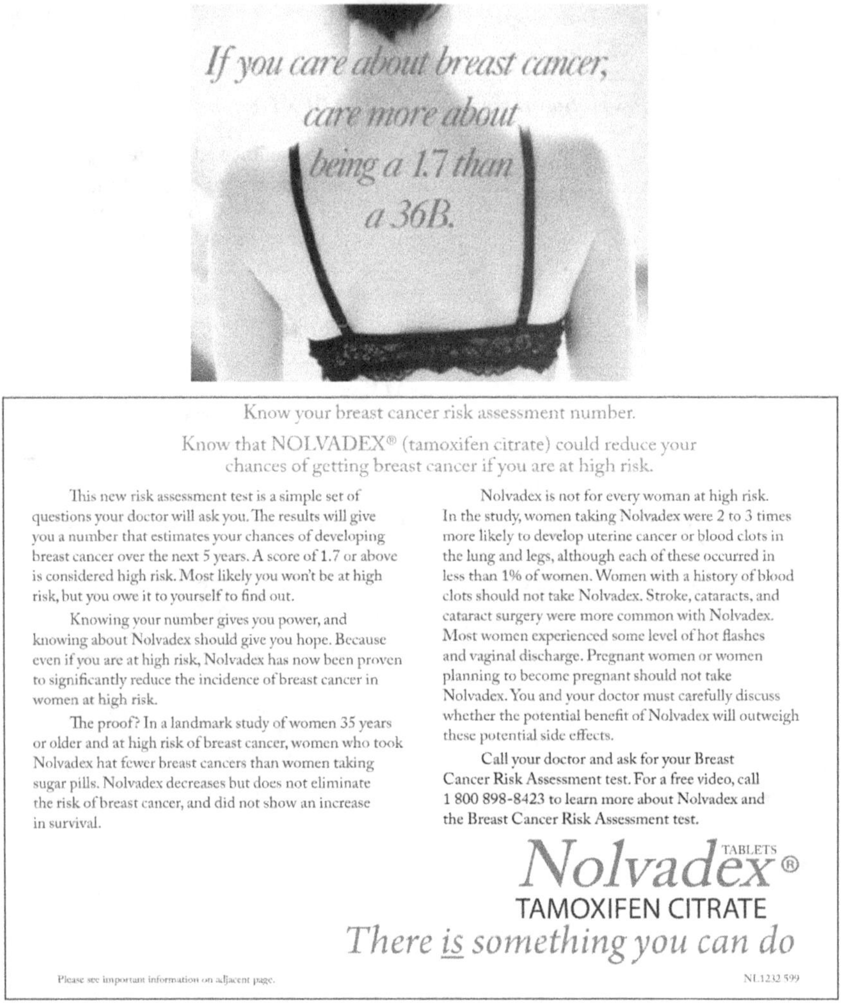

Figure 5.16: The original image and text of the Nolvadex (tamoxifen) advertisement (compare to the drug facts box Table 5.10).

those with lower educational attainment) like it, think the data are valuable, and, most importantly, can understand information presented (Woloshin, Schwartz, & Welch, 2004; Schwartz, Woloshin, & Welch, 2007). We hope that such tables can become a routine element in communicating data to the public.

Transparent Numbers In our final section, we summarize transparent and nontransparent ways to communicate health statistics (Table 5.11). They are arranged in pairs, with definitions and examples provided. In the literature, one sometimes finds a general distinction between probability

Table 5.11: Some Confusing and Transparent Representations of Health Statistics

Confusing Representation	Transparent Representation
Single-event probabilities	*Frequency statements*
Definition: A probability that refers to an individual event or person, as opposed to a class of events or people, is called a single-event probability. In practice, single-event probabilities are often expressed in percentages, and occasionally as "X chances out of 100," rather than as a probability ranging between 0 and 1.	Definition: A frequency states the risk in relation to a specified reference class.
Example: "If you take Prozac, the probability that you will experience sexual problems is 30% to 50% (or: 30 to 50 chances out of 100)."	Example: "Out of every 10 of my patients who take Prozac, 3 to 5 experience a sexual problem."
Relative risks	*Absolute risks*
Definition: A relative risk is a ratio of the probabilities of the event occurring in one group (usually the treatment group) versus another group (usually the control group). The relative risk reduction of the treatment is calculated as 1 minus the relative risk: $\text{Relative risk reduction} = 1 - \frac{P_{\text{treatment}}}{P_{\text{control}}}$	Definition: The absolute risk in both the treatment and the control groups is simply the corresponding baseline risk. The absolute risk reduction is calculated by subtracting the absolute risk in the treatment group from the absolute risk in the control group: $\text{Absolute risk reduction} = P_{\text{control}} - P_{\text{treatment}}$
Example: "Mammography screening reduces the risk of dying from breast cancer by about 20%."	Example: "Mammography screening reduces the risk of dying from breast cancer by about 1 in 1,000, from about 5 in 1,000 to about 4 in 1,000."
Survival rates	*Mortality rates*
Definition: The survival rate is the number of patients alive at a specified time *following diagnosis* (such as after 5 years) divided by the number of patients diagnosed.	Definition: The mortality rate is the number of people in a group who die annually from a disease, divided by the total number of people in the group.
Example: "The 5-year survival rate for people diagnosed with prostate cancer is 98% in the USA vs. 71% in Britain."	Example: "There are 26 prostate cancer deaths per 100,000 American men vs. 27 per 100,000 men in Britain."
Conditional probabilities	*Natural frequencies*
Definition: A conditional probability $p(A \mid B)$ is the probability of an event A given an event B.	Definition: A class of N events (persons) is subdivided into groups by two binary variables. The four resulting joint frequencies are called natural frequencies. Note that these are "raw counts" that sum up to N, unlike relative frequencies or conditional probabilities that are normalized with respect to the base rates of the event in question. Generalization to more than two variables and variable values are straightforward.
Example: See Figures 5.3 and 5.8.	Example: See Figures 5.3 and 5.8.

format and frequency format. Yet there are different kinds of probabilities and frequencies, and some are less confusing than others (Brase, 2002, 2008; Gigerenzer & Hoffrage, 1995). For instance, an unconditional probability statement that specifies a reference class is clear ("The probability that 50-year-old American women will die of colon cancer in the next 10 years is 2 in 1,000"), whereas conditional probabilities tend to confuse ("the probability of colon cancer given a positive screening test" is often mistaken for "the probability of a positive screening test given colon cancer"). Table 5.11 distinguishes various kinds of probability and frequency representations.

Use frequency statements, not single-event probabilities. One non-transparent representation we have not discussed so far is a single-event probability statement. It is defined as a statement in which a probability refers to a singular person or event rather than to a class. A good illustration is weather prediction: "There is a 30% probability of rain tomorrow" is a single-event probability. By definition, no reference class is mentioned, but since people tend to think in terms of classes, misunderstanding is inevitable. Some citizens believe the statement to mean that it will rain tomorrow 30% *of the time,* others that it will rain in 30% *of the area*, or that it will rain on 30% *of the days* for which the announcement was made (Gigerenzer, Hertwig, van den Broek, Fasolo, & Katsikopoulos, 2005). The ambiguity of the reference class—time, area, or days—can be avoided by making a frequency statement, such as "it will rain on 30% of the days."

Similarly, when in clinical practice a physician tells a patient: "If you take Prozac, you have a 30 to 50% chance of developing a sexual problem, such as impotence or loss of interest," this single-event statement invites misunderstanding. As in the case of probabilities of rain, confusion will mostly go unnoticed. After learning of this problem, one psychiatrist changed the way he communicated the risk to his patients from single-event statements to frequency statements: "Out of every 10 patients who take Prozac, 3 to 5 experience a sexual problem." Psychologically that made a difference: Patients who were informed in terms of frequencies were less anxious about taking Prozac. When the psychiatrist asked his patients how they had understood the single-event statement, it turned out that many had thought that something would go awry in 30 to 50% of their sexual encounters (Gigerenzer, 2002). The psychiatrist had been thinking of all his patients who take Prozac, whereas his patients thought of themselves alone. Several studies have shown systematic differences in the interpretation of single-event and frequency statements (Gigerenzer, Hoffrage, & Kleinbölting, 1991; Slovic, Monahan, & MacGregor, 2000; Tan et al., 2005).

Use absolute risks, not relative risks. There exist several reviews of studies comparing relative risks with absolute risks (Covey, 2007; Edwards, Elwyn, Covey, Matthews, & Pill, 2001; McGettigan, Sly,

O'Connell, Hill, & Henry, 1999; Moxey, O'Connell, McGettigan, & Henry, 2003). The common finding is that relative risk reductions lead people to systematically overestimate treatment effects. Why are relative risks confusing for many people? As mentioned before, this statistic is mute about the baseline risks (in Table 5.11: from 5 to 4 in 1,000) and the absolute effect size (1 in 1,000). Moreover, when patients hear about a 20% risk reduction, they are likely to think that this percentage refers to people like themselves, such as people who participate in cancer screening. Yet it refers to the baseline of people who do not participate in screening and die of cancer.

Use mortality rates, not survival rates. Until recently, there were no studies on whether patients or physicians understand that in screening, higher survival rates do not necessarily mean longer life. Two such studies now exist. About 70–80% of a representative sample of U.S. primary care physicians and an equal proportion of German physicians wrongly confused survival with mortality, just as Giuliani did (Wegwarth, Gaissmaier, & Gigerenzer, 2011; Wegwarth, Schwartz, Woloshin, Gaissmaier, & Gigerenzer, 2012).

Use natural frequencies, not conditional probabilities. Estimating the probability of disease given a positive test (or any other posterior probability) is much easier with natural frequencies than with conditional probabilities (sensitivities and specificities). Note that this distinction refers to situations where *two* variables are considered: Natural frequencies are *joint* frequencies, as shown in Figures 5.3 and 5.8. Gigerenzer and Hoffrage (1995, 1999) showed that natural frequencies—but not relative frequencies—facilitate judgments. This fact has been repeatedly misrepresented in the literature, where our thesis is often held to be that *all* frequency representations improve judgments (see Hoffrage et al., 2002).

Caution It should be noted that providing people with accurate, balanced, accessible data on disease risk and treatment benefit could have an untoward side effect. People may be very surprised about how small many of the risks and benefits are. Consequently, they may dismiss as unimportant interventions that physicians see as extremely valuable. For example, in one of our studies (Woloshin, Schwartz, & Welch, 2004), participants were very optimistic about the effectiveness of three different drugs; in each case, these perceptions dropped substantially after seeing the actual data. The effect, however, was similar for all drugs. This is concerning, since one of the drugs, a statin used to treat men with high cholesterol but no prior myocardial infarction, showed a reduction of overall mortality over 5 years from 4 in 100 to 3 in 100 patients. We suspect that many respondents did not appreciate the real magnitude of this effect: Few drugs now being manufactured can match this reduction in all-cause mortality among relatively healthy outpatients. To truly judge how well a drug (or other intervention) works, people need a context—that is, some sense of the magnitude of the benefit of other interventions. Undoubtedly, most people lack such knowledge and overestimate the benefits of drugs. We believe that reactions to benefit data will change as people have more exposure to them; that is, as consumers become better calibrated to effect sizes, they will be better able to discriminate among drugs and interventions. It is important to provide this context to make sure consumers do not discount small but important effects.

Reference Class and Transparency Much of the mental confusion that defines nontransparency seems to be caused by the reference class to which a health statistic applies (Gigerenzer & Edwards, 2003). Single-event probabilities specify by definition no class of events, and relative risks often refer to a reference class that is different from the class people are thinking of. Sensitivities and specificities are conditional on two different reference classes (patients with disease and patients without disease), whereas natural frequencies all refer to the same reference class (all patients). And survival and mortality rates crucially differ in their denominator—that is, the class of events they refer to. Clarity about the reference class to which a health statistic refers is one of the central tools in attaining health literacy.

VII. The Dream of Statistical Literacy

Two millennia separated the Athens of Aristotle and the Paris of Claude Bernard, but the two men shared one article of faith: Science is about causes, not chances. Not until 1654, when the French mathematicians Blaise Pascal and Pierre Fermat exchanged letters on gambling problems, did mathematical probability arrive on the scene. This curiously late appearance was christened "the scandal of philosophy" by philosopher Ian Hacking (1975). In the following centuries, the "probabilistic revolution" (Krüger, Gigerenzer, & Morgan, 1987) changed science and everyday life, beginning slowly but resulting in enormous transformations. It turned deterministic physics into statistical mechanics and quantum theory, changed biology by introducing Darwinian variation and random drift, and redefined the nature of scientific experiments by introducing repetition and randomization. Yet this revolution in thought has not yet reached patients and physicians in their understanding of health statistics.

We hope that this monograph stimulates researchers to contribute to solving the problem of collective statistical illiteracy and to develop and implement efficient and transparent representations of health statistics. Nonetheless, the dream of statistical literacy is of a broader scope and is fundamental to a functioning democracy. It embodies the Enlightenment ideal of people's emergence from their self-imposed immaturity. In Kant's (1784) words, "Dare to know!"

Acknowledgments

We are grateful to Adrian Barton, Klaus Eichler, Mirta Galesic, Ulrich Hoffrage, Julian Marewski, Jutta Mata, Ingrid Mühlhauser, and Odette Wegwarth for their comments.

Chapter 6

Public Knowledge of Benefits of Breast and Prostate Cancer Screening in Europe

Women and men in countries with modern health systems are confronted with the question of whether to participate in screening for breast cancer and prostate cancer. Yet, because screening can also lead to harms such as overtreatment, they need to understand the potential benefits of these screening programs before they can make informed decisions about participating. Ideally, physicians, health pamphlets, and other information sources should assist in clarifying the actual size of benefits.

Screening for breast cancer with mammography is widely encouraged by governmental programs in both the European Union (EU) and the United States under the assumption that the screening programs save lives. In the case of breast cancer, an analysis of randomized trials with some 247,000 women aged 40–74 years showed that for every 1,000 women who participated in screening, 3.9 diagnosed with breast cancer died, compared with 5.0 among those who did not participate (Nyström, 2002). The follow-up time ranged between 5.8 and 20.2 years. Thus, the absolute risk reduction was on the order of one in 1,000 (Humphrey, Helfand, Chan, & Woolf, 2002). The authors of a review of six trials involving half a million women estimated the absolute risk reduction to be approximately one in 2,000 (Gøtzsche & Jørgensen, 2013). Note that this benefit relates to fewer breast cancer deaths; no reduction in mortality from all cancers or other causes was found. Whether the potential of screening to reduce breast cancer mortality outweighs the harms of overdiagnosis and overtreatment is still under discussion (Elmore, Barton, Moceri, Polk, Arena, & Fletcher, 1998; Schwartz, Woloshin, Fowler Jr, & Welch, 2004; Welch, 2004).

Screening for prostate cancer with prostate-specific antigen (PSA) tests, although often encouraged by physicians and health information pamphlets,

Originally published as Gigerenzer, G., Mata, J., & Frank, R. (2009). Public knowledge of benefits of breast and prostate cancer screening in Europe. *Journal of the National Cancer Institute, 101*, 1216–1220. This chapter has been slightly updated.

is not part of governmental screening programs and is recommended by few medical organizations. The evidence for any benefit of screening is limited. The U.S. Preventive Services Task Force (2012) reviewed the available studies and recommended against PSA screening for prostate cancer. A European randomized trial reported a prostate cancer-specific mortality reduction of one in 781 after 13 years (Schröder, Hugosson, Roobol et al., 2014), but a randomized trial in the United States found no reduction after 7 or 10 years (Andriole, Grubb, Buys, et al., 2009). Thus, the best estimate seems to be a reduction of death from prostate cancer of zero or one for every 1,000 men screened, and the evidence is insufficient to determine whether the benefits outweigh the harms, such as incontinence through overtreatment of nonprogressive cancers (Andriole, Grubb, Buys, et al., 2009; Concato, Wells, Horowitz, et al., 2006).

This study addresses two main questions: 1) Do women and men have realistic knowledge about the benefits of mammography and PSA screening, respectively? and 2) What information sources do they rely on? Here, we also addressed a related question: Does the frequency of consulting a given source improve understanding of benefits? To our knowledge, this is the first European survey of women's and men's perceptions of the benefits of mammography and PSA screening, and the information sources that they rely on, with representative samples of the general population.

We conducted a survey of the public's knowledge of the benefits of screening in eight countries of the EU and the European part of Russia. The eight EU countries include about 75% of people in the 27 EU countries and have a total population of about 500 million. The European part of Russia has a population of about 106 million out of a total of 143 million Russians. The percentage of women who have had mammography is 57 in Germany, 78 in France, 76 in Austria, 85 in the Netherlands, 66 in Italy, 75 in the United Kingdom, 52 in Spain, 47 in Poland (for women aged 45–54 years), and 19 in Russia (Binkowska & Debski, 2005; World Health Organization, 2008). PSA screening programs do not exist in the nine countries, apart from a regional state-funded program in Tyrol, Austria. National health systems are predominantly financed by taxes in the United Kingdom, Italy, and Poland and by contributions to social health insurance in Austria, France, Germany, and the Netherlands.

The data were collected as part of the European Consumer Study 2007 conducted between September and December 2006 by the Gesellschaft für Konsumforschung (GfK)-Nürnberg Group (GfK-Nürnberg & Frank, 2007). Participants within each country were selected according to a quota method based on the official statistics concerning five variables: region, size of household, sex, profession, and age (Särndal, Swensson, & Wretman, 1992). The population in each country was first segmented into subgroups based on these five criteria, and within each subgroup, subjects were sampled in proportion to their distribution in the entire country. Initial contacts were made by telephone; the interviews were conducted in the participants' homes. Consistent with earlier representative quota sampling surveys conducted by the GfK Group, across all countries, about 60% of initial phone contacts resulted in a complete interview; in the remaining cases, sampling was continued until the quotas were met. Across all countries, the age distribution of participants was as follows: 14–19 years (8.4%), 20–29 years (16.6%), 30–39 years (18.0%), 40–49 years (18.4%), 50–59 years (15.2%), 60–69 years (11.8%), and 70 years and older (11.5%). The total number of interviews was 10,228, with 2,054 in Germany and 2,019 in Russia (the countries with the largest populations); 1,005 in France, 1,042 in the United Kingdom, 1,007 in Italy, 1,019 in Poland, and 1,024 in Spain; and 501 in Austria and 557 in the Netherlands (the two countries with the smallest populations). Participants were questioned in face-to-face personal interviews with computer assistance, except in Russia, where for security reasons, interviewers used paper and pencil. Using personal interviews avoided some of the problems of telephone interview methods, such as excluding poorer households without telephones and hence introducing a bias in comparisons between countries.

As a measure of the perceived benefit of mammography screening, we focused on cancer-specific mortality reduction because this is the end point typically communicated to the public (as opposed to total mortality reduction, for example). Women were questioned as follows:

> "1,000 women age 40 and older from the general population participate every two years in screening for breast cancer with mammography. After 10 years, the benefit is measured. Please estimate how many fewer women die from breast cancer in the group who participate in screening compared to women who do not participate in screening."

The response alternatives were 0, 1, 10, 50, 100, 200 (out of 1,000), and "I don't know." For the perceived benefit of PSA screening, men were questioned similarly:

> "1,000 men age 50 and older from the general population participate every two years in screening for prostate cancer with PSA tests. After 10 years, the benefit is measured. Please estimate how many fewer men die from prostate cancer in the group who participate in screening compared to men who do not participate in screening."

The response alternatives were the same as those used for breast cancer screening.

To measure the frequency of information sources used, we asked participants how often they used each of 14 sources that were divided into four categories as follows: family and/or friends (considered both a source and a category), experts (general practitioner and pharmacist), general media (television, popular magazines, daily newspapers, and radio), and health-specific sources (pamphlets by health organizations, reference books, health insurance, Internet, consumer counseling, patient counseling, and self-help organizations). The response alternatives were never, rarely, sometimes, frequently, and don't know.

We calculated the proportion of best estimates of screening benefits for all countries, all age groups, and for the group of citizens aged 50–69 years who are targeted by the screening campaigns. The proportion of participants reporting use of sources of health information was calculated for all countries, all age groups, and all of the 14 sources. Correlation coefficients between frequency of use of particular sources of health information and estimates of screening benefits were calculated. For mammography screening, overestimation of benefit was defined as the difference between the estimated benefit (expressed in X out of 1,000 women) and one out of 1,000. For instance, if the estimate was 50 in 1,000, the overestimation was 49 in 1,000. A positive correlation means the higher the reported frequency of use, the larger the overestimation. For PSA screening, the same procedure was used except that estimates of 0 were not scored as underestimation, but 0 and 1 in 1,000 were considered equally accurate. The correlations between overestimation and frequency of use of particular sources did not include participants who answered the question concerning the benefit of screening with "don't know" (Table 6.1 shows the frequency of these responses).

Box 6.1 Context and Caveats

Prior Knowledge

Given the harms that can ensue from cancer screening procedures, people's decisions as to whether to undergo cancer screening should be based on a realistic knowledge of its benefits.

Study Design

Face-to-face-interviews were conducted among a representative sample of men and women in nine European countries, who were asked to choose among estimates of the number of fewer cancer-specific deaths (per 1,000 individuals screened) by prostate-specific antigen and mammography screening, respectively. Participants were also queried as to their sources of medical information.

Contribution

This study found dramatic (by an order of magnitude or more) overestimation of the benefits (absolute cancer-specific mortality reduction) of mammography and prostate-specific antigen testing in the vast majority of women and men, respectively, in all countries surveyed. Frequent consultation of sources of medical information (including physicians) was not associated with more realistic knowledge of the benefits of screening.

Implications

A basis for informed decisions by people about participation in screening for breast and prostate cancer is largely nonexistent in Europe, suggesting inadequacies in the information made available to the public.

Limitations

The influence of the public's overestimation of screening benefits on actual participation in screening was not addressed in this study, and the work was restricted to European countries.

From the Editors

Table 6.1: Estimated Reduction of Breast Cancer Mortality through Regular Participation in Mammography Screening (Women Only)*

Reduction Out of 1,000	Percentage of Responders									
	Mean	Germany	France	Austria	The Netherlands	Italy	United Kingdom	Spain	Poland	Russia
None	6.4	1.4	0.8	2.4	0.7	5.3	2.0	3.9	4.2	16.1
1	1.5	0.8	1.3	2.9	1.4	1.3	1.9	2.7	0.8	1.7
10	11.7	12.8	15.7	11.0	10.7	10.6	10.3	6.9	9.7	12.4
50	18.9	21.3	21.7	22.1	22.6	17.4	13.9	11.7	20.5	20.1
100	15.0	16.8	21.5	20.8	22.5	13.9	17.0	11.3	14.8	10.8
200	15.2	13.7	23.7	11.0	20.1	15.2	26.9	15.7	17.1	6.8
Don't know	31.4	33.1	15.3	29.8	22.1	36.3	28.0	48.0	32.9	32.1

*Question: How many fewer women die from breast cancer in the group who participate in screening, compared to women who do not participate in screening? Mean across all nine countries is weighted by sample size.

Among all participants, only 1.5% of women (range across different countries: 0.8%–2.9%) chose the best estimate for reduction in mortality due to breast cancer screening, that is, one woman for every 1,000 screened (Table 6.1). Four times as many women answered that the benefit was zero, and 92.1% overestimated the benefit by at least one order of magnitude or answered that they did not know; this proportion was higher (95.9%) in the eight EU countries due to the large proportion of no-benefit estimates in Russia. The greatest overestimation was observed in France, the Netherlands, and the United Kingdom, where more than 40% of the women answered that the reduction in mortality was 100 or 200 women per 1,000 screened; in the United Kingdom, almost 27% chose the highest figure. These three countries also had high participation rates in mammography screening. In Russia, where the availability of mammography equipment is limited (Rozhkova & Kochtetova, 2005), the percentage of women who exhibited overestimation or did not know was the lowest of the countries surveyed, 82%.

Some of the women included in our study were younger than women targeted by screening programs and may have had little motivation to inform themselves about screening. However, in every country, the percentage of women who gave the best estimate was lower among those aged 50–69 years and thus targeted by screening programs than among women younger than 50 years, and in every country but Russia, the proportion of 50- to 69-year-old women giving the best estimate was smaller than in all other age groups.

In all countries surveyed, only 10.7% of men made reasonable estimates of the benefits of prostate cancer screening (ie, deaths from prostate cancer prevented for every 1,000 men screened were less than or equal to one, Table 6.2); 89.3% overestimated or answered that they did not know. Like their female counterparts, more than 40% of the French men estimated that screening would save 100 or 200 men from dying from prostate cancer per 1,000 screened. Men in Austria, the Netherlands, Spain, and the United Kingdom made similar overestimates. As observed for women, the percentage of Russian men who overestimated the benefits or did not know was the lowest among the nine countries surveyed, 77%.

Similar to what was observed in women, the distribution of estimates made by men between the ages of 50 and 69 years was not more accurate than what was observed overall. The percentage of men who estimated zero and one life saved decreased from 8.3% and 2.4%, respectively, in all age groups to 7.3% and 1.9%, respectively, among men aged 50–69 years.

Most (59%) women reported using one or more sources frequently, compared with 47% of men (data not shown). In every country, older citizens searched for more information than younger ones (data not shown).

Within the general categories of health information sources, family and friends, experts, general media, and health-specific sources, the correlations between the frequencies of use of two sources were consistently high (correlation coefficients >.5), whereas the correlations between sources from different categories were consistently lower (data not shown). The sources of health-related information reported most often were family and/or friends,

Table 6.2: Estimated Reduction of Prostate Cancer Mortality through Regular Participation in Prostate-Specific Antigen Screening (Men Only)*

Reduction Out of 1,000	Percentage of Responders									
	Mean	Germany	France	Austria	The Netherlands	Italy	United Kingdom	Spain	Poland	Russia
None	8.3	3.8	1.6	4.1	3.0	5.7	0.5	9.3	5.0	20.3
1	2.4	2.3	2.7	3.5	2.2	1.8	0.9	4.3	0.7	2.9
10	14.4	17.7	16.9	24.4	11.5	11.9	15.9	17.0	13.9	10.7
50	19.3	23.0	21.6	27.1	20.2	18.5	17.3	25.1	17.9	15.0
100	14.0	17.2	21.1	20.8	20.3	9.2	15.6	18.8	14.5	7.3
200	11.8	9.7	20.2	14.2	14.2	12.2	19.5	17.9	11.3	3.4
Don't know	29.8	26.3	15.9	5.9	28.5	40.6	30.2	7.6	36.7	40.4

*Question: How many fewer men die from prostate cancer in the group who participate in screening, compared to men who do not participate in screening? Mean across all nine countries is weighted by sample size.

followed in descending order by experts (general practitioner and pharmacist), general media (television was the most reported source in this category), and health-specific sources (among all participants, the seven sources in this category were the least used among the 14 sources).

Individual trends according to country were observed with respect to sources of health information (Table 6.3). In Poland and Russia, family and/or friends were by far the most often reported source of information. In Austria, France, Germany, Italy, and Spain, the general practitioner was the primary source of information, and, except for family and friends, little use was made of other sources in these countries. The Netherlands had the most even distribution of reported information sources. In the United Kingdom, the frequency of reported consultation of most sources of information was generally low. For only two sources did British citizens report higher than average frequencies.

Frequent consulting of sources was not associated with an increase in understanding of the benefits of screening, but instead was often associated with overestimation. For the women in Austria, France, Germany, Poland, Russia, Spain, and the United Kingdom, there was no single source of information whose frequent use was associated with more accurate understanding of the benefits. By contrast, German women who more often consulted leaflets and pamphlets from medical organizations (41% of Germans use this source; Table 6.3) tended to overestimate the benefit of mammography screening ($r = .15$, 95% confidence interval [CI] = 0.07 to 0.23), as did French women ($r = .12$, 95% CI = 0.04 to 0.29). The German women who more often consulted a general practitioner ($r = .10$, 95% CI = 0.02 to 0.18) or a pharmacist ($r = .11$, 95% CI = 0.03 to 0.19) for health information also had less accurate understanding of benefits.

The only sources associated with improved knowledge of the benefits of breast cancer screening were consumer counseling in the Netherlands ($r = -.18$, 95% CI = −0.35 to −0.01) and in Italy ($r = -0.17$, 95% CI = −0.27 to −0.07) and patient counseling ($r = -.16$, 95% CI = −0.26 to −0.06) and self-help groups ($r = -.12$, 95% CI = −0.22 to −0.02) in Italy alone.

The results for PSA screening confirmed the general conclusion that consultation of sources of medical information is not associated with knowledge of the benefits of screening. For men in Austria, Germany, the Netherlands, Russia, and Spain, there was no single source whose frequent use was associated with better understanding of benefits. Information from health insurances was associated with less overestimation in France ($r = -.11$, 95% CI = −0.20 to −0.02), Poland ($r = -.13$, 95% CI = −0.25 to −0.01), and Italy ($r = -.18$, 95% CI = −0.29 to −0.08), and information from radio with less overestimation in the United Kingdom ($r = -.11$, 95% CI = −0.21 to −0.01).

For both mammography and PSA screening, there was no single country in which frequent consulting of general practitioners and health pamphlets improved understanding of benefits. The overall effect across all nine countries was a slight positive correlation between overestimation and frequency

Table 6.3: Percentage of Participants Reporting That They Use Specific Sources of Health Information Sometimes or Frequently*

Source	Mean†	Germany	France	Austria	The Netherlands	Italy	United Kingdom	Spain	Poland	Russia
Family/friends	62	65	60	61	50	62	53	47	67	69‡
General practitioner	59	68	69	68	50	79‡	53	72	43	44
Pharmacist	54	56	62	59	54	70‡	49	66	49	43
Television	43	45	57‡	43	51	38	35	32	42	42
Popular magazines	26	36	39‡	33	33	20	22	21	30	18
Daily newspaper	25	29	38‡	38	30	19	25	24	25	20
Radio	23	20	36‡	34	28	12	22	21	30	23
Leaflets and pamphlets by health organizations	21	41‡	36	23	30	13	14	17	12	14
Reference books about health topics	20	20	23	23	27‡	15	25	15	15	22
Health insurance company	17	19	27	20	44	3	9	54‡	21	4
Internet (e.g., health portals)	15	17	21	17	42‡	11	26	16	14	7
Consumer counseling	6	3	8	4	20‡	4	3	9	4	6
Patient counseling	6	2	3	3	20‡	6	5	8	9	5
Self-help organizations	4	3	5	2	8‡	2	4	6	3	4

*Response alternatives were never, rarely, sometimes, frequently, and don't know.
†Mean across all nine countries was weighted by sample size.
‡Highest value for each source.

of consultation for general practitioners ($r = .07$, 95% CI = 0.05 to 0.09) and health pamphlets ($r = .06$, 95% CI = 0.04 to 0.08).

In this survey of more than 10,000 people in nine European countries, 92% of women and 89% of men overestimated the benefits of mammography and PSA screening, respectively, by an order of magnitude or more, or stated that they did not know what the benefits were. This percentage was the lowest in Russia, with 82% for women and 77% for men. Consulting general practitioners, health pamphlets, and other information sources generally did not increase accurate knowledge of benefits; the only major exception was information from health insurances about PSA screening.

Our use of a numerical response scale with particular categories (0, 1, 10, 50, 100, 200) may have influenced participants' estimates and may have contributed to the large amount of overestimation observed. However, we have indirect evidence that an open response format might not reduce the degree of overestimation. At the time of this study (December 2006), we conducted an independent survey with a different polling institute (TNS Emnid) in Germany and with a new representative sample of 1,018 citizens, in which we included the question: "Early detection with mammography reduces the risk of dying from breast cancer by 25%. Assume that 1,000 women aged 40 and older participate regularly in screening. How many fewer would die of breast cancer?" No response categories were used. The proportion of correct answers was equally low, and overestimation was even larger, with a median estimate of 500 lives saved for every 1,000 women screened by mammography (Chapter 5).

This study did not assess perceived harms and economic costs, or whether the degree of overestimation of benefit translates into higher participation in screening. An association between overestimation and participation has been demonstrated in other studies, although this association was not observed for African American women (Miller & Champion, 1997; Price, Desmond, Slenker, Smith, & Stewart, 1992). We also do not know whether the results are generalizable to other countries. Domenighetti et al. found similar overestimation of mammography in telephone interviews conducted with women in Switzerland and the United States and also reported overestimation for women in the United Kingdom and Italy, but we are not aware of any surveys of the perceived benefit of PSA tests that were conducted simultaneously in different countries (Domenighetti, D'Avanzo, Egger et al., 2003). Nor are we aware of any representative nationwide survey of the perceived quantitative benefit of mammography or PSA screening in the United States. A study with 145 American women with above-average education reported an average perceived breast cancer-specific mortality reduction of 60 in 1,000 (Black, Nease Jr, Tosteson, 1995), and a study of 207 women attending general internal medicine clinics in Wisconsin reported that 76% overestimated the relative risk reduction (Haggstrom & Schapira, 2006).

We do not know why women and men overestimate the benefits of screening, but the results in Table 6.3 may indicate potential reasons. After family and friends, whose information might actually derive from the other sources in Table 6.3, the most frequently mentioned sources were general practitioner and pharmacist. Studies on physicians' lack of knowledge about the benefits of screening and conflicts of interest support the possibility that these professionals contribute to overestimation (Chapter 5; Steurer, Held, Schmidt, Gigerenzer, Tag, & Bachmann, 2009; Welch, 2004). The observation that health-specific sources rarely improve understanding of screening (except for health insurance in several countries) also implicates these sources as a further potential cause, a hypothesis that is consistent with the findings that few pamphlets, letters of invitation, and websites explain the size of the benefit. If they do, the explanation is almost always in terms of a relative risk reduction rather than in the more transparent form of an absolute risk reduction (see Chapters 4 and 5).

In conclusion, this study documents that information about the benefits of mammography and PSA screening has not reached the general public in nine European countries, including the age group targeted by screening programs. Knowing the benefit of a treatment is a necessary condition for informed consent and rational decision making. At present, however, the available information sources are not designed to communicate benefits clearly. As a consequence, preconditions for informed decisions about participation in screening are largely nonexistent in Europe.

Part III

Smart Heuristics

Chapter 7

Heuristic Decision Making

Introduction

How are decisions made? Three major answers have been proposed: The mind applies logic, statistics, or heuristics. Yet these mental tools have not been treated as equals, each suited to a particular kind of problem, as we believe they should be. Rather, rules of logic and statistics have been linked to rational reasoning and heuristics linked to error-prone intuitions or even irrationality. Since the 1970s, this opposition has been entrenched in psychological research, from the heuristics-and-biases program (Tversky & Kahneman, 1974) to various two-system theories of reasoning (Evans, 2008). Deviations from logical or statistical principles became routinely interpreted as judgmental biases and attributed to cognitive heuristics such as "representativeness" or to an intuitive "System 1." The bottom line was that people often rely on heuristics, but they would be better off in terms of accuracy if they did not. As Kahneman (2003) explained in his Nobel Memorial Lecture: "Our research attempted to obtain a map of bounded rationality, by exploring the systematic biases that separate the beliefs that people have and the choices they make from the optimal beliefs and choices assumed in rational-agent models" (p. 1449). In this research, it is assumed that the conditions for rational models hold and can thus define optimal reasoning. The "father" of bounded rationality, Simon (1989), however, asked a fundamentally different question, leading to a different research program.

> Simon's question: "How do human beings reason when the conditions for rationality postulated by the model of neoclassical economics are not met?" (p. 377)

As Simon (1979, p. 500) stressed in his Nobel Memorial Lecture, the classical model of rationality requires knowledge of all the relevant alternatives,

Originally published as Gigerenzer, G., & Gaissmaier, W. (2011). Heuristic decision making. *Annual Review of Psychology, 62*, 451–482. The chapter has been slightly edited.

their consequences and probabilities, and a predictable world without surprises. These conditions, however, are rarely met for the problems that individuals and organizations face. The economist Knight (1921) distinguished between worlds of risk, in which probabilities can be measured in terms of frequencies or propensities, and worlds of uncertainty, where that is not the case. In uncertain worlds, part of the relevant information is unknown or has to be estimated from small samples, so that the conditions for rational decision theory are not met, making it an inappropriate norm for optimal reasoning (Binmore, 2009). In an uncertain world, one can no longer assume that "rational" models automatically provide the correct answer. Even small deviations from the model conditions can matter. In fact, models of risk can lead to disaster when applied to an uncertain world, as Stiglitz (2010) noted with respect to the financial crash of 2008: "It simply wasn't true that a world with *almost* perfect information was very similar to one in which there was perfect information" (p. 243, emphasis added). And Soros (2009) concluded that "rational expectations theory is no longer taken seriously outside academic circles" (p. 6).

In recent years, research has moved beyond small worlds such as the ultimatum game and choice between monetary gambles. To test how well heuristics perform in uncertain worlds, one needs formal models of heuristics. Such tests are not possible as long as heuristics are only vaguely characterized by general labels, because labels cannot make the precise predictions that statistical techniques can.

When heuristics were formalized, a surprising discovery was made. In a number of uncertain worlds, simple heuristics were more accurate than standard statistical methods that have the same or more information. These results became known as less-is-more effects: There is an inverse U-shaped relation between level of accuracy and amount of information, computation, or time. In other words, there is a point where more is not better, but harmful. Starting in the late 1990s, it was shown for the first time that relying on one good reason (and ignoring the rest) can lead to higher predictive accuracy than achieved by a linear multiple regression (Czerlinski et al., 1999; Gigerenzer & Goldstein, 1996b) and a three-layer feed-forward connectionist network trained using the back propagation algorithm (Brighton, 2006; Chater et al., 2003; Gigerenzer & Brighton, 2009). These results put heuristics on par with standard statistical models of "rational" cognition (see Gigerenzer, 2008). Simon (1999) spoke of a "revolution in cognitive science, striking a great blow for sanity in the approach to human rationality."

The revolution Simon referred to could not have happened without formal models and the power of modern computers. Moreover, it is a "revolution" in the original sense of the term, building on earlier demonstrations of the robust beauty of simple models. These include Dawes and Corrigan (1974) and Einhorn and Hogarth (1975), who showed that simple equal weights predict about as well as—and sometimes better than—multiple regression with "optimal" beta weights. Their important work has not received the recognition it deserves and is not even mentioned in standard textbooks in econometrics (Hogarth, 2012).

Although the study of heuristics has been typically considered as purely descriptive, less-is-more effects open up a prescriptive role for heuristics, resulting in two research questions:

Description: Which heuristics do people use in which situations?
Prescription: When should people rely on a given heuristic rather than a complex strategy to make more accurate judgments?

Scope of Review

We review a field that is in a fundamental transition, focusing on the major new ideas. The literature on heuristics does not speak with one voice, and we do not attempt to cover it exhaustively. Rather than presenting a patchwork of ideas to the reader, we organize this review within a theoretical framework and restrict it to (*a*) formal models of heuristics and (*b*) inferences rather than preferences.

The first restriction excludes explanation by mere labels but also by verbally stated processes that have not been formalized, such as the tools-to-theories heuristic in scientific discovery (Gigerenzer, 1991). Formal models allow rigorous tests of both descriptive and prescriptive questions. "Inference" refers to tasks for which a unique criterion exists, whereas "preference" (or preferential choice) refers to tasks where no such criteria exist, as in matters of taste. The advantage of studying inference is that the accuracy of a strategy can be determined. At the same time, we agree with Weber and Johnson (2009) that inferences and preferences draw on the same cognitive processes; in fact, most heuristics covered in this review can be used for preferential choice as well, as illustrated with examples from consumer choice and health. Note that the general term "decision making" is used here to cover both inferences and preferences.

Heuristics: strategies that ignore information to make decisions faster, more frugally, and/or more accurately than more complex methods

Risk: a situation in which all relevant alternatives, their consequences, and probabilities are known, and where the future is certain, so that the optimal solution to a problem can be determined

Uncertainty: a situation in which some relevant information is unknown or must be estimated from samples, and the future is uncertain, violating the conditions for rational decision theory

Less-is-more effects: when less information or computation leads to more accurate judgments than more information or computation

We begin with a brief, incomplete history (for more, see Groner, Groner, & Bischof, 1983), define the term heuristic, and provide an illustration of the use of heuristics in organizations, including an empirical demonstration of a less-is-more effect.

What Is a Heuristic?

The term heuristic is of Greek origin and means, "serving to find out or discover." Einstein included the term in the title of his Nobel prize-winning paper from 1905 on quantum physics, indicating that the view he presented was incomplete but highly useful (Holton, 1988, pp. 360–361). Max Wertheimer, who was a close friend of Einstein, and his fellow Gestalt psychologists spoke of heuristic methods such as "looking around" to guide search for information. The mathematician George Polya distinguished heuristics from analytical methods: For instance, heuristics are needed to find a proof, whereas analysis is for checking a proof. Simon and Allen Newell, a student of Polya, developed formal models of heuristics to limit large search spaces. Luce (1956), Tversky (1972), Dawes (1979), and others studied models of heuristics, such as lexicographic rules, elimination-by-aspect, and equal-weight rules. Payne and colleagues (1993) provided evidence for the adaptive use of these and other heuristics in their seminal research. Similarly, behavioral biologists studied experimentally the rules of thumb (their term for heuristics) that animals use for choosing food sites, nest sites, or mates (Hutchinson & Gigerenzer, 2005). After an initial phase dominated by logic, researchers in artificial intelligence (AI) began to study heuristics that can solve problems that logic and probability cannot, such as NP-complete (computationally intractable) problems. While AI researchers began to study how heuristics make computers smart, psychologists in the 1970s became interested in demonstrating human reasoning errors, and they used the term heuristic to explain why people make errors. This change in the evaluation of heuristics went hand-in-hand with replacing models of heuristics with vague labels, such as "availability" and, later, "affect." Unlike in biology and AI, heuristics became tied to biases, whereas the content-free laws of logic and probability became identified with the principles of sound thinking (Kahneman, 2003; Tversky & Kahneman, 1974). The resulting heuristics-and-biases program has had immense influence, contributing to the emergence of behavioral economics and behavioral law and economics.

Definition

Many definitions of heuristics exist. Kahneman and Frederick (2002) proposed that a heuristic assesses a target attribute by another property (attribute substitution) that comes more readily to mind. Shah and Oppenheimer (2008) proposed that all heuristics rely on effort reduction by one or more of the following: (*a*) examining fewer cues, (*b*) reducing the effort of retrieving cue values, (*c*) simplifying the weighting of cues, (*d*) integrating less information, and (*e*) examining fewer alternatives. Although both attribute substitution and effort reduction are involved, attribute substitution is less specific because most inference methods, including multiple regression, entail it: An unknown criterion is estimated by cues. For the purpose of this review, we adopt the following definition:

> A heuristic is a strategy that ignores part of the information, with the goal of making decisions more quickly, frugally, and/or accurately than more complex methods.

Let us explain the terms. Heuristics are a subset of strategies; strategies also include complex regression or Bayesian models. The part of the information that is ignored is covered by Shah and Oppenheimer's list of five aspects. The goal of making judgments more quickly and frugally is consistent with the goal of effort reduction, where "frugal" is often measured by the number of cues that a heuristic searches. Of course, there is no strict dichotomy between heuristic and nonheuristic, as strategies can ignore more or less information. The goal of making judgments more accurately by ignoring information is new. It goes beyond the classical assumption that a heuristic trades off some accuracy for less effort. Unlike the two-system models of reasoning that link heuristics to unconscious, associative, and error-prone processes, no such link is made in this review. Every heuristic reviewed in this article can also be relied upon consciously and is defined as a rule. The amount of error it generates can be measured and compared to other strategies.

Consider the following illustration of decision making in an uncertain world.

Less-Can-Be-More: Managers' One-Good-Reason Decisions

Commercial retailers need to distinguish those customers who are likely to purchase again in a given time frame (active customers) from those who are not (inactive customers). These companies have a large database containing the amount, kind, and date of every customer's previous purchases. Based on this information, how can an executive predict which customers will be active in the future?

Statistically sophisticated academics might opt for a Bayesian analysis, regression analysis, or some other optimizing strategy to predict the probability that a customer with a given purchase history is active at some future time. Researchers in business share this vision, and the state-of-the-art approach is the Pareto/NBD model (negative binomial distribution; Schmittlein & Peterson, 1994). This model assumes that purchases follow a Poisson process with a purchase parameter λ, that customer lifetimes follow an exponential distribution with a dropout rate parameter μ, and that, across customers, purchase and dropout rates are distributed according to a gamma distribution. However, most managers in Europe, North America, Japan, Brazil, and India rely on "intuitive" heuristics rather than on this or similar statistical forecasting methods (Parikh, 1994). Wübben and Wangenheim (2008) reported that experienced managers use a simple recency-of-last-purchase rule:

> Hiatus heuristic: If a customer has not purchased within a certain number of months (the hiatus), the customer is classified as inactive; otherwise, the customer is classified as active.

The managers of an apparel retailer and an airline relied on nine months as the hiatus, whereas the hiatus of an online CD retailer was six months. Note that by relying on recency only, the managers ignore information such as the frequency and the spacing of previous purchases. Yet how accurate is the heuristic compared to the Pareto/NBD model? To investigate this question, the Pareto/NBD model was allowed to estimate its parameters from 40 weeks of data and was tested over the following 40 weeks. The hiatus heuristic does not need to estimate any parameters. For the apparel retailer, the hiatus heuristic correctly classified 83% of customers, whereas the Pareto/NBD model classified only 75% correctly. For the airline, the score was 77% versus 74%, and for the online CD business, the two methods tied at 77% (Wübben & Wangenheim, 2008). Similar results were found for forecasting future best customers and for a second complex statistical model.

This study demonstrated empirically a less-is-more effect: The complex model had all the information the simple heuristic used and more, performed extensive estimations and computations, but nevertheless made more errors. In this situation, "big data" analysis proved to be inferior to a smart heuristic. The study also showed how important it is to formalize a heuristic so that its predictions can be tested and compared to competing models.

Adaptive toolbox: the cognitive heuristics, their building blocks (e.g., rules for search, stopping, decision), and the core capacities (e.g., recognition memory) they exploit

Ecological rationality: the study of ecological rationality investigates in which environments a given strategy is better than other strategies (better—not best—because in situations of uncertainty the optimal strategy is unknown)

Accuracy–effort trade-off: the traditional explanation why people use heuristics, assuming that effort is traded against accuracy. Not generally true (see less-is-more effects)

The Adaptive Toolbox

Formal models of heuristics represent progress over labels, but precision alone is not enough to build a science of heuristics. For instance, behavioral biology has experimentally identified various rules of thumb that animals use, which often look like curiosities in the absence of an overarching theory (Hutchinson & Gigerenzer, 2005). Further progress requires a theoretical framework that reaches beyond a list of heuristics. One step toward such a theory is to look for common building blocks, from which the various heuristics are constructed as an organizing principle. This would allow reducing the larger number of heuristics to a smaller number of components, similar to how the number of chemical elements in the periodic table is built from a small number of particles. Three building blocks have been proposed (Gigerenzer et al., 1999):

1. Search rules specify in what direction the search extends in the search space.
2. Stopping rules specify when the search is stopped.
3. Decision rules specify how the final decision is reached.

For instance, the hiatus heuristic searches for recency-of-last-purchase information; stops when it is found, ignoring further information; and uses a nine-month threshold to make the decision. Similarly, Simon's (1955) satisficing heuristic searches through options in any order, stops as soon the first option exceeds an aspiration level, and chooses this option. Many but not all heuristics are composed of these three building blocks; thus, the list of building blocks is incomplete.

The collection of heuristics and building blocks an individual or a species has at its disposal for constructing heuristics, together with the core mental capacities that building blocks exploit, has been called the adaptive toolbox (Gigerenzer et al., 1999). Core capacities include recognition memory, frequency monitoring, object tracking, and the ability to imitate. These vary systematically between species and individuals. Heuristics can be fast and frugal only because the core capacities are already in place.

How are heuristics selected for a given problem? Although some authors implied that the selection problem is unique to heuristics (Glöckner et al., 2010; Newell, 2005), it equally applies to statistical models of mind. There are many such models. Even if one proposes that the mind has only one tool in its statistical toolbox, such as Bayes, regression, or neural network, the strategy selection problem translates into the question of how parameter values are selected for each new problem (Marewski, 2010).

Several principles appear to guide learning which strategy to select. First, heuristics and their underlying core capacities can be (partly) hardwired by evolution, as it appears to be in bees' collective decision about the location of a new hive (Seeley, 2001) and in perceptual mechanisms for inferring the extension of objects in three-dimensional space (Kleffner & Ramachandran, 1992). The second selection principle is based on individual learning; a formal model is Rieskamp and Otto's (2006) strategy selection learning theory. Third, heuristics can be selected and learned by social processes, as in imitation and explicit teaching of heuristics (e.g., Snook et al., 2004). Finally, the content of individual memory determines in the first place which heuristics can be used, and some heuristics' very applicability appears to be correlated with their "ecological rationality" (see below). For instance, the fluency heuristic is most likely to be applicable in situations where it is also likely to succeed (Marewski & Schooler, 2011).

Why Heuristics?

Two answers have been proposed to the question of why heuristics are useful: the accuracy–effort trade-off, and the ecological rationality of heuristics.

Accuracy–Effort Trade-off

The classical explanation is that people save effort with heuristics, but at the cost of accuracy (Payne et al., 1993; Shah & Oppenheimer, 2008). In this view, humans and other animals rely on heuristics because information search and computation cost time and effort; heuristics trade off some loss in accuracy for faster and more frugal cognition.

There are two interpretations of this trade-off: (*a*) Rational trade-offs. Not every decision is important enough to warrant spending the time to find the best course of action; thus, people choose shortcuts that save effort. The program on the adaptive decision maker (Payne et al., 1993) is built on the assumption that heuristics achieve a beneficial trade-off between accuracy and effort. Here, relying on heuristics can be rational in the sense that costs of effort are higher than the gain in accuracy. (*b*) Cognitive limitations. Capacity limitations prevent us from acting rationally and force us to rely on heuristics, which are considered a source of judgmental errors.

The accuracy–effort trade-off is regularly touted as a potentially universal law of cognition. Yet the study on the hiatus heuristic illustrated that this assumption is not generally correct. The hiatus heuristic saves effort compared to the sophisticated Pareto/NBD model, but is also more accurate: a less-is-more effect.

Ecological Rationality

Less-is-more effects require a new conception of why people rely on heuristics. The study of the ecological rationality of heuristics, or strategies in general, is such a new framework: "A heuristic is ecologically rational to the degree that it is adapted to the structure of the environment" (Gigerenzer et al., 1999, p. 13). Vernon L. Smith (2003) used this definition in his Nobel lecture and generalized it from heuristics to markets and institutions. The study of ecological rationality fleshes out Simon's scissors analogy: "Human rational behavior (and the rational behavior of all physical symbol systems) is shaped by a scissors whose two blades are the structure of task environments and the computational capabilities of the actor" (Simon, 1990, p. 7). If one looks only at one blade, cognition, one cannot understand why and when it succeeds or fails. The study of ecological rationality addresses two related questions: How does cognition exploit environmental structures, and how does it deal with error?

Exploiting environmental structure. In which environments will a given heuristic succeed, and in which will it fail? Environmental structures that have been identified include (Todd et al., 2012):

1. Predictability: how well a criterion can be predicted.
2. Redundancy: the correlation between cues.
3. Sample size: number of observations (relative to number of cues).
4. Variability in weights: the distribution of the cue weights (e.g., skewed or uniform).

For instance, heuristics that rely on only one reason, such as the hiatus heuristic and take-the-best heuristic (see below), tend to succeed (relative to strategies that rely on more reasons) in environments with (*a*) moderate to high uncertainty (Hogarth & Karelaia, 2007) and (*b*) moderate to high redundancy (Dieckmann & Rieskamp, 2007). For customer activity, low predictability means that it is difficult to predict future purchases, and redundancy might be reflected in a high correlation between length of hiatus and spacing of previous purchases. The study of ecological rationality results in comparative statements of the kind "strategy *X* is more accurate (frugal, fast) than *Y* in environment *E*" or in quantitative relations between the performance of strategy *X* when the structure of an environment changes (e.g., Baucells, Carrasco, & Hogarth, 2008; Karelaia, 2006; Martignon & Hoffrage, 2002). Specific findings are introduced below.

Dealing with error. In much research on reasoning, a bias typically refers to ignoring part of the information, as in the base rate fallacy. This can be captured by the equation:

$$\text{Error} = \text{bias} + \varepsilon, \quad (1)$$

where ε is an irreducible random error. In this view, if the bias is eliminated, good inferences are obtained. In statistical theory (Geman, Bienenstock, & Doursat, 1992), however, there are three sources of errors:

$$\text{Error} = \text{bias}^2 + \text{variance} + \varepsilon, \quad (2)$$

where bias refers to a systematic deviation between a model and the true state, as in Equation 1. To define the meaning of variance, consider 100 people who rely on the same strategy, but each one has a different sample of observations from the same population to estimate its parameters. Because of sampling error, the 100 inferences may not be the same. Across samples, bias is the difference between the mean prediction and the true state of nature, and variance is the expected squared deviation around this mean. To illustrate, the nine-month hiatus heuristic has a bias but zero variance, because it has no free parameters to adjust to specific samples. In contrast, the Pareto/NBD model has free parameters and is likely to suffer from both variance and bias. Variance decreases with increasing sample size, but also with simpler strategies that have fewer free parameters (and less flexible functional forms; Pitt et al., 2002). Thus, a cognitive system needs to draw a balance between being biased and flexible (variance) rather than simply trying to eliminate bias. In the extreme, as illustrated by the nine-month hiatus, the total elimination of variance at the price of higher bias can lead to better inferences. This "bias–variance dilemma" helps to explicate the rationality of simple heuristics and how less can be more (Brighton & Gigerenzer, 2008; Gigerenzer & Brighton, 2009).

The study of ecological rationality is related to the view that human cognition is adapted to its past environment (Cosmides & Tooby, 2006), yet it should not be confused with the biological concept of adaptation. A match between a heuristic and an environmental structure does not imply that the heuristic evolved because of that environment (Hutchinson & Gigerenzer, 2005). The distinction between ecological and logical rationality is linked to that between correspondence and coherence (Hammond, 2007), but it is not identical. If correspondence means achieving a goal in the world rather than cohering to a rule of logic, correspondence and ecological rationality refer to similar goals—although the study of the latter adds a mathematical analysis of the relation between heuristic and environment. If correspondence, however, means that the mental representation corresponds to the world, as in a fairly accurate mental model or in Shepard's (2001) view of the mind as a mirror reflecting the world, then ecological rationality is different. A heuristic is functional, not a veridical copy of the world.

Ecological rationality does not mean that all people are perfectly adapted to their environment. As Simon (1992) noted, if that were the case, one would only need to study the environment to predict behavior; the study of heuristics would be obsolete.

Methodological Principles

Formal models of heuristics are indispensable for progress, yet remain the exception in psychology. Much of the research first documents an error of judgment and thereafter attributes it to a heuristic. In a widely cited experiment, Tversky and Kahneman (1973) reported that certain letters were falsely judged to occur more frequently in the first than the third position in

English words. They attributed this error to the availability heuristic: Words with a letter in the first position come to mind more easily. Note that availability was introduced after the fact, without any independent measure or test. Once the heuristic is formalized, conclusions change. Sedlmeier and colleagues (1998) defined and modeled the two most common meanings of availability—the speed of retrieval of the first word and the number of retrieved words within a constant time period. Neither version of the availability heuristic could predict participants' frequency estimates. Instead, estimated frequencies were best predicted by actual frequencies, consistent with the classical findings by Attneave (1953). Formal models protect against the seductive power of general labels.

We are concerned about the replacement of formal models by general labels in parts of psychology. For instance, Tversky's (1977) seminal model of similarity makes testable predictions (e.g., the asymmetry of similarity), whereas the widely cited label "representativeness" can predict little but is so flexible that it is consistent with many judgments, including opposite intuitions (Ayton & Fischer, 2004). Similarly, research on the adaptive decision maker (Payne et al., 1993) and the adaptive toolbox (Gigerenzer et al., 1999) has studied formal models of heuristics, which have been ignored in two-system theories of reasoning in favor of a "System 1" (Evans, 2008). The problem with two-system theories "is the lack of any predictive power and the tendency to employ them as an after-the-fact explanation" (Keren & Schul, 2009, p. 544). Moving backward from existing models to labels is a rare event in science, which typically proceeds in the opposite direction. The study of formal models entails four methodological principles.

Comparative Versus Singular Tests

All models are wrong. But some predict better than others and lead to novel questions. Therefore, tests of cognitive strategies need to be comparative, that is, test several models. This differs from the widespread practice of null hypothesis testing, where only one model (the null) is specified.

Test of Individuals Versus Group Means

Numerous studies have documented systematic individual differences in the use of heuristics (e.g., Lee & Cummins, 2004; Nosofsky & Bergert, 2007), including in old age (Mata et al., 2007). In the presence of individual differences, tests of group mean differences can be highly misleading (see Pachur et al., 2008).

Testing the Adaptive Versus Universal Use of Heuristics

Research has shifted from asking whether people use one heuristic in all situations to asking whether heuristics are applied in situations where these are ecologically rational. For instance, Bröder began by asking whether all people use the take-the-best heuristic all the time, but soon asked whether people rely on take-the-best in situations where it is ecologically rational, for instance, when cue validities are highly skewed (Bröder & Schiffer, 2003, 2006).

Prediction Versus Fitting

Prediction takes place when the data have not yet been observed and a model with fixed parameter values is used to predict them; fitting takes place when the data have already been observed and the parameters of a model are chosen so that they maximize the fit (such as R^2). In general, the more free parameters a model has, the better the fit, but this does not hold for predictions. In an uncertain world where parameters need to be estimated from small or unreliable samples, the function between predictive accuracy and the flexibility of a model (e.g., number of free parameters) is typically inversely U-shaped. Both too few and too many parameters can hurt performance (Pitt et al., 2002). Competing models of strategies should be tested for their predictive ability, not their ability to fit already known data.

In the next sections, we review four classes of heuristics. The first class exploits recognition memory, the second relies on one good reason only (and ignores all other reasons), the third weights all cues or alternatives equally, and the fourth relies on social information. As mentioned in the introduction, formal models of heuristics allow asking two questions: whether they can describe decisions, and whether they can prescribe how to make better decisions than, say, a complex statistical method. The prescriptive question is particularly relevant for organizations, from business to health care. Organizations seem ideally suited to the application of heuristics because of the inherent uncertainty and the pressure to act quickly. One might therefore presume that plenty of studies have investigated fast-and-frugal heuristics in organizations. Not so, as Hodgkinson and Healey (2008) pointed out. Our working hypothesis is that heuristic decision making in individuals and organizations can be modeled by the same cognitive building blocks: rules of search, stopping, and decision.

Recognition-Based Decision Making

The recognition memory literature indicates that a sense of recognition (often called familiarity) appears in consciousness earlier than recollection (Ratcliff & McKoon, 1989). The first class of heuristics exploits this core capacity.

Recognition Heuristic

The goal is to make inferences about a criterion that is not directly accessible to the decision maker, based on recognition retrieved from memory. This is possible in an environment (reference class) R where the recognition of alternatives $a, b \in R$ positively correlates with their criterion values. For two alternatives, the heuristic is defined as (Goldstein & Gigerenzer, 2002, p. 76):

> Recognition heuristic: If one of two alternatives is recognized and the other is not, then infer that the recognized alternative has the higher value with respect to the criterion.

The higher the recognition validity α for a given criterion, the more ecologically rational it is to rely on this heuristic and the more likely people will rely on it. For each individual, α can be computed by

$$\alpha = C/(C+W),$$

where C is the number of correct inferences the recognition heuristic would make, computed across all pairs in which one alternative is recognized and the other is not, and W is the number of wrong inferences.

A number of studies addressed the question of whether people rely on the recognition heuristic in an ecologically rational way. For instance, name recognition of Swiss cities is a valid predictor of their population ($\alpha = 0.86$) but not their distance from the center of Switzerland ($\alpha = 0.51$). Pohl (2006) reported that 89% of inferences accorded with the model in judgments of population, compared to only 54% in judgments of the distance. More generally, there is a positive correlation of $r = 0.64$ between the recognition validity and the proportion of judgments consistent with the recognition heuristic across 11 studies (Pachur et al., 2012; see also Chapter 8). Similarly, old and young people alike adjust their reliance on the recognition heuristic between environments with high versus low recognition validities, even though old people have poorer recognition memory (Pachur et al., 2009).

The recognition heuristic is a model that relies on recognition only. This leads to the testable prediction that people who rely on it will ignore strong, contradicting cues (so-called noncompensatory inferences). Several studies that taught participants between one and three contradicting cues, typically of higher validity than α (Newell & Fernandez, 2006; Pachur et al., 2008; Richter & Späth, 2006, experiment 3), reported that mean accordance rates decreased. A reanalysis of these studies at an individual level, however, showed that typically about half of the participants consistently followed the recognition heuristic in every single trial, even in the presence of up to three contradicting cues (Pachur et al., 2008).

The model of the recognition heuristic does not distinguish between pairs where the model leads to a correct inference and pairs where it leads to a wrong inference. However, the mean accordance rates were 90% and 74%, respectively (Hilbig & Pohl, 2008; Pohl, 2006). Together with the effect of contradicting cues, this result indicated that some people did not follow the recognition heuristic, although the overall accordance rates remain high. Various authors concluded that people relied on a compensatory strategy, such as weighting and adding of all cues (e.g., Hilbig & Pohl, 2008;

Oppenheimer, 2003). None of the studies above, however, formulated and tested a compensatory strategy against the recognition heuristic, leaving the strategies that participants relied on unknown. One study since tested five compensatory models and found that none could predict judgments better than the simple model of the recognition heuristic (Marewski et al., 2010).

The recognition heuristic model also makes another bold prediction:

> If $\alpha > \beta$, and α, β are independent of n, then a less-is-more effect will be observed.

Here, β is the knowledge validity, measured as $C/(C + W)$ for all pairs in which both alternatives are recognized, and n is the number of alternatives an individual recognizes. A less-is-more effect means that the function between accuracy and n is inversely U-shaped rather than monotonically increasing. Some studies reported less-is-more effects empirically among two, three, or four alternatives (Frosch et al., 2007; Goldstein & Gigerenzer, 2002) and in group decisions (Reimer & Katsikopoulos, 2004), whereas others failed to do so (Pachur & Biele, 2007; Pohl, 2006), possibly because the effect is predicted to be small [see Katsikopoulos (2010) for an excellent analysis of the evidence]. Using a signal detection analysis, Pleskac (2007) showed how the less-is-more effect depends on the false alarms and miss rates in the recognition judgments.

Dougherty et al. (2008) criticized the model of the recognition heuristic for treating recognition as binary input (threshold model) rather than continuously. In contrast, Bröder and Schütz (2009) argued that the widespread critique of threshold models is largely invalid. In a reanalysis of 59 published studies, they concluded that threshold models in fact fit the data better in about half of the cases.

Predicting Wimbledon. Although much of the work has addressed the descriptive question of what proportion of people rely on the heuristic when it is ecologically rational, the prescriptive question is how well the heuristic can compete with well-established forecasting instruments (Goldstein & Gigerenzer, 2009). For instance, Serwe and Frings (2006) reported that collective recognition of amateur players (who knew only half of the contestants) turned out to be a better predictor of the 2004 Wimbledon tennis match outcomes (72% correct) than did the Association of Tennis Professionals (ATP) Entry Ranking (66%), ATP Champions Race (68%), and the seeding of the Wimbledon experts (69%). Scheibehenne and Bröder (2007) found the same surprising result for Wimbledon 2006.

Predicting elections. Gaissmaier and Marewski (2011) put the recognition heuristic to a test in predicting federal and state elections in Germany. Surprisingly, forecasts based on name recognition were as accurate as interviewing voters about their voting intentions. This particularly holds true when predicting the success of small parties, for which no polls are usually available because those polls would require huge samples. In contrast to surveys of voting intentions, recognition-based forecasts can be computed from small, "lousy" samples.

Investment. In three studies on predicting the stock market, Ortmann et al. (2008) reported that recognition-based portfolios (the set of most-recognized options), on average, outperformed managed funds such as the Fidelity Growth Fund, the market (Dow or Dax), chance portfolios, and stock experts. In contrast, Boyd (2001) found no such advantage when he used college students' recognition of stocks rather than that of the general public. It is imperative to understand why and under what conditions this simple heuristic can survive in financial markets without making a systematic market analysis. This remains an open question.

Consumer choice. The recognition heuristic could be a first step in consideration set formation (Marewski et al., 2010), as it allows the choice set to be quickly reduced. This idea is consistent with research that suggests that priming a familiar brand increases the probability that it will be considered for purchase (e.g., Coates et al., 2004). Brand recognition can be even more important than attributes that are a more direct reflection of quality. For instance, in a blind test, most people preferred a jar of high-quality peanut butter to two alternative jars of low-quality peanut butter. Yet when a familiar brand label was attached to one of the low-quality jars, the preferences changed. Most (73%) now preferred the jar with the label they recognized, and only 20% preferred the unlabeled jar with the high-quality peanut butter (Hoyer & Brown, 1990). Brand recognition may well dominate the taste cues, or the taste cues themselves might even be changed by brand recognition—people "taste" the brand name.

Fluency Heuristic

The recognition heuristic is mute about the underlying recognition process, just as Bayes rule is mute about source of prior probabilities. Dougherty et al. (2008) argued that it needs to be embedded in a theory of the recognition process. Schooler and Hertwig (2005) implemented the heuristic based on the ACT-R (Adaptive Control of Thought–Rational) model of memory, which showed how forgetting—a process often seen as nuisance and handicap—can be functional in the context of inference, generating less-is-more effects. In this same work, the fluency heuristic was formulated for situations when both alternatives are recognized, that is, when the recognition heuristic cannot be applied:

> Fluency heuristic: If both alternatives are recognized but one is recognized faster, then infer that this alternative has the higher value with respect to the criterion.

The fluency heuristic builds on earlier work on fluency (Jacoby & Dallas, 1981). For instance, fluent processing that stems from previous exposure can increase the perceived truth of repeated assertions (Hertwig et al., 1997) and the perceived fame of names (Jacoby et al., 1989), and it is related to the mere exposure effect (Zajonc, 1968). People's sense of fluency has been reported to predict the performance of stocks (Alter & Oppenheimer, 2006).

By formalizing the fluency heuristic, Schooler and Hertwig (2005) clearly defined the difference between the recognition and fluency heuristics and contributed to the progress in replacing verbal labels with computational models. The fluency heuristic is ecologically rational if the speed of recognition is correlated with the criterion, that is, the fluency validity >0.5. Hertwig et al. (2008) reported that the validity of fluency for predicting variables such as sales figures and wealth was always lower than recognition validity, although always above chance. Subsequently, they showed that people can accurately tell the difference between two recognition latencies if the difference exceeded 100 ms, and that across three environments, the mean proportions of inferences consistent with the fluency heuristic were 74%, 63%, and 68%, respectively. Accordance rates were as high as 82% when differences in recognition latencies were large. Deriving the fluency heuristic's prediction for individual people and individual items is a strong test. Yet it is not how the impact of fluency is commonly tested in social and cognitive psychology, where researchers tend to manipulate fluency experimentally and observe the consequences.

Fluency also plays a role when alternatives are not given (as in a two-alternative choice) but need to be generated from memory. Johnson and Raab (2003) proposed a variant of the fluency heuristic when alternatives are sequentially retrieved rather than simultaneously perceived:

> Take-the-first heuristic: Choose the first alternative that comes to mind.

Johnson and Raab (2003) showed experienced handball players video sequences from a professional game and asked what they would have done—e.g., pass the ball to the player at the left or take a shot. On average, the first option that came to mind was better than later options and when more time was given to inspect the situation. This result was replicated for basketball players (Hepler, 2008). Klein's (2004) recognition-primed decision model for expertise appears to be closely related to the take-the-first heuristic.

Neural Basis of Recognition and Evaluation

Although a number of studies have shown that people do not automatically use the recognition heuristic when it can be applied, it is less clear how this evaluation process can be modeled. A functional magnetic resonance imaging study tested whether the two processes, recognition and evaluation, can be separated on a neural basis (Volz et al., 2006). Participants were given two tasks: The first involved only a recognition judgment ("Have you ever heard of Modena? Milan?"), while the second involved an inference in which participants could rely on the recognition heuristic ("Which city has the larger population: Milan or Modena?"). For mere recognition judgments, activation in the precuneus, an area that is known from independent studies to respond to recognition confidence

(Yonelinas, Otten, Shaw, & Rugg, 2005), was reported. In the inference task, precuneus activation was also observed, as predicted, and activation was detected in the anterior frontomedian cortex (aFMC), which has been linked in earlier studies to evaluative judgments and self-referential processing. The aFMC activation could represent the neural basis of this evaluation of ecological rationality. Furthermore, the neural evidence suggests that the recognition heuristic may be relied upon by default, consistent with the finding that response times were considerably faster when participants' inferences followed the recognition heuristic than when they did not (Pachur & Hertwig, 2006; Volz et al., 2006).

One-Reason Decision Making

Whereas the recognition and fluency heuristics base decisions on recognition information, other heuristics rely on recall. One class looks for only one "clever" cue and bases its decision on that cue alone. The hiatus heuristic is one example. A second class involves sequential search through cues, and it may search for more than one cue but also bases its decision on only one. Examples include lexicographic rules (Fishburn, 1974; Luce, 1956) and elimination-by-aspect (Tversky, 1972). These heuristics were originally developed for preferences; here, we focus on models of inferences.

One-Clever-Cue Heuristics

Many animal species appear to rely on a single "clever" cue for locating food, nest sites, or mates. For instance, in order to pursue a prey or a mate, bats, birds, and fish do not compute trajectories in three-dimensional space, but simply maintain a constant optical angle between their target and themselves—a strategy called the gaze heuristic (Gigerenzer, 2007; Shaffer et al., 2004). In order to catch a fly ball, baseball outfielders and cricket players rely on the same kind of heuristics rather than trying to compute the ball's trajectory (McLeod & Dienes, 1996). Similarly, to choose a mate, a peahen investigates only three or four of the peacocks displaying in a lek and chooses the one with the largest number of eyespots (Petrie & Halliday, 1994).

When are one-clever-cue heuristics ecologically rational? In general, in situations where (i) the variability of cue weights and redundancy is high (e.g., noncompensatory weights; see Martignon & Hoffrage, 2002), (ii) dominance or cumulative dominance holds for cue values (Baucells et al., 2008), and (iii) redundancy is large and sample size is small (see Hogarth & Karelaia, 2007; Katsikopoulos et al., 2010; McGrath, 2008).

Geographic profiling. The task of geographic profiling is to predict where a serial criminal is most likely to live given the sites of the crimes. Typically, geographic profiling is performed by sophisticated statistical software programs, such as CrimeStat, that calculate a probability distribution across possible locations. Snook and colleagues (2005) were among the first to challenge the "complexity equals accuracy" assumptions in the field of profiling. They tested the circle heuristic, which predicts the criminal's most likely location in the center of a circle drawn through the two most distant sites of crime. It relies on one cue only, the largest distance. In a comparison with 10 other profiling strategies, the heuristic predicted the locations best. Complex profiling strategies appear to become more accurate if the sample size is large, that is, when the number of crime locations known for an offender is nine or higher. Snook et al. (2004) taught two heuristics (including the circle heuristic) to laypeople in criminology and reported that after a single session, laypeople became about as accurate in predicting offender locations as the CrimeStat algorithm. These results led to a heated debate with proponents of commercial optimization algorithms in profiling (e.g., Rossmo, 2005).

One-reason decisions: a class of heuristics that bases judgments on one good reason only, ignoring other cues (e.g., take-the-best and hiatus heuristic)

Take-the-Best

The take-the-best heuristic is a model of how people infer which of two alternatives has a higher value on a criterion, based on binary cue values retrieved from memory. For convenience, the cue value that signals a higher criterion value is 1, and the other cue value is 0. Take-the-best consists of three building blocks:

1. Search rule: Search through cues in order of their validity.
2. Stopping rule: Stop on finding the first cue that discriminates between the alternatives (i.e., cue values are 1 and 0).
3. Decision rule: Infer that the alternative with the positive cue value (1) has the higher criterion value.

Take-the-best simplifies decision making by stopping after the first cue and ordering cues unconditionally according to validity *v*, which is given by:

$$v = C/(C+W),$$

where *C* is the number of correct inferences when a cue discriminates, and *W* is the number of wrong inferences. Alternative search rules such as success (Martignon & Hoffrage, 2002; Newell et al., 2004) and discrimination (Gigerenzer & Goldstein, 1996b) have been investigated. Todd and Dieckmann (2005) studied alternative simple principles for learning cue orders. Karelaia (2006) showed that a "confirmatory" stopping rule—stop after two cues are found that point to the same alternative—leads to remarkably robust results across varying cue orders, which is ecologically rational in situations where the decision maker knows little about the validity of the cues.

A striking discovery was that take-the-best can predict more accurately than linear multiple regression models (Czerlinski et al., 1999). It can even predict more accurately than complex nonlinear strategies. Figure 7.1 shows the predictive accuracy of an exemplar-based model (nearest-neighbor classifier), Quinlan's decision-tree induction algorithm C4.5, and classification and regression trees (CARTs), compared to take-the-best. In both tasks, and across most sample sizes, take-the-best achieves higher predictive accuracy than each of the three complex strategies (Brighton & Gigerenzer, 2011). This

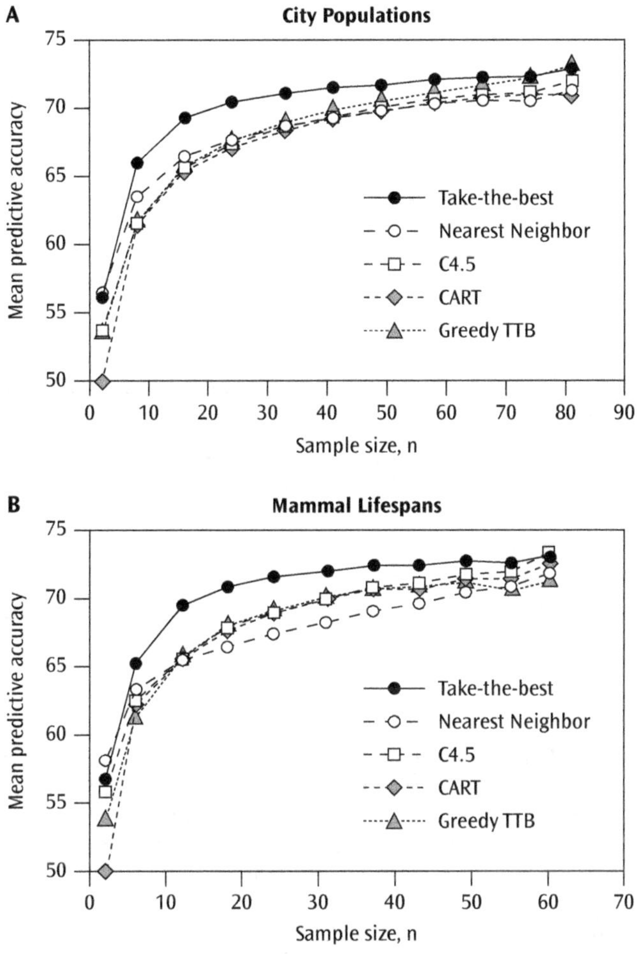

Figure 7.1: Less can be more. A competition between take-the-best and three well-known learning algorithms [nearest neighbor classifier, Quinlan's decision-tree induction algorithm C4.5, and classification and regression tree (CART)], also including a greedy version of take-the-best (TTB) that orders cues by conditional validity instead of unconditional validity. Mean predictive accuracy in cross-validation is plotted as a function of the sample size of the training set. The two tasks were deciding (*A*) which of two German cities has more inhabitants, and (*B*) which of two mammal species lives longer on average (Brighton & Gigerenzer, 2011).

is not to say that relying on one good reason is always better, but the result in Figure 7.1 is the most frequently obtained in a total of 20 environments. Note that CARTs have been designed to be robust against estimation error (variance) due to small samples and other factors. These complex algorithms can mimic the outcome of take-the-best in the sense that they are models that include take-the-best as a special case. Yet, although their greater flexibility leads to better fit of known data, more general models do not necessarily lead to better predictions of unknown data.

As noted above, take-the-best orders cues unconditionally, unlike the other models in Figure 7.1. Ordering cues conditionally, that is, taking their interdependencies into account, may seem a more rational strategy. In fact, in a small world where all cue validities are perfectly known, conditional validity leads to higher (or at least equal) accuracy than unconditional validity (Schmitt & Martignon, 2006). However, in uncertain worlds as in Figure 7.1, where the cue order needs to be estimated from samples, this no longer holds. If one makes take-the-best more sophisticated by ordering cues conditionally (greedy take-the-best), the predictive accuracy drops to the level of the complex strategies (Figure 7.1). This suggests that the predictive power of take-the-best stems mostly from the search rule rather than the stopping rule.

The ecological rationality of take-the-best has been studied in three different situations: (*a*) when the cue order is known (Katsikopoulos & Martignon, 2006; Martignon & Hoffrage, 2002), (*b*) when error is introduced in that knowledge (Hogarth & Karelaia, 2007), and (*c*) when the order of cues needs to be inferred from samples (Brighton, 2006; Gigerenzer & Brighton, 2009). This research has led to important results that explain when and why less is more. For instance, why is take-the-best often as accurate as linear decision rules? One result is that take-the-best produces exactly the same inferences as a linear rule if any of three conditions holds:

1. *Dominance* (the cue values of one object dominate those of the other).
2. *Cumulative dominance* (the cue values of one object cumulatively dominate those of the other; see Baucells et al., 2008).
3. *Noncompensatoriness* (the weight of each higher ranked cue is larger than the sum of the weights of all lower-ranked cues; see Martignon & Hoffrage, 2002, and Chapter 8).

How often are these conditions met? An investigation of 51 natural environments (such as predicting which house will be sold at a higher price; which party dining in a restaurant will give a higher tip) showed that these conditions are the rule rather than the exception. In 90% of all inferences (97% when cues are binary or dichotimized), at least one of the three conditions was met in natural environments (Şimşek, 2013). Here, relying on one-reason decision-making generates the same accuracy as when computing the weights of many cues in a linear decision rule.

Many experimental studies asked the descriptive question whether take-the-best can predict people's inferences (e.g., Bröder, 2003; Bröder & Gaissmaier, 2007; Bröder & Schiffer, 2006; Newell & Shanks, 2003; Rieskamp &

Hoffrage, 1999). Dieckmann and Rieskamp (2007) first showed that in environments with high redundancy, take-the-best is as accurate as and more frugal than naïve Bayes (a strategy that integrates all cues), and then experimentally demonstrated that in high-redundancy environments, take-the-best predicted participants' judgments best, whereas in low-redundancy environments, compensatory strategies predicted best, indicating adaptive strategy selection. Rieskamp and Otto (2006) showed that in an environment with high variability of cue validities, judgments consistent with take-the-best increased over experimental trials from 28% to 71%, whereas in an environment with low variability, they decreased to 12%. Bröder (2003) reported similar selection of take-the-best dependent on the variability or cue validities. In several experiments, individuals classified as take-the-best users for tasks where the heuristic is ecologically rational showed higher IQs than those who were classified as compensatory decision makers, suggesting that cognitive capacity as measured by IQ "is not consumed by strategy execution, but rather by strategy selection" (Bröder & Newell, 2008, p. 209).

Bergert and Nosofsky (2007) formulated a stochastic version of take-the-best, tested it against a weighted additive model at the individual level, and concluded that the vast majority of participants adopted the take-the-best heuristic. Comparing take-the-best with both weighted additive and exemplar models of categorization, Nosofsky and Bergert (2007) found that most participants did not use an exemplar-based strategy but instead followed the response time predictions of take-the-best. Bröder and Gaissmaier (2007) analyzed five published experiments and one new experiment, and reported that in all instances when decision outcomes indicated the use of take-the-best, decision times increased monotonically with the number of cues that had to be searched in memory, as predicted by take-the-best's search and stopping rules. Taken together, these studies indicate systematic individual differences in strategy use and adaptive use of take-the-best.

García-Retamero and Dhami (2009) tested how policemen, professional burglars, and laypeople infer which of two residential properties is more likely to be burgled. Both expert groups' inferences were best modeled by take-the-best, and laypeople's inferences by a weighted additive rule. The latter may reflect that laypeople need to explore all the information, whereas experts know what is relevant, consistent with findings of the literature on expertise (Ericsson et al., 2007; Reyna & Lloyd, 2006; Shanteau, 1992).

Concerns were raised by Juslin and Persson (2002) that take-the-best is not so simple after all but requires complex computations for ordering the cues; Dougherty et al. (2008) and Newell (2005) voiced similar concerns. First, it is true that estimating validity order can sometimes be nontrivial, yet it is simpler than estimating other kinds of weights such as regression weights. Second, people estimate order from samples rather than by calculating the "true" order from perfect knowledge about the entire population, as Juslin and Persson assumed. Even with minute sample sizes of two to ten—resulting in estimated orders that deviate from the true order—take-the-best predicted more accurately than multiple regression when both were provided with continuous cue values (Katsikopoulos et al., 2010). Finally, a person does not need to learn cue orders individually but instead can learn from others, as through teaching and imitation (Gigerenzer et al., 2008).

Consumer choice. How do consumers decide which product to buy among an ever-increasing assortment on the Internet or on supermarket shelves? The classical methodology to answer this question has been conjoint analysis, which assumes a weighted linear combination of features or cues. When John Hauser, a proponent of conjoint analysis, began to test models of heuristics, he found to his surprise that sequential heuristics predict consumer choices well (Hauser et al., 2009). Examples are decisions between computers (Kohli & Jedidi, 2007) and smartphones (Yee et al., 2007). In particular, heuristics are important early in the decision process to form a consideration set, which consists of eliminating most products from further consideration. Once the consideration set is formed, consumers evaluate the remaining options more carefully (Gaskin et al., 2007; see also Reisen et al., 2008). Within their consideration set of potential suppliers, they then appear to trade off price and reliability to reach their final choice.

Literature search. How should an organization design a search algorithm for prioritizing literature searches from the PsycINFO database? Lee and colleagues (2002) engineered two methods for identifying articles relevant to a given topic of interest (e.g., eyewitness testimony), one a variant of take-the-best, the other a Bayesian model using all available information. Lee et al. tested both methods on ten actual literature searches and measured the methods' performances against effort (i.e., the proportion of the articles read by the user) and accuracy (i.e., proportion of relevant articles found). The variant of take-the-best was as good as or better than the Bayesian model, particularly in searches in which the proportion of relevant articles was small.

Fast-and-Frugal Trees

One way to model classification is in terms of trees. For instance, Bayes rule can be represented as a tree with 2^m leaves, where m is the number of binary cues or attributes. Natural frequencies provide such a representation (see Chapter 3). Yet when the number of cues grows, a Bayesian analysis—with or without natural frequencies—becomes computationally intractable or fraught with estimation error because one typically has too few data points for the thousands of leaves of such a gigantic tree. A fast-and-frugal tree has only $m + 1$ leaves and thus is likely more robust. It has building blocks similar to take-the-best (Martignon et al., 2003):

1. Search rule: Search through cues in a predetermined order.
2. Stopping rule: Stop search as soon as a cue leads to an exit.
3. Decision rule: Classify the object accordingly.

Fast-and-frugal trees are used by experts in many fields, from cancer screening to bail decisions (see Figure 7.2). Martignon et al. (2008) tested the accuracy of fast-and-frugal trees in 30 classification problems from fields such as medicine, sports, and economics. They reported that complex

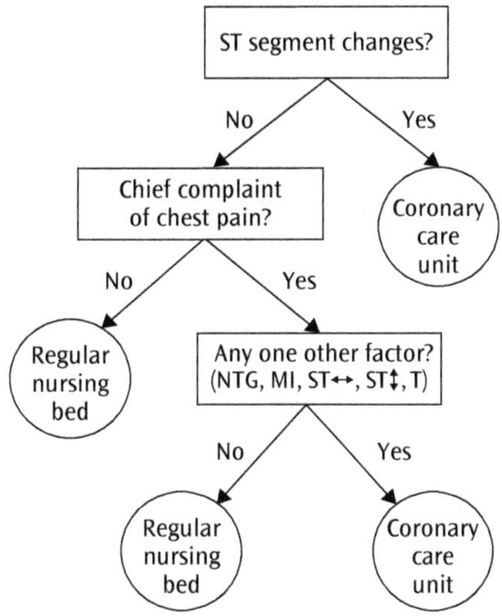

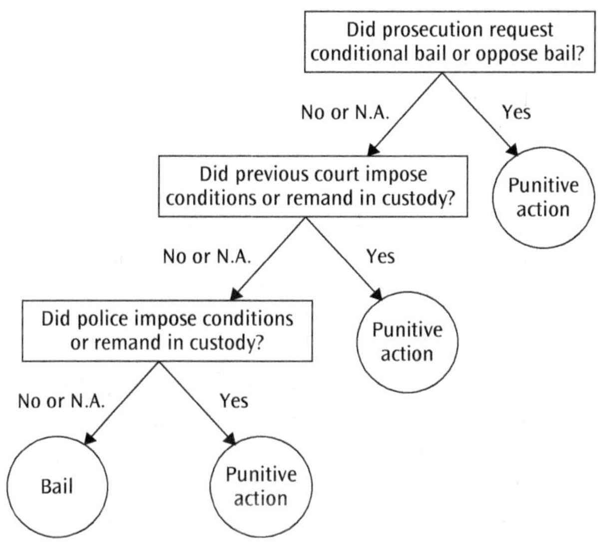

Figure 7.2: Fast-and-frugal trees for medical and legal decisions. The tree on the top prescribes how emergency physicians can detect acute ischemic heart disease. It only asks up to three yes/no questions, namely whether the patient's electrocardiogram shows a certain anomaly ("ST segment changes"), whether chest pain is the patient's primary complaint, and whether there is any other factor (Green & Mehr, 1997). The tree on the bottom describes how magistrates at a London court decided whether to bail a defendant or to react punitively by imposing conditions such as curfew or imprisonment. The logic is defensive and "passes the buck." The tree predicted 92% of bail decisions correctly (Dhami, 2003). Abbreviations: MI, myocardial infarction; N.A., not applicable; NTG, nitroglycerin; T, T-waves with peaking or inversion.

benchmark strategies including logistic regression excelled in data fitting, but fast-and-frugal trees were close or identical to these strategies in their predictive accuracy.

Emergency medicine. When patients arrive at the hospital with severe chest pain, emergency physicians have to decide quickly whether they suffer from acute ischemic heart disease and should be assigned to the intensive coronary care unit (ICU). In a Michigan hospital, doctors preferred to err on what they believed was the safe side by sending about 90% of the patients to the ICU, although only about 25% of these actually had a myocardial infarction (Green & Mehr, 1997). The result was an overly crowded ICU, a decrease in quality of care, an increase in cost, and a risk of serious infection among those who were incorrectly assigned. Something had to be done. Green and Mehr (1997) tried two solutions: (*a*) a logistic regression, the Heart Disease Predictive Instrument (HDPI), and (*b*) a fast-and-frugal tree. To use the HDPI, doctors received a chart with some 50 probabilities, checked the presence and absence of symptoms, and inserted the relevant probabilities into a pocket calculator. The fast-and-frugal tree ignored all probabilities and asked only a few yes-or-no questions (Figure 7.2, top). Ultimately, the tree was more accurate in predicting actual heart attacks than the HDPI: It sent fewer patients who suffered from a heart attack wrongly into a regular bed and also nearly halved physicians' high false alarm rate. Last but not least, the tree was transparent, easy to memorize, and easy to modify, and was accepted by physicians who disliked relying on a logistic regression they barely understood.

Easy memorization is an important feature of fast-and-frugal trees, particularly in emergency situations. After the terrorist attacks on September 11, 2001, START (Simple Triage and Rapid Treatment; Cook, 2001) helped paramedics to quickly classify victims into two major categories: those who needed medical treatment immediately and those whose treatment could be delayed. A tree with only two cues—age and duration of fever—was developed to decide upon macrolide prescription in young children with community-acquired pneumonia (Fischer et al., 2002). This tree was slightly less accurate than a scoring system based on logistic regression (72% versus 75%), but it does not require any expensive technology and thus can be applied to millions of children worldwide who would otherwise not have access to health care.

How to model physicians' thinking? Taking for granted that physicians use heuristics for diagnosing patients, the medical community quickly adopted the heuristics-and-biases view and has left it mainly unrevised as of today (Croskerry, 2009). For instance, Elstein (1999) described heuristics as "mental shortcuts commonly used in decision making that can lead to faulty reasoning or conclusions" (p. 791) and blamed them for many errors in clinical reasoning. Some researchers, however, recognize their potential to improve decisions. McDonald (1996), for one, wrote, "admitting the role of heuristics confers no shame" (p. 56). Rather, the goal should be to formalize and understand heuristics so that their use can be effectively taught, which could lead to less practice variation and more efficient medical care.

"The next frontier will involve fast and frugal heuristics; rules for patients and clinicians alike" (Elwyn et al., 2001, p. 358).

For diagnosis, which is a form of classification, fast-and-frugal trees potentially model how physicians make decisions. For treatment choice, all heuristics described above are potential models. Both fast-and-frugal trees and other heuristics differ from traditional models of medical decision making, such as logistic regression for classification and expected utility maximization for choice. Dhami and Harries (2001) compared a fast-and-frugal tree ("matching heuristic") to a linear regression model on general practitioners' decisions to prescribe lipid-lowering drugs for a set of hypothetical patients. Both models fitted prescription decisions equally well, but the simple tree relied on less information. Similar results were obtained by Smith and Gilhooly (2006) and Backlund et al. (2009). These studies reported only fitting—not predicting—physicians' judgments, which is a limitation. More direct evidence comes from the routine use of fast-and-frugal trees by physicians in cancer screening and HIV tests.

Bail decisions. Heuristics matter in the law in multiple respects. They play a role in the making of law (Haidt et al., 2006) as well as in litigation (Hastie & Wittenbrink, 2006). In both domains, there has been debate whether heuristics are a problem or a solution (Gigerenzer & Engel, 2006).

One of the initial decisions of the legal system is whether to bail a defendant unconditionally or to react punitively by conditions such as curfew or imprisonment. In England and Wales, around two million bail decisions are made every year by benches of two or three magistrates, 99.9% of whom are members of the local community without legal training. How do they make these decisions? When magistrates were interviewed, they generally responded that they thoroughly examined and weighed all information in a complex way (Dhami & Ayton, 2001). However, when Dhami (2003) observed several hundreds of trials in two London courts, she found that the average time magistrates spent on a case was 6 to 10 minutes and that their decisions could be predicted better with a fast-and-frugal tree ("matching heuristic") than with weighting and adding all information (Figure 7.2, bottom). The logic of the tree appears to be to "pass the buck," because it copies the punitive decisions of the prosecution, a previous court, or the police. It violates due process because it ignores relevant information about the defendant. In the two courts, the fast-and-frugal trees predicted 92% and 85% of all decisions correctly (cross-validation), compared to 86% and 73% by a weighted additive model that would correspond to due process and what magistrates responded in the interviews.

Trade-off Heuristics

Unlike recognition-based and one-reason decisions, the third class of heuristics weights cues or alternatives equally and thus makes trade-offs (compensatory strategies).

Tallying

Whereas take-the-best ignores cues (but includes a simple form of weighting cues by ordering them), tallying ignores weights, weighting all cues equally. It entails simply counting the number of cues favoring one alternative in comparison to others.

1. Search rule: Search through cues in any order.
2. Stopping rule: Stop search after m out of a total of M cues (with $1 < m \leq M$). If the number of positive cues is the same for both alternatives, search for another cue. If no more cues are found, guess.
3. Decision rule: Decide for the alternative that is favored by more cues.

Dawes (1979; Dawes & Corrigan, 1974) showed that tallying was about as accurate as multiple regression and sometimes even better. In a more extensive test across 20 environments, Czerlinski et al. (1999) demonstrated that tallying had, on average, a higher predictive accuracy. The challenge is to figure out *when* this is the case. Einhorn and Hogarth (1975) found that unit-weight models were successful in comparison to multiple regression when the ratio of alternatives to cues was 10 or smaller, the linear predictability of the criterion was small ($R^2 \leq 0.5$), and cues were highly redundant. Relatively few studies have identified conditions under which people would use a tallying strategy. Interestingly, it seems that more people prefer to dispense with particular cues (as in one-reason decision making) than with cue order or weights (Bröder & Schiffer, 2003; Rieskamp & Hoffrage, 2008; but see Wang, 2008). One reason for the relatively low prevalence of tallying could be that these studies used only few cues, typically four or five. Below we provide two illustrations of the prescriptive use of tallying in institutions (for more, see Astebro & Elhedhli, 2006; Graefe & Armstrong, 2013; Lichtman, 2008; Wang, 2008).

Magnetic resonance imaging (MRI) or simple bedside rules? There are about 2.6 million emergency room visits for dizziness or vertigo in the United States every year (Kattah et al., 2009). The challenging task for the emergency physician is to detect the rare cases where dizziness is due to a dangerous brainstem or cerebellar stroke. Frontline misdiagnosis of strokes happens in about 35% of the cases. One solution to this challenge could be technology. Getting an early MRI with diffusion-weighted imaging takes 5 to 10 minutes plus several hours of waiting time, costs more than $1,000, and is not readily available everywhere. However, Kattah et al. (2009) developed a simple bedside eye exam that actually outperforms MRI and takes only about one minute: It consists of three tests and raises an alarm if at least one indicates a stroke. This simple tallying rule correctly detected 100% of those patients who actually had a stroke (sensitivity), whereas an early MRI detected only 88%. Out of 25 patients who did not have a stroke, the bedside exam raised a false alarm in only one case (i.e., 4% false positive rate = 96% specificity). Even though the MRI did not raise any false alarms, the bedside exam seems preferable in total, given that misses are more severe than false alarms and that it is faster, cheaper, and universally applicable.

Avoiding avalanche accidents. Hikers and skiers need to know when avalanches could occur. The obvious clues method is a tallying heuristic that checks how many out of seven cues have been observed en route or on the slope that is evaluated (McCammon & Hägeli, 2007). These cues include whether there has been an avalanche in the past 48 hours and whether there is liquid water present on the snow surface as a result of recent sudden warming. When more than three of these cues are present on a given slope, the situation should be considered dangerous. With this simple tallying strategy, 92% of the historical accidents (where the method would have been applicable) could have been prevented.

> **Trade-offs:** a class of heuristics that weights all cues or alternatives equally and thus makes trade-offs (e.g., tallying and 1/N)

Mapping Model

How do people arrive at quantitative estimates based on cues? The mapping model assumes that people tally the number of relevant cues with an object's positive values (von Helversen & Rieskamp, 2008). The estimate is the median criterion value of objects with the same number of positive cues. The mapping model captured people's judgment better than a linear regression and an exemplar model when the criterion values followed a skewed distribution.

Sentencing decision. In the adversarial U.S. legal system, the vast majority of cases are closed by plea bargaining, where the prosecution and defense negotiate a sentence, which is then ratified by a judge. In contrast, in Germany and many other countries, plea bargaining before a case goes to court is an exception rather than the rule. Here, the judge has to determine an appropriate sentence proportional to the offender's guilt, within the range of the minimum and maximum sentence specified for each offense. The single most important factor influencing judges' decisions is the prosecution's sentencing recommendation. How should the prosecution make its recommendation? The German penal code lists over 20 factors to consider. The legal literature recommends a three-step strategy: Determine first all relevant factors and the direction of their effect on the sentence (aggravating or mitigating), then weight these by their importance, and add them up to determine the sentence. Von Helversen and Rieskamp (2009) analyzed trial records of sentencing and tested five models of how sentencing decisions have been made in theft, fraud, and forgery, including a linear regression model. The best predictions of actual sentences were obtained by the mapping model, a heuristic model of quantitative estimation, based on a simple tallying rule described above. As von Helversen and Rieskamp (2009) pointed out, this result "provides further evidence that legal decision makers rely heavily on simple decision heuristics . . . and suggests that eliciting these employed heuristics is an important step in understanding and improving legal decision making" (pp. 389–390).

1/N Rule

Another variant of the equal weighting principle is the 1/*N* rule, which is a simple heuristic for the allocation of resources (time, money) to *N* alternatives:

1/*N* rule: Allocate resources equally to each of *N* alternatives.

This rule is also known as the equality heuristic (Messick, 1993). Sharing an amount of money equally is the modal response in the one-shot ultimatum game for adults and also the most frequent split in children's group decisions, contrary to the predictions of selfish behavior (Takezawa et al., 2006).

Investment. When deciding how to allocate financial resources among *N* options, some individuals rely on the 1/*N* rule (Benartzi & Thaler, 2001), which allocates financial resources equally across all alternatives. The 1/*N* rule was compared to 14 optimizing models, including a Nobel Prize–winning model, Markowitz's mean variance portfolio, in seven investment problems (DeMiguel et al., 2009). To estimate the models' parameters, each optimizing strategy received 10 years of stock data and then had to predict the next month's performance on this basis. The same procedure was repeated, with a moving window, for the next month, and so forth, until no data were left. Note that 1/*N* does not have any free parameters that need to be estimated (and thus has no error due to "variance"). Nevertheless, it came out first on certainty equivalent returns, second on turnover, and fifth on the Sharpe ratio. None of the complex optimizing models could consistently beat it.

Social Intelligence

According to the social intelligence hypothesis, also called the Machiavellian intelligence hypothesis (Whiten & Byrne, 1997), highly social species such as humans and other social primates should be intellectually superior to less social ones because the social environment is more complex, less predictable, and more intellectually challenging. In Humphrey's (1976/1988) words, social primates "must be able to calculate the consequences of their own behavior, to calculate the likely behaviours of others, to calculate the balance of advantage and loss" (p. 19). For the sake of argument, let us assume that the social world is indeed more complex and unpredictable than the nonsocial one. Would social intelligence therefore require more complex cognition? Not necessarily, according to the following two hypotheses (Hertwig & Herzog, 2009):

1. Social intelligence does not require complex mental calculation; it also works with heuristics.
2. The same heuristics that underlie nonsocial decision making also apply to social decisions (but not vice versa).

The justification for hypothesis 1 is the same as for complex nonsocial problems: The more unpredictable a situation is, the more information needs to be ignored to predict the future. One reason for hypothesis 2 is that the distinction between social and nonsocial is a common oversimplification in the first place. Nevertheless, for the purpose of this review, we distinguish two meanings of social: whether the input into a strategy is social information (e.g., when imitating the behavior of a peer) or not (e.g., features of digital cameras), and whether the task is a game against nature or a social game involving other humans (Hertwig et al., 2012b). The goals of social intelligence go beyond accuracy, frugality, and making fast decisions. They include transparency, group loyalty, and accountability (Lerner & Tetlock, 1999). Consistent with hypothesis 2, heuristics from all three classes discussed above have been investigated in social situations. Below are a few examples.

Recognition-Based Decisions

Reimer and Katsikopoulos (2004) first showed analytically that less-is-more effects are larger in group decisions than in individual decisions and subsequently demonstrated this effect empirically in an experiment in which another fascinating phenomenon emerged. Consider a group of three in which one member recognized only city *a* while the other two members recognized both cities *a* and *b* and individually chose *b* as the larger one. The majority rule predicts that *b* would always be selected, yet in 59% of the cases, the final group decision was *a*, following the one who had not heard of *b*.

One-Reason Decision Making

As mentioned above, the behavior of most people in the one-shot ultimatum game is inconsistent with the classical economic predictions. Most researchers nevertheless retained the utility-maximizing framework and added free parameters for other-regarding dispositions (e.g., Fehr & Schmidt, 1999). In contrast, Rubinstein (2003) called for a radical change, "to open the black box of decision making, and come up with some completely new and fresh modeling devices" (p. 1215). Hertwig, Fischbacher, and Bruhin (2012) did so and modeled the individual differences observed in the ultimatum game by fast-and-frugal trees of different sizes, involving one to four cues. The number of cues predicted how long decisions took.

Trade-off Heuristics

Many parents try to divide their time every day between their *N* children equally by 1/*N*. If parents have only two children, 1/*N* will attain the long-term goal of providing each child with as much time as the other. But if there are three or more children (excepting multiple births), only the first-born and last-born have exclusive time with the parents, while the middle-borns have to share with their siblings throughout their childhood and thus end up receiving less time in total. The simple 1/*N* rule predicts a complex pattern of care time for each child, a pattern observed in a survey of 1,296 families (Hertwig et al., 2002). This result illustrates that a heuristic and its goal (fair division during childhood) are not the same—the environment has the last word. The majority rule is a second example of a tallying rule applied to group decisions; it also defines democratic voting systems (Hastie & Kameda, 2005).

Social Heuristics

Although the heuristics discussed so far can be fed with both social and nonsocial information, there are genuinely social heuristics designed exclusively for social information. Examples include imitation heuristics, tit-for-tat, the social-circle heuristic, and averaging the judgments of others to exploit the "wisdom of crowds" (Hertwig & Herzog, 2009). Imitate-the-successful, for instance, speeds up learning of cue orders and can find orders that excel take-the-best's validity order (García-Retamero, Takezawa, & Gigerenzer 2009). Social heuristics prove particularly helpful in situations in which the actor has little knowledge. The classic example is that of Francis Galton, who visited a livestock fair where villagers estimated the weight of an ox and was surprised to find that their median and mean average estimates were only 9 and 1 pounds, respectively, off the actual weight of 1198 pounds (Galton, 1907).

A peculiar social rule is the default heuristic: "If there is a default, do nothing about it." Defaults are set by institutions and act as implicit recommendations (Johnson & Goldstein, 2003). Every year, an estimated 5,000 Americans and 1,000 Germans die while waiting for a suitable organ donor. Although most citizens profess that they approve of organ donation, relatively few sign a donor card: only about 28% and 12% in the United States and Germany, respectively. In contrast, 99.9% of the French and Austrians are potential donors. These striking differences can be explained by the default heuristic. In explicit-consent societies such as Germany, the law prescribes that nobody is a donor unless one opts in. In presumed-consent societies such as France, the default is that everyone is a donor unless one opts out. Although most people appear to follow the same heuristic, the result is drastically different because the legal environment differs.

Very few studies use large-scale demographic data to test social heuristics. For instance, marriage patterns are studied by demographers without much attention to the social heuristics that generate these, and vice versa. Todd and colleagues (2005) had the ingenious methodological insight that the aggregate demographic data rule out certain heuristics and can be used to test various satisficing strategies for mate choice.

Moral Behavior

Although moral behavior has long been attributed to conscious reflection, Haidt and Bjorklund (2008) argued that reasons are typically used to justify behavior after the fact and that the causes are mostly unconscious or intuitive. Gigerenzer (2010) proposed that these unconscious causes are often social heuristics, such as imitating the behavior of peers in order to gain acceptance by the group. Note that one and the same social heuristic can lead to behavior evaluated as moral or immoral, such as when imitating benevolent or malevolent peer behavior. This perspective on moral behavior is different from assuming that people have internalized specific moral rules such as don't steal and don't kill.

Moral behavior has been related to "sacred values" (Fiske & Tetlock, 1997). If one understands sacred values as top cues in lexicographic heuristics, decisions between alternatives where a sacred value conflicts with a secular value (e.g., life versus money) should be faster and easier than when two sacred values (e.g., one life versus another) conflict with each other, as reported by Hanselmann and Tanner (2008). Baron and Ritov (2009) argued that, from a utilitarian perspective, this form of one-reason decision making can cause great problems for policy decisions as it could prevent trade-offs for the greater good. In the same vein, Sunstein (2005) asserted that moral heuristics can lead to great error, but added that we would not necessarily "be better off without them. On the contrary, such heuristics might well produce better results, from the moral point of view, than the feasible alternatives" (p. 535). Cosmides and Tooby (2006) located the origins of moral heuristics in our ancestral world of tiny bands of individuals. An example of a moral heuristic is an intuitive search rule that looks for information that could reveal whether one has been cheated in a social contract. This heuristic correctly predicts when information search in the Wason selection task contradicts propositional logic (Gigerenzer, 2000).

Conclusions

We began this review with the observation that the three major tools for modeling decision making—logic, statistics, and heuristics—have not been treated equally, with each suited to a particular kind of problem. Instead, in psychology, heuristics became associated with errors and contrasted with logical and statistical rules that were believed to define rational thinking in all situations. Yet this view has been questioned for uncertain worlds where the assumptions of rational models are not met. We reviewed studies on decisions by individuals and institutions, including business, medical, and legal decision making, that show that heuristics can often be more accurate than complex "rational" strategies. This puts heuristics on a par with statistical methods and emphasizes a new ecological question: In what environment does a given strategy (heuristic or otherwise) succeed? This insight adds a prescriptive research program to that of the existing descriptive research program on heuristics. Pioneers such as Dawes, Hogarth, and Makridakis demonstrated years ago that simple forecasting methods can often predict better than standard statistical procedures; as James March, one of the most influential researchers in organizational decision making, put it more than 30 years ago, "If behavior that apparently deviates from standard procedures of calculated rationality can be shown to be intelligent, then it can plausibly be argued that models of calculated rationality are deficient not only as descriptors of human behavior but also as guides to intelligent choice" (1978, p. 593).

Nonetheless, a large community continues to routinely model behavior with complex statistical procedures without testing these against simple rules. Yet a fundamental change in thinking about human and animal behavior seems to be occurring. Mathematical biologists McNamara and Houston (2009) described this shift: "Although behavioral ecologists have built complex models of optimal behavior in simple environments, we argue that they need to focus on simple mechanisms that perform well in complex environments" (p. 670).

Formal models also help to answer the descriptive question of when people rely on what heuristic. As for the prescriptive question, a similar conflict is waged between those who argue in favor of classical statistical techniques as models of the mind (typically weighting and adding of all information) and those who argue that many people consistently rely on heuristics. The best way to decide is comparative testing; the difficulty is to understand the individual differences reported in most experiments.

With all these new insights, we are left with big challenges. How should we develop a systematic theory of the building blocks of heuristics and the core capacities as well as environmental structures that these exploit? To what extent can the emerging science of heuristic decision making provide a unifying framework for the study of the mind? One way to proceed is theory integration, that is, to connect the simple heuristics framework with other theoretical frameworks in psychology. This is already happening with ACT-R (Adaptive Control of Thought-Rational) theory (Schooler & Hertwig, 2005), signal detection theory (Luan et al., 2011, Pleskac, 2007), and the heuristics-and-biases program (Read & Grushka-Cockayne, 2011). In physics, theory integration, such as quantum theory and relativity theory, is a primary goal. In psychology, theory integration is not accorded the importance it deserves; instead, the field still resembles a colorful loose patchwork. We envision that the study of cognitive heuristics may help to sew some of the pieces together.

Summary Points

1. Heuristics can be more accurate than more complex strategies even though they process less information (less-is-more effects).
2. A heuristic is not good or bad, rational or irrational; its accuracy depends on the structure of the environment (ecological rationality).

3. Heuristics are embodied and situated in the sense that they exploit core capacities of the brain and their success depends on the structure of the environment. They provide an alternative to stable traits, attitudes, preferences, and other internal explanations of behavior.
4. With sufficient experience, people learn to select proper heuristics from their adaptive toolbox.
5. Usually, the same heuristic can be used both consciously and unconsciously, for inferences and preferences, and underlies social as well as nonsocial intelligence.
6. Decision making in organizations typically involves heuristics because the conditions for rational models rarely hold in an uncertain world.

Future Issues

1. How do people learn, individually or socially, to use heuristics in an adaptive way? And what prevents them from doing so? (For a start: Rieskamp & Otto, 2006)
2. Does intelligence mean knowing when to select which strategy from the adaptive toolbox? (For a start: Bröder & Newell, 2008)
3. Are gut feelings based on heuristics, and if so, on which? (For a start: Gigerenzer, 2007)
4. To what extent is moral (and immoral) behavior guided by social heuristics? (For a start: Gigerenzer, 2010; Sunstein, 2005)
5. How does the content of the adaptive toolbox change over the life span and between cultures? (For a start: Mata et al., 2007)
6. Can people adapt the use of heuristics to idiosyncrasies in their core capacities, such as differences in memory span, but also differences in knowledge? (For a start: Bröder & Gaissmaier, 2007)
7. Which heuristics do humans share with which animals, and why? (For a start: Hutchinson & Gigerenzer, 2005 and commentaries)
8. Finally, the overarching goal: Develop a systematic theory of the building blocks of cognitive heuristics (such as search, stopping, and decision rules) anchored in core capacities and the social and physical structures they exploit.

Acknowledgments

We thank Mirta Galesic, Sebastian Hafenbrädl, Ralph Hertwig, Ulrich Hoffrage, Konstantinos Katsikopoulos, Julian N. Marewski, and Lael J. Schooler for helpful comments.

Chapter 8

The Recognition Heuristic

A Decade of Research

Introduction

With Herbert Simon's (1990) emphasis on recognition memory and limited search as a starting point, it was only a small logical step toward the recognition heuristic, which exploits the potential information in a lack of recognition. In accordance with Simon's emphasis on computational models, the recognition principle (as it was first called) was formulated as a building block of take-the-best and other heuristics, in order to model inferences from memory (Gigerenzer & Goldstein, 1996b). Subsequently, it was realized that this initial building block could function as a stand-alone model for the same type of inferences, and it was named the recognition heuristic (Goldstein & Gigerenzer, 1999, 2002).

In reality, the recognition heuristic was not derived in such a logical manner. Serendipity, the luck of finding something one was not looking for, assisted its birth. Gigerenzer, Hoffrage, and Kleinbölting (1991, Prediction 4) had deduced from probabilistic mental models theory a situation in which the "hard–easy" effect would disappear. In his dissertation, Ulrich Hoffrage (1995; described in Hoffrage, 2011) set out to test this prediction, for which he needed two sets of questions, one hard, one easy. Hoffrage chose questions concerning the populations of American cities and German cities, which are respectively hard and easy for German students—or so everyone thought. Surprisingly, the students scored slightly higher when tested on a representative sample of American cities than on German ones. The result ruined the experiment. How could people score more correct answers in a domain in which they knew less? For days, our research group failed to think of a cognitive process that makes more out of less. Finally, Anton Kühberger pointed out that the explanation was tucked

Originally published as Gigerenzer, G., & Goldstein, D. G. (2011). The recognition heuristic: A decade of research. *Judgment and Decision Making*, 6, 100–121 away in the Gigerenzer et al. (1991) article, which mentioned "familiarity" as a probabilistic cue. If a person has heard of one city but not the other, this lack of recognition can be informative, indicating that the recognized city probably has the larger population. For the German cities, the participants could not rely on the recognition heuristic—they knew too much. This serendipitous discovery also revealed a crucial condition for the successful reliance on recognition: a substantial correlation between recognition and population (the recognition validity), and a representative sampling of the cities. We return to this condition later.

One possible reason why it took us so long to find the answer was our training in classical statistical models. In a weighted linear model, adding a cue or predictor can never decrease its fit, such as unadjusted R^2, and the same is true for Bayes rule (McGrath, 2008). This more-is-better principle holds for fitting parameters to known data, but not necessarily for predicting what one does not already know, as the German students had to do. A good cognitive heuristic, however, should excel in foresight as well as in hindsight.

The possibility that people could sometimes do better with less knowledge has generated much interest and controversy in the social sciences and in the media. In May 2009, the BBC, intrigued by the idea of less being more, decided to test the effect on their Radio 4 "More or less" program. Listeners in New York and London were asked whether Detroit or Milwaukee has the larger population. In exploratory studies for his dissertation, one of us (Goldstein, 1997) had found that about 60% of American students answered correctly ("Detroit"), compared to 90% of a corresponding group of German participants. The BBC is not known for running tightly controlled studies, and so we were somewhat uneasy about whether they could reproduce a less-is-more effect. But they did. In New York, 65% of the listeners got the answer right, whereas in London, 82% did so—as close as one can hope for an informal replication.

Our initial work on the recognition heuristic has stimulated dozens of articles comprising theoretical advancements, critique, and above all, progress. This work has contributed much to understanding the potential and limits of this simple model, but we also believe that its broad reception represents a larger shift in research practice. This change occurs in three directions:

1. *From labels to models of heuristics.* It is an interesting feature of the recent history of psychology that vague labels such as *availability* had remained largely unquestioned for three decades (for an exception, see Wallsten, 1983), whereas the precise model of the recognition heuristic immediately led to heated debates.
2. *From preferences to inferences.* While formal models such as elimination-by-aspects and lexicographic rules have been studied for preferences (e.g., Payne, Bettman, & Johnson, 1993; Tversky, 1972), their accuracy was typically measured against the gold standard of adding and weighting all information. In this earlier

research, a heuristic could not—by definition—be more accurate, only more frugal. For inferences, in contrast, there exists a clear-cut criterion, and thus it was possible to show that cognition can actually achieve more accuracy with simple heuristics than with weighted additive rules. This leads to the third shift in research.

3. *From logical to ecological rationality.* For decades, human rationality was studied in psychology by proposing a logical or statistical rule (e.g., truth table logic; Bayes rule) as normative in all situations, and then constructing an artificial problem in which this rule could be followed, such as the Wason Selection Task (Wason, 1971) or the Linda Problem (Kahneman & Tversky, 1982). In contrast, the question of ecological rationality asks in which environment a given strategy (heuristic or otherwise) excels and in which it fails. No rule is known that is rational per se, or best in all tasks. Parts of the psychological research community have resisted the asking of questions about ecological as opposed to logical rationality.

We begin our review of the progress made in the last decade with the two key processes that govern the use of the recognition heuristic: recognition and evaluation, the latter of which corresponds to a judgment of its ecological rationality.

The Recognition Process

The recognition heuristic makes inferences about criteria that are not directly accessible to the decision maker. When the criterion is known or can be logically deduced, *inferential* heuristics like the recognition heuristic do not apply. Relying on the heuristic is ecologically rational in an environment R where the recognition of objects $a, b \in R$ is strong and positively correlates with their criterion values. For two objects, the heuristic is:

> If one of two objects is recognized and the other is not, then infer that the recognized object has the higher value with respect to the criterion.

In our original work (Gigerenzer & Goldstein, 1996b, pp. 651–652; Goldstein & Gigerenzer, 2002, pp. 76–78), we assumed that the recognition heuristic will model inferences when three conditions hold:

(i) there is a substantial recognition validity;
(ii) inferences are made from memory, rather than from tables of information ("inferences from givens"), meaning that cue values for unrecognized objects are missing values; and
(iii) recognition stems from a person's natural environment (i.e., before entering the laboratory), as opposed to experimentally induced recognition.

We return to these characteristics below. We would like to emphasize that such a definition of the domain is essential, just as in game theory, where rational strategies are defined (and restricted) to game features such as anonymity, a fixed number of repetitions, and no reputation effects. This is not to say that studies that test predictions outside the domain are useless; on the contrary, they help to map out the boundary conditions more clearly, as we ourselves and other researchers have done. For example, we conducted a long-run experiment that subtly induced recognition over several weeks to investigate the effect of exogenous recognition on choice (Goldstein & Gigerenzer, 2002, pp. 84–85).

The recognition heuristic stands on the shoulders of the core psychological capacity of recognition memory; without it, the heuristic could not do its job. However, the recognition heuristic is mute about the nature of the recognition process, just as Bayes rule is mute about where the prior probabilities come from. Heuristics exploit core capacities in order to make fast and frugal judgments. Examples include recall memory (e.g., take-the-best), frequency monitoring (e.g., fast-and-frugal trees), mimicry (e.g., imitate-the-majority), and object tracking (e.g., gaze heuristic), with some heuristics taking advantage of several capacities (Gigerenzer & Brighton, 2009).

Connecting the Recognition Heuristic with the Recognition Process

In our original work, we did not investigate the link between the recognition heuristic and theories of the underlying recognition process. Since then, progress has been made toward this goal of theory integration, a topic that is of utmost importance in fields such as physics but is given little attention in psychology. As Walter Mischel (2006) put it, many psychologists still tend to treat theories like toothbrushes—no self-respecting person wants to use anyone else's. In one step toward theory integration, the recognition heuristic has been implemented based on the ACT-R model of memory (Anderson & Lebiere, 1998), which showed in some detail how forgetting—a process often seen as a nuisance and a handicap—can be functional in the context of inference (Schooler & Hertwig, 2005). In this same work, the fluency heuristic (Table 8.1) was formulated for the situation when both alternatives are recognized, that is, when the recognition heuristic cannot be applied. This work also integrated earlier work on fluency (e.g., Jacoby & Dallas, 1981) into the simple heuristics framework, defined the difference between the recognition and fluency heuristics, and thus contributed toward replacing verbal labels with computational models. Moreover, Schooler and Hertwig's analysis challenges the common belief that cognitive limits, such as forgetting or a limited working memory, inevitably pose liabilities for the human mind. Some cognitive limits foster specific cognitive processes, and at the same time some cognitive processes exploit specific cognitive limits—as may be the case in the interplay of forgetting and heuristic inference.

Table 8.1: Four Heuristics from the Adaptive Toolbox. Which to use for a given task? The content of individual memory determines whether an individual can apply the recognition heuristic (or other heuristics), and an evaluation process determines whether it should be applied.

Heuristic	Definition	Ecologically Rational If:	Predictions
Recognition heuristic (Goldstein & Gigerenzer, 2002)	If one of two alternatives is recognized, infer that it has the higher value on the criterion.	Recognition validity >.5	Contradicting information about the recognized object is ignored; less-is-more effect; forgetting is beneficial.
Fluency heuristic (Schooler & Hertwig, 2005)	If one alternative is recognized faster than another, infer that it has the higher value on the criterion.	Fluency validity >.5	Less-is-more effect; forgetting is beneficial (Hertwig et al., 2008).
Take-the-best (Gigerenzer & Goldstein, 1996b)	To infer which of two alternatives has the higher value: (1) search through cues in order of validity, (2) stop search as soon as a cue discriminates, (3) choose the alternative this cue favors.	Cue weights vary highly; moderate to high redundancy; scarce information (Hogarth & Karelaia, 2005, 2006; Katsikopoulos & Martignon, 2006; Martignon & Hoffrage, 1999, 2002).	Can predict as or more accurately as linear regression (Czerlinski et al., 1999), neural networks, exemplar models, and CARTs (Brighton, 2006).
Tallying/ unit-weight linear model (Dawes, 1979)	To estimate a criterion, do not estimate weights but simply count the number of favoring cues.	Cue weights vary little; low redundancy (Hogarth & Karelaia, 2005, 2006).	Can predict as or more accurately than multiple regression (Czerlinski, Gigerenzer, & Goldstein, 1999).

A second theoretical integration has combined a signal detection model of recognition memory with the recognition heuristic (Pleskac, 2007). In our original work, we had not separately analyzed how the recognition validity changes depending on what proportion of recognition judgments are correct. When recognizing an object, people can go wrong by erroneously recognizing something that they have never encountered before ("false alarms") and by failing to recognize something that they have previously encountered ("misses"). Pleskac showed that, as the error rate of recognition increases, the accuracy of the recognition heuristic declines, and that the less-is-more effect is more likely when participants' sensitivity (d') is high, whereas low sensitivities lead to "more-is-more." Furthermore, when people are cognizant of their level of recognition knowledge, they can increase their inferential accuracy by adjusting their decision criterion accordingly. When the amount of knowledge is very low, it may be prudent to be conservative in judging something as previously encountered; with increasing knowledge, however, it is better to become more liberal and classify something as previously encountered, even if one is not absolutely certain.

Is Recognition Binary?

In our original work, we modeled the input for the recognition heuristic, the recognition judgment, as binary. For instance, the brand name *Sony* would be modeled as either recognized or not. This simplifying assumption has been criticized (e.g., Hilbig & Pohl, 2009). However, it is consistent with theories that distinguish between a continuous process of recognition (or familiarity) and a binary judgment, such as signal detection theory, where an underlying continuous sensory process is transformed by a decision criterion into a binary recognition judgment. Moreover, there is now evidence that not only the recognition judgment, but the recognition process itself may be binary or threshold-like in nature. Bröder and Schütz (2009) argued that the widespread critique of threshold models is largely invalid, because it is, for the most part, based on confidence ratings, which are nondiagnostic for rejecting threshold models. In a reanalysis of 59 published studies, they concluded that threshold models fit the data better in about half of the cases. Thus, our assumption of a binary input into the recognition heuristic is a simplification, but not an unreasonable one, as it is supported by evidence and theories of the nature of recognition. (But see Hoffrage, 2011, sec. 3.3.5, for some evidence against a simple threshold.) Note that a model that assumes binary recognition judgments does not imply that organisms are unable to assess the degree to which something is familiar or frequent in the environment (Malmberg, 2002). In fact, models such as the fluency heuristic exploit such information (Schooler & Hertwig, 2005).

Individual Recognition Memory Constrains the Selection of Heuristics

No heuristic is applied indiscriminately to all situations (Payne, Bettman, & Johnson, 1993; Goldstein et al., 2001), and the recognition heuristic is no exception. How are heuristics selected from the adaptive toolbox? Marewski and Schooler (2011) have developed an ACT-R model of how memory can constrain the set of applicable heuristics. Consider the following set of strategies: the recognition heuristic, the fluency heuristic, take-the-best, and tallying (Table 8.1), in connection with the task of betting money on which tennis player, Andy Roddick or Tommy Robredo, will win against the other. Each of the four heuristics is *potentially* applicable for this task (the gaze heuristic, for instance, would be inapplicable). Whether a strategy is *actually* applicable for a given individual, however, depends on the state of individual memory. First, if an individual is ignorant about tennis and has heard of neither of the players, none of the heuristics can be applied and that person might simply guess. Second, if a person has heard of Roddick but not of Robredo, this state of memory restricts the choice set to the recognition heuristic; the bet would be on Roddick. As it turns out, Roddick and Robredo have played 25 sets against each other so far (by 2010) and Roddick has won 24 of them. The person who uses the recognition heuristic will, by definition, not be able to recall this fact from memory, having never heard of Robredo, but can nevertheless bet correctly. Third, consider an individual who has heard of both players, but recalls nothing else about them. This state of memory excludes the recognition heuristic, as well as take-the-best and tallying, and limits the choice set to the fluency heuristic: If both players are recognized, but one was recognized more quickly than the other, predict that the more quickly recognized player will win the game.

Finally, consider an individual more knowledgeable about tennis who has heard of both players, and can also recall the values of both on relevant cues, such as their current ATP Champions Race ranking, their ATP Entry ranking, their seeding by the Wimbledon experts, and the results of their previous matches. This state of memory again excludes the recognition heuristic, but leaves the other three heuristics in the choice set. To choose between these, principles of ecological rationality come into play. For instance, if cues are moderately to highly redundant, take-the-best has an advantage over tallying, and participants in experiments tend to prefer take-the-best after simply observing the structure of the environment (such as the degree to which cues were intercorrelated): No feedback about accuracy appears to be necessary (Dieckmann & Rieskamp, 2007). When feedback is available, Strategy Selection Learning theory (SSL theory) provides a quantitative model of heuristic selection (Rieskamp & Otto, 2006). SSL theory makes predictions about the probability that a person selects one heuristic within a defined set and shows how learning by feedback leads to adaptive strategy selection.

To summarize: In the decade after our initial work, we now have a sharp distinction between the recognition heuristic and the fluency heuristic. The effects of misses and false alarms on the accuracy of the recognition heuristic are better understood. Our grossly simplifying assumption of modeling recognition judgments as binary turned out to be largely consistent with a body of empirical evidence, although this issue is far from being settled. We postulate that individual recognition memory is the basis for the first of two steps by which an individual decides whether to rely on the recognition heuristic for solving a given task. The state of recognition memory determines whether it can be applied, while an evaluation process, our next topic, determines whether it should be applied.

The Evaluation Process

If the recognition heuristic satisfies the individual memory constraint (to recognize one of two objects), then an evaluation process is needed to determine whether relying on the recognition heuristic is ecologically rational for the particular inference being made. We titled our 2002 article "Models of *ecological* rationality: The recognition heuristic," emphasizing that the heuristic is not general-purpose, but selected in an adaptive way that depends on the environment (i.e., ecology). In our original work, we had specified one condition for the ecological rationality of the recognition heuristic (Goldstein & Gigerenzer, 2002, p. 87):

Substantial recognition validity. The recognition validity for a given criterion must be substantially higher than chance (α >.5).

Evaluating the recognition validity requires the existence of reference class R of objects (Goldstein & Gigerenzer, 2002, p. 78; Gigerenzer & Goldstein, 1996b, p. 654). We take this opportunity to clarify:

Precondition 1: Existence of a reference class. Without a reference class R (such as the class of all 128 contestants in Wimbledon Gentleman Singles), neither the experimenter nor the participant can estimate whether there is a substantial recognition validity. In other words, the more uncertain one is about the identity of the reference class, the less one can know about whether relying on the recognition heuristic is ecologically rational.

Precondition 2: Representative sampling. Assuming a substantial recognition validity, the successful use of the recognition heuristic for a specific pair $a, b \in R$ presupposes that it has been representatively sampled from R, rather than selectively sampled in a biased way (e.g., such that a high recognition validity in R is misleading for the specific task). For instance, when we asked international audiences during talks we have given outside the United States whether Detroit or Milwaukee has the larger population, some answered "Milwaukee" despite never having heard of it; they explained that they thought it was a trick question, that is, one selectively sampled for being counterintuitive. This suspicion reflects the widespread view that psychologists routinely deceive their participants (Hertwig & Ortmann, 2001), but is not the only reason why people may suspect biased sampling and overrule the recognition heuristic. Yet, whereas biased sampling can be hard for a participant to judge, the absence of a meaningful reference class can easily be noticed. Thus, we assume that, in the presence of a substantial recognition validity, people will consider applying the recognition heuristic by default, that is, unless there is reason to assume biased sampling of objects. The issue of representative sampling of questions is described in detail in Gigerenzer et al. (1991) and Hoffrage (2011).

Table 8.2 (in the Appendix) includes all studies we know of that report correct predictions of judgments by the recognition heuristic ("accordance rates"). It reports the reference class, the criterion, and whether the three conditions that define the domain of the recognition heuristic were in place. It also reports two methodological features: whether the recognition heuristic was tested comparatively against alternative models, and whether individual analyses were performed, as opposed to means only (see below). The last column shows the recognition validity and the mean correct predictions of judgments by the recognition heuristic.

Does the Strength of Recognition Validity Relate to the Predictive Accuracy of the Recognition Heuristic?

In our 2002 article, we had not systematically investigated this question. The research of the last years suggests that the answer may be affirmative. For instance, name recognition of Swiss cities is a valid predictor of their population ($\alpha = .86$), but not for their distance from the center of Switzerland ($\alpha = .51$). Pohl (2006) reported that 89% of inferences accorded with the recognition heuristic model in judgments of population, compared to only 54% in judgments of the distance. Similarly, the first study on aging indicates that old and young people alike adjust their reliance on the recognition heuristic between environments with high versus low recognition validities, even though old people have worse recognition memory (Pachur, Mata, & Schooler, 2009).

In Figure 8.1, we plotted the recognition validities against the proportion of correct predictions by the recognition heuristic (accordance rates) for all study conditions from Table 8.2. Included are all studies that reported recognition validities and correct predictions, even when the objects were not representatively sampled, as well as studies that tested the recognition heuristic outside its domain. Figure 8.1 shows that when participants were provided with up to three negative cues (black triangles and squares), the results still fall into the general pattern, while studies that tested the recognition heuristic outside its domain appear to result in lower correct predictions. Studies that used inferences from givens or experimentally induced recognition are shown by white and gray diamonds. Across all study conditions, the correct predictions is $r = .57$.

Note that there are two interpretations of this correlation. One is that most individuals engage in probability matching, that is, they rely on the heuristic in a proportion of trials that corresponds to the recognition validity. However, the evidence does not support this hypothesis (Pachur, Bröder, & Marewski, 2008, figure 5; Pachur & Hertwig, 2006). The second assumes differences among individuals and tasks, for instance that people tend to rely on the recognition heuristic consistently when the validity for them is high, but when the validity decreases, some people suspend the default and follow some other strategy. Analyses of individual differences, such as in Figure 8.2 below, indicate that the latter may be the rule.

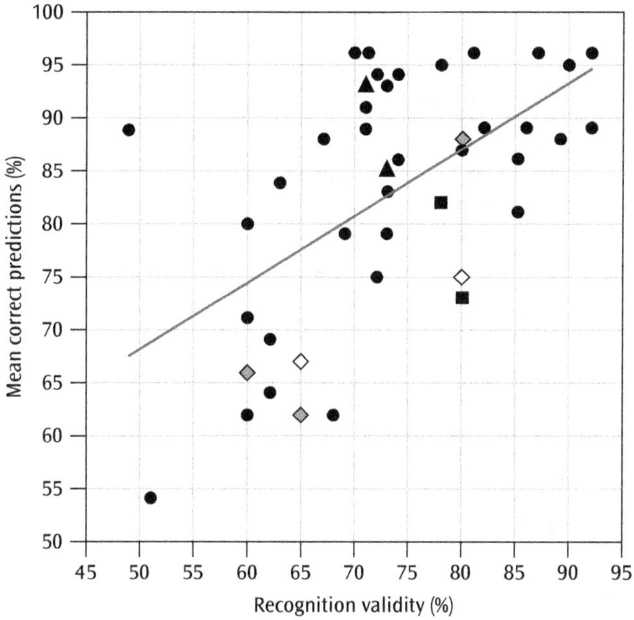

Figure 8.1: Relationship between recognition validity and mean percentage of correct predictions of the recognition heuristic (accordance rate). Included are all 43 experiments or conditions in Table 8.2 where alpha and accordance rates were reported, inside and outside the domain of the recognition heuristic. Black symbols represent experiments/ conditions with natural recognition and inferences from memory. Black triangles = 3 negative (contradicting) cues; black squares = 1 negative (contradicting) cue. White diamonds = repetition during the experiment rather than natural recognition (Bröder & Eichler, 2006); gray diamonds = repetition and inferences from givens (Newell & Shanks, 2004). Here, repetition validity is reported instead of recognition validity. Richter and Späth (2006, Exp. 3) reported a rank correlation instead of alpha, which we transformed into an estimate of alpha using Equation 2 in Martignon and Hoffrage (1999). Mixtures of positive and negative cues (Pachur, Bröder, & Marewski, 2008, Exp. 1, all accordance rates >.96) are not included. The best fitting linear relation is shown; the Pearson correlation is $r = .57$.

Is Reliance on the Recognition Heuristic Sensitive to the Specification of a Reference Class?

Substantial recognition validities have been reported in various environments, including predicting the winners in the Wimbledon Gentlemen Singles matches in the class of final contestants (Scheibehenne & Bröder, 2007; Serwe & Frings, 2006), inferring the career points of professional hockey players (Snook & Cullen, 2006), and forecasting the election results of political parties and candidates (Marewski et al., 2010). In each of these studies, a large majority of participants consistently made inferences that were predicted by the recognition heuristic. In contrast, accordance appears to be low in studies where no reference class was specified or neither researchers nor participants could possibly estimate the recognition validity. For instance, Oppenheimer (2003, Experiment 2) asked students at Stanford University in a questionnaire to make paired comparison judgments of city population. There were six key pairs, each of which consisted of a nonexistent, fictional city, such as "Heingjing," and a real city, selected for specific reasons, such as Chernobyl (nuclear disaster), Cupertino (close proximity), and Nantucket (popular limerick). Because no real reference class exists, it is not possible to determine the recognition validity in this study. In the end, in less than half of the cases were the recognized cities judged to be larger (Table 8.2). The study concluded that it "found no evidence for the recognition heuristic despite using *the same task* as in the original studies" (p. B7, italics in the original). However, it was not the same task. In the original studies, participants knew the reference class, knew it was real, and were tested on its identity (Goldstein & Gigerenzer, 2002).

Let us make it clear. We do not assume that people follow the recognition heuristic unconditionally, for example independently of recognition validity, as a few researchers have implied. Sensitivity to the specification of reference classes (or lack thereof) has been documented in research by ourselves and others, and is of general importance for understanding human judgment. For instance, single-event probabilities by definition do not specify a class, which results in confusion and large individual differences in interpretation (see Chapter 2), as in probabilistic weather forecasts (Gigerenzer et al., 2005) and clinical judgments of the chances of a patient harming someone (Slovic, Monahan, & MacGregor, 2000).

Are People Sensitive to the Presence or Lack of Representative Sampling?

As far as we can see, in environments with substantial recognition validity, a well-defined reference class, and representative sampling, a substantial proportion of participants act in accordance with the recognition heuristic when making inferences from memory. Moreover, if there is a reference class with a substantial recognition validity, such as the height of the largest mountains, but the objects are selected so that the sample recognition validity is close to chance level, recognition heuristic accordance can still be quite high. The outlier in the left upper left corner of Figure 8.1 is such a case of selected sampling (Pohl 2006, Experiment 4; see Table 8.2). As mentioned above, although it is easy to detect whether there is a meaningful reference class, it is sometimes difficult to detect whether objects are randomly or selectively sampled from this class. Pohl's experiment with a selected sample of mountains suggests that people might assume random selection in the absence of any red flags. Except for Hoffrage (1995, 2011), we are not aware of any systematic study that varied representative and biased sampling in inferences from recognition; however, studies on Bayesian judgments suggest sensitivity to random versus biased sampling if the sampling is performed or witnessed by the participant (Gigerenzer, Hell, & Blank, 1988).

In summary, two processes, recognition and evaluation, do much to guide the adaptive selection of the recognition heuristic. They can be formulated as two questions: "Do I recognize one object but not the other?" "If so, is it reasonable to rely on the recognition heuristic in this situation?" The first constrains the set of applicable heuristics. The second process provides an evaluation check, judging the ecological rationality of the heuristic in a given task. Experimental results indicate that participants were sensitive to differences in recognition validity between tasks (Figure 8.1) and that they were sensitive to the existence of a meaningful reference class *R* (Table 8.2). Sensitivity to sampling from *R* needs to be studied. How this sensitivity and the associated evaluation process works is not yet well understood; however, the research in the following section provides some progress and hypotheses.

The Neural Basis of the Recognition and Evaluation Processes

An fMRI study tested whether the two processes, recognition and evaluation, can be separated on a neural basis (Volz et al., 2006). Participants were given two kinds of tasks; the first involved only a recognition judgment ("Have you ever heard of Modena? of Milan?"), while the second involved an inference in which participants could rely on the recognition heuristic ("Which city has the larger population: Milan or Modena?"). For mere recognition judgments, activation in the precuneus, an area that is known from independent studies to respond to recognition confidence, was observed (Yonelinas et al., 2005). In the inference task, precuneus activation was also observed, as expected, and in addition activation was detected in the anterior frontomedian cortex (aFMC), which has been linked in earlier studies to evaluative judgments and self-referential processing. These results indicate that the neural processes elicited by the two tasks of recognition and evaluation are not identical, as an automatic interpretation of the use of the heuristic would imply, but suggest a separate evaluation process that determines whether to select the recognition heuristic for a given task. The aFMC activation could represent the neural basis of this evaluation of ecological rationality.

The neural evidence furthermore suggests that the recognition heuristic may be relied upon by default, as opposed to being just one of many strategies. The default can be overthrown by information indicating that it is not ecologically rational to apply the heuristic in a particular task because recognition is not predictive of the criterion (Volz et al., 2006).

The default interpretation is also supported by behavioral data. Response time data from Pachur and Hertwig (2006) as well as Volz et al. suggest that recognition judgments are made before other knowledge can be recalled. Consistent with this hypothesis, these authors show that response times were considerably faster when participants' inferences accorded with the recognition heuristic than when they did not. Similarly, participants' inferences accorded with the recognition heuristic more often when they were put under time pressure. Moreover, even though older people have slower reaction times, they also reacted faster when choosing the recognized object (Pachur et al., 2009). These findings are consistent with the recognition memory literature, indicating that a sense of recognition (often called *familiarity*) arrives in consciousness earlier than recollection (e.g., Ratcliff & McKoon, 1989). Recognition judgments are made very quickly, and the recognition heuristic appears to be a default strategy that can be overthrown by information contradicting its perceived ecological rationality.

Correcting Misconceptions

We have seen three misconceptions about the nature of the recognition heuristic. The first was already briefly mentioned, the second concerns the meaning of a noncompensatory strategy, and the third the original domain of the heuristic.

Misunderstanding #1: All people rely indiscriminately on the recognition heuristic in all situations. Some researchers have ascribed to us the view that the recognition heuristic is "universally applied" or that people "rely on recognition blindly" (e.g., Richter & Späth, 2006, p. 160). Others used multinominal models to test the null hypothesis that people rely on the recognition heuristic 100% (or 96%) of the time, and found that only some people exhibit this level of consistency (see Brighton & Gigerenzer, 2011). We know of no model of judgment that predicts 96% correctly, and in all situations. In contrast, our view was and is that the recognition heuristic—like other heuristics—is likely to be applied when it is ecologically valid, not in all situations. This is implied by the very notion of the adaptive toolbox. Furthermore, different individuals select different heuristics, as we shall discuss.

Misunderstanding #2: A noncompensatory strategy ignores all other information, not just other cue values. Consider an ordered set of M binary cues, $C_1, \ldots C_M$. These cues are noncompensatory for a given strategy if every cue C_j outweighs any possible combination of cues after C_j, that is, C_{j+1} to C_M. In the special case of a weighted linear model with a set of weights $W = \{w_1, \ldots, w_M\}$, a strategy is noncompensatory if (Martignon & Hoffrage, 1999):

$$w_j > \sum_{k>j} w_k \quad \text{for every } i \leq j \leq M \tag{1}$$

In words, a linear model is noncompensatory if, for a given ordering of the weights, each weight is larger than the sum of all subsequent weights. A simple example is the set {1, 1/2, 1/4, 1/8, 1/16}. Noncompensatory models include lexicographic rules, conjunctive rules, disjunctive rules, and elimination-by-aspects (Hogarth, 1980; Tversky, 1972).

The definition shows that *noncompensatory* refers to a relationship between one cue and other cues, not a relationship between one cue and the criterion (e.g., knowing a city is small). We clarify this here, because in our original article we used the terms *further cue values* and *further information* interchangeably, assuming the technical meaning of *noncompensatory* to be known. But that was not always the case, and thus we may have contributed to the misunderstanding. For instance, people recognize the name *Madoff* as a money manager of the last decade but do not infer him to have been reputable because they have direct knowledge about him on this criterion. With certain knowledge about the criterion, no inference may be needed—one might simply make deductions using a local mental model (see Gigerenzer et al., 1991, figure 2). If there is no inference to be made, the recognition heuristic does not apply, and the proportion of inferences consistent with the recognition heuristic is likely to be at or below chance level (Pachur & Hertwig, 2006; see also Hilbig, Pohl, & Bröder, 2009). Proper tests of noncompensatory processing introduce cues for the recognized object (but not the unrecognized object) that contradict the inference the recognition heuristic would make (see below). In sum, a noncompensatory process ignores cues, but not information in general, such as information concerning criterion values or the recognition validity.

Misunderstanding #3: The recognition heuristic is a model of inference in general, rather than of inference from memory. The recognition heuristic, like take-the-best, was explicitly proposed as a model of inferences made from memory, that is, inferences in which each object's cue values are retrieved from memory, as opposed to inferences from givens, in which the cue values are provided by the experimenter (Gigerenzer & Goldstein, 1996b, pp. 651–652). Inferences from memory are logically different from inferences based on external information. If one has not heard of an object, its cue values cannot be recalled from memory (although the name itself may, quite rarely, impart cue values, much like "80 proof whiskey" reveals its alcohol content). Thus, in inferences from memory, recognition is *not* like other cues. Rather, recognition can be seen as a prior condition for being able to recall further cue values from memory. In inferences from givens, in contrast, this logical relation does not hold, and recognition could, but need not, be treated as just another cue. Note that proponents of evidence accumulation models or parallel constraint satisfaction models tend to neglect this fundamental distinction when they consider "recognition as one cue among others" (Hilbig & Pohl, 2009, p. 1297; see also Glöckner & Bröder, 2011).

In addition to this logical difference, memory-based inferences are also psychologically different. Memory-based inferences require search in memory for cue values, whereas inferences from givens do not. Importantly, search in memory has been shown to elicit more noncompensatory processing

(Bröder & Schiffer, 2006). Nevertheless, some tests of the recognition heuristic focused on inferences from given information, even about unrecognized objects (e.g., Newell & Shanks, 2004; Glöckner & Bröder, 2011). Moreover, in some studies, recognition was induced experimentally by repetition within a session rather than arising naturally over time (e.g., Bröder & Eichler, 2006; Newell & Shanks, 2004). These studies went beyond the domain of the recognition heuristic and mostly show lower levels of correct predictions (Figure 8.1). We are very supportive of testing precise models for a variety of tasks, such as making inferences about an unrecognized product whose attributes values are listed on the box, but we emphasize that this task environment is outside that which is modeled by the recognition heuristic.

To summarize: No strategy is, or should be, universally applied. Assuming an automatic use of the recognition heuristic is the first misunderstanding. The proper question is: When do people rely on the heuristic, and when should they? The terms *noncompensatory* and *compensatory* refer to how cues are processed when making inferences about a criterion; they do not refer to ignoring or using information about the criterion or about the ecological rationality of a strategy. Finally, the recognition heuristic is a model of inference from memory, not from givens, with recognition and its validity learned in a person's natural environment.

Testing Noncompensatory Inferences

Although some models of heuristics are compensatory (for instance, unit-weight models or tallying in Table 8.1, and the compensatory "recognition cue" models in Gigerenzer & Goldstein, 1996), many process information without making trade-offs, that is, in a noncompensatory way. For instance, the availability heuristic (Tversky & Kahneman, 1973) predicts that judgments of likelihood are based on the speed (or number, since definitions vary) with which instances come to mind. It appears to process speed (or number) in a noncompensatory way; no integration of other cues is mentioned. Similarly, the affect heuristic (Slovic, Finucane, Peters, & MacGregor, 2002) captures the notion that judgments are based on the affective tag associated with an object. Both heuristics appear to entail that people make judgments based on only a single piece of information—ease of retrieval and affect, respectively—and ignore further cue values. Being described in general terms rather than as an explicit computational model, however, the assumption of noncompensatory processing is not made explicit and may not even be intended by some authors. Perhaps this lack of clarity is one reason why the various apparent examples of one-reason decision making postulated by the heuristics-and-biases program have not sparked debate over noncompensatory processing.

Not so with the recognition heuristic. When we spelled out that the recognition heuristic is a model that relies on recognition and does not incorporate further probabilistic cues, this modeling assumption drew heavy fire.

The intense reaction continues to puzzle us, given that noncompensatory processes have been frequently reported. Over 20 years ago, a classic review of 45 *process-tracing* (as opposed to outcome) studies of decision making concluded, "the results firmly demonstrate that noncompensatory strategies were the dominant mode used by decision makers" (Ford et al., 1989, p. 75). Today, we know of several structures of environments in which not making trade-offs leads to faster, more accurate, and more robust inferences than one can achieve with compensatory processes, and vice versa (e.g., higher cue redundancy favors noncompensatory processing whereas higher independence between cues favors compensatory processing; see Gigerenzer & Brighton, 2009; Hogarth & Karelaia, 2006; Katsikopoulos & Martignon, 2006; Martignon & Hoffrage, 2002).

Ideally, research proceeds by first identifying environments in which a noncompensatory heuristic is ecologically rational, and then testing whether people rely on that heuristic in this environment or switch to compensatory strategies when the environment is changed accordingly (e.g., Dieckmann & Rieskamp, 2007; Rieskamp & Otto, 2006). Tests of whether and when people process recognition in a noncompensatory way fall mostly into two groups. One group did not test the recognition heuristic in its domain, that is, with substantial recognition validity, inferences from memory, and natural recognition (Table 8.2). Instead, tests were performed in situations where the recognition validity was unknown or could not be determined (e.g., Richter & Späth, 2006, Experiment 1; Oppenheimer, 2003, Experiments 1 and 2), in which recognition was not natural but induced by the experimenter (e.g., Bröder & Eichler, 2006; Newell & Shanks, 2004, Experiments 1 and 2), in which inferences were made from givens (e.g., Newell & Shanks, 2004, Experiments 1 and 2) or cue values were provided for unrecognized objects (Glöckner & Bröder, 2011). The second group tested noncompensatory processing of recognition in its proper domain. One of the first was an experiment by Richter and Späth (2006, Experiment 3), which we briefly review here given that it has been incorrectly presented as evidence against noncompensatory processing.

Richter and Späth asked whether the recognition heuristic would predict inferences in the presence of a strong, contradicting cue. German participants were taught whether certain recognized American cities have international airports or not. The airport cue was chosen as being the most valid (mean subjective validity = .82) among six cues tested in a pilot study. Moreover, the biserial rank correlation between population rank and airport was larger than that between population rank and recognition, –.71 versus –.56. There were three memory states for recognized cities: positive cue (with international airport), no cue (unknown), and negative cue (no international airport). Richter and Späth reported that in these three states, 98%, 95%, and 82% of the inferences were in accordance with the recognition heuristic, respectively, and they concluded that "no evidence was found in favor of a noncompensatory use of recognition" (p. 159). Puzzled by that conclusion, which was based on averages, we asked the authors for

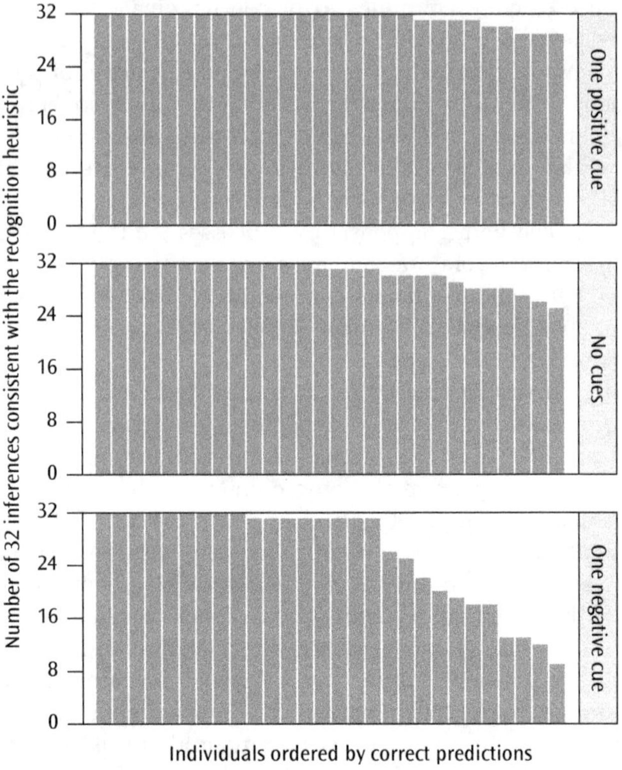

Figure 8.2: A reanalysis of Richter and Späth's (2006) Experiment 3, which tested the noncompensatory use of recognition in inferences from memory with substantial recognition validity. Each bar represents one participant, and its height the number of inferences (out of a total of 32) consistent with the recognition heuristic. The upper panel shows how often each participant judged a recognized city as larger than an unrecognized one when they were told that the recognized city had an international airport (positive cue). The middle panel shows the same when the participants had no information about whether the city had an international airport (no cue). The lower panel shows the critical test in which participants were told that the recognized city had no such airport (negative cue). Even in this critical test, the majority of participants made nearly every inference in accordance with the recognition heuristic. In contrast to this reanalysis, Richter and Späth (2006) did not report their individual data and concluded from the group means (98%, 95%, and 82% of the choices consistent with the recognition heuristic) that there is "no evidence was found in favor of a noncompensatory use of recognition" (see text).

the individual data, which they cordially provided and which are shown in Figure 8.2. These data show that, in the presence of a strong contradicting cue (lower panel), the *majority of people* chose the recognized objects 97% to 100% of the time, as predicted by the recognition heuristic, while the others appeared to guess or follow some other strategy. This pattern was intraindividually highly consistent, with zero or one deviations out of 32 judgments per participant, a degree of consistency rarely obtained in judgment and decision-making research.

Pachur et al. (2008) reviewed the literature and found similar results to those in Figure 8.2. They concluded that, when the recognition validity is high, inferences from memory are frequently consistent with a noncompensatory use of recognition, even in the presence of conflicting cues. In the authors' own study, participants had knowledge of three conflicting (negative) cues indicating that the recognized object should have a small criterion value; nevertheless, about half of the participants chose the recognized object in every single trial. Individual differences were similar to those in Figure 8.2 (lower panel). Note that if it were true that most people consistently made trade-offs between recognition and opposing-valued cues or sets of cues *that have a higher validity than the recognition validity*, then in such situations most people should exhibit about 0% accordance. However, such individuals are not observed in Figure 8.2. Only a few were observed in Pachur et al.'s experiments (2008, p. 195) and in their reanalysis of Newell and Fernandez (2006).

Similarly, in studies on the role of name recognition in political forecasts, most voters always behaved in accordance with the recognition heuristic, whether or not there was a strong conflicting cue present (Marewski et al., 2010). As Table 8.2 shows, individual analyses reveal that a large proportion of participants consistently made inferences in accordance with the recognition heuristic, even with up to three conflicting cues.

How to model the people who deviate from the predictions of the recognition heuristic? A common proposal in the literature has been that these people integrate recognition information with other cues in a compensatory fashion. But, to our knowledge, in none of these articles was a compensatory model formulated and tested against the recognition heuristic. Testing such models is essential to theorizing for several reasons. First, if some individuals do not accord with the recognition heuristic, this does not logically imply that they rely on a compensatory process. They might simply guess, or rely on the best cue beyond recognition, as in a lexicographic rule, and thus adopt a different noncompensatory process. Second, because no model can explain all behavior, one needs to show that there are others that can explain more.

We know of only one study that has formulated compensatory models and tested them against the recognition heuristic (Marewski et al., 2010).[1] The five alternatives integrate recognition with further cues for the recognized object (Table 8.2). The alternative models had free parameters that allowed them to both mimic the recognition heuristic *and* predict the opposite pattern, depending on the parameter tuning. That is, they included the recognition heuristic as a special case. Because these alternatives use free parameters and the recognition heuristic uses none, it is important to test

1. As mentioned, Glöckner and Bröder (2011) work outside the domain of the recognition heuristic in which people know the cue values of unrecognized objects.

how well the models predict (rather than fit) judgments. None of the five compensatory models could predict judgments better than the recognition heuristic, which performed the best overall. The study showed that although the recognition heuristic cannot predict with 100% accuracy, particularly in the presence of contradicting cues, this by itself does not imply that compensatory models can actually predict better.

To summarize: The recognition heuristic is a simple, noncompensatory model of inference from memory. We personally have no doubts that recognition is sometimes dealt with in a compensatory way, especially when the ecology favors doing so. A number of studies have conducted critical tests in which recognized objects with negative cue values were compared with unknown objects. The typical results were that (i) the mean accordance rates decreased when one or more negative cue values were introduced, while (ii) a large proportion of participants' judgments nevertheless accorded consistently with the recognition heuristic's predictions. Result (i) has been interpreted as implying compensatory decision making, but no compensatory models were put forth to test this claim. In contrast, the first test of five compensatory models showed that in fact none could predict people's inferences as well as the noncompensatory use of recognition.

Methodological Principles

The previous section suggests a methodology to be followed. We summarize here two relevant principles.

Principle 1: Test heuristics against competing models; do not claim victory for a model that was neither specified nor tested. This principle seems obvious, but it has been routinely neglected in the study of the recognition heuristic. Among the studies that claimed to have found evidence for compensatory models, we are not aware of a single one that has actually tested such a model. Hilbig and Pohl (2009) attempted to do so, and we applaud the direction they took. They used as alternatives two large model classes, evidence-accumulation models and neural nets, which they also treated as one. Since these model classes can mimic the outcomes of the recognition heuristic, multiple regression models, as well as virtually any inferential strategy ever proposed in cognitive psychology, it is not clear to us how they derived their specific predictions from such flexible models. We ourselves have proposed and tested linear "recognition cue" models that treat recognition in a compensatory way (Gigerenzer & Goldstein, 1996b). We suspect that the origin of this methodological flaw is deeply rooted in generations of researchers who have been taught that hypotheses testing amounts to null hypothesis testing, that is, rejecting a precisely stated null hypothesis in favor of an unspecified alternative hypothesis. This biased procedure is not a swift route to scientific progress (Gigerenzer et al., 1989).

Principle 2: Analyze individual data; do not base conclusions on averages only. This principle is necessary because there are systematic individual differences in cognitive strategies. These differences have been reported across the entire life span, from children's arithmetical reasoning (e.g., Shrager & Siegler, 1998), judgments of area (Gigerenzer & Richter, 1990), and Bayesian inferences (Zhu & Gigerenzer, 2006) to decision making in old age (Mata, Schooler, & Rieskamp, 2007). If individual differences exist, analyses based only on means (across individuals) do not allow conclusions about underlying processes. One simple solution is to always analyze data on the individual level, as in Figure 8.2, which can reveal the existence of individual differences. A more theoretically guided approach would be to specify competing models, and test what proportion of participants can be predicted by each model (Marewski et al., 2010).

Results We Had Not Anticipated a Decade Ago

Thanks to the researchers who set out to study the recognition heuristic by means of analysis, computer simulation, and experiment, we have more than once been taught lessons by unexpected results. We cannot list all here, but briefly mention three of the many surprises.

Less-Is-More Effects Are Theoretically Stronger in Group Decision Making

Reimer and Katsikopoulos (2004) extended the role of name recognition from individual to collective decision making. If one member in a group recognizes only one of two alternatives, but the others recognize all and have some further cue knowledge, should the group follow the most ignorant member who can rely on the recognition heuristic? The authors first deduced analytically that less-is-more effects can emerge in a group context and that these effects are stronger in magnitude than in individual decisions. The conditions are similar to those we had arrived at for individual decisions: If the recognition validity is higher than the knowledge validity, both are independent of the number n of objects recognized, and some further assumptions concerning the homogeneity of the groups hold, then the relationship between accuracy and n is inversely U-shaped. That is, there should exist groups whose members recognize fewer objects but reach a higher accuracy than do groups who recognize more objects. The authors reported less-is-more-effects in group decision making in an experiment, where also a fascinating new phenomenon emerged. Consider a group of three in which one member recognized only city a, while the other two members recognized both cities and individually chose b as the larger one. What would the group decide after consulting with one another? The majority rule predicts b, yet in most cases, the final group decision was a. This result suggests that a lack of recognition has a special status not only in individual decisions, as originally proposed, but in group decisions as well.

Less-Is-More Effects Are Stronger with >2 Alternatives and Positive Framing

In our original work, we relied on tasks with two alternatives to deduce the size of less-is-more effects analytically. Recently, the recognition heuristic has been generalized to more than two alternatives (McCloy, Beaman, & Smith, 2008; Marewski et al., 2010). Does the less-is-more effect also hold in choices involving multiple objects? It does. An experiment on inferring who is the richest citizen in a set demonstrated less-is-more effects irrespective of whether the task was to choose among two, three, or four alternatives (Frosch, Beaman, & McCloy, 2007). Moreover, one can analytically show that this generalization of the recognition heuristic implies that the size of the effect increases when more alternatives are involved. Surprisingly, for three and more alternatives, the model implies a framing effect. If the question is framed positively, such as "Which of the people is richest?" the less-is-more effect is more pronounced than for the question "Which of the people is poorest?" (McCloy et al., 2008). This work illustrates the importance of analytically deriving predictions from the recognition heuristic, in order to see what the model implies and what it does not.

The Power of Laypersons' Recognition for Prediction

A widely entrenched view about heuristics is that yes, people rely on them because of their limited cognitive capacities, but no, they cannot often lead to good inferences. Skeptical of the power of the recognition heuristic to yield good decisions, Serwe and Frings (2006) set out to test it in a task in which they were confident that it would fail: predicting the winners of the 127 Gentleman Singles Wimbledon tennis matches. They were skeptical for good reasons. First, tennis heroes rise and fall quickly; by the time their names have finally found a place in collective recognition memory, their prowess may already be fading. Second, athletes are best known within their home country, even if they do not perform particularly well in the international arena. Recognition of an athlete should thus be a poor guide to predicting whether he or she will win an international match. To demonstrate these suspected Achilles' heels of the recognition heuristic, Serwe and Frings needed semi-ignorant people, ideally, those who recognized about half of the contestants. Among others, they contacted German amateur tennis players, who indeed recognized on average only about half of the contestants in the 2004 Wimbledon Gentlemen's Singles tennis tournament. Next, all Wimbledon players were ranked according to the number of participants who had heard of them. How well would this "collective recognition" predict the winners of the matches? Recognition turned out to be a better predictor (72% correct) than the ATP Entry Ranking (66%), the ATP Champions Race (68%), and the seeding of the Wimbledon experts (69%). These unexpected results took the authors by surprise. When they presented their results to the ABC Research Group, the surprise was on both sides. Could it have been a lucky strike, ripe for publication in the *Journal of Irreproducible Results*? Scheibehenne and Bröder (2007) set out to test whether the findings would replicate for Wimbledon 2005—and found basically the same result. In addition, when asked to predict the match winners, the amateur tennis players predicted in around 90% of the cases that the recognized player would win. Thus, there can be powerful "wisdom of the crowd" in laypeople's collective recognition.

Collective recognition has also been used for investment in the stock market, which is reviewed in Ortmann, Gigerenzer, Borges, and Goldstein (2008), and for forecasting sport events, as reviewed in Goldstein and Gigerenzer (2009).

Open Questions and Future Research

In this chapter, we have addressed a number of research directions that we think are important to pursue, such as integration with theories of recognition memory and deeper understanding of the evaluation process. We close with a selection of open questions and issues.

Is the Noncompensatory Process Implemented in the Stopping Rule or in the Decision Rule?

The definition of the recognition heuristic allows both interpretations. The classic definition of compensatory and noncompensatory processes locates the difference in the decision rule: Cues for the recognized object may or may not come to mind; the relevant question is whether they are used when making the decision. In contrast, we had made a stronger modeling assumption, namely that search for cue information in memory is stopped if only one alternative is recognized, which locates the absence of trade-offs already in the stopping rule. It is not easy to decide between these alternatives. For instance, Hilbig and Pohl (2009) reported that mean decision times were shorter when participants had further knowledge about a city as opposed to when they had none, and interpreted this difference against our interpretation of the process and in favor of an unspecified "evidence-based model." Decision time predictions, however, cannot be derived from our simple model without making additional assumptions, and highly specific ones. We do not deal with decision times extensively in the limited space of this review, but elaborate on one point here. Decision time predictions as well as recognition time (fluency) are best derived from a model of the core cognitive capacities involved (Marewski & Schooler, 2011). To illustrate, it is not correct that our interpretation implies no difference in decision times; this prediction would, among others, require that the speed of recognition (fluency) be uncorrelated with the size of the object or, in general, the criterion value. Recognition, however, tends to be faster for larger objects (Hertwig, Herzog, Schooler, & Reimer, 2008; Schooler & Hertwig, 2005). Thus, if speed of recognition is correlated with the actual size and if objects that people know more about are larger, mean decision times are likely to be shorter when additional knowledge is available (Marewski et al., 2010; Hilbig & Pohl, 2009, Experiment 3, did address this issue). No cue integration is needed to explain this result. The question whether the noncompensatory process is located in the stopping or in the decision rule is still an open issue. The answer to this question does not concern the outcome prediction of the recognition heuristic, only the process that leads to this outcome.

How Do People Adapt Their Use of Recognition to Changes in Recognition Validity?

Two striking observations have been reported. First, whereas accordance rates are correlated with the recognition validities across tasks (see Figure 8.1), the individual accordance rates within a task appear to be unrelated to the individual recognition validities (Pachur & Hertwig, 2006; Pohl, 2006). This result may be due to the low mean recognition validity (.60) or the low variability in individual recognition validities (most were between .55 and .65) in Pachur and Hertwig's study, or to the use of selected rather than representative samples in some of Pohl's sets (mountains, rivers, and islands). Whatever the reasons, this observation deserves a closer investigation. Although it suggests limits to the adaptive use of recognition, a second observation suggests an enhanced adaptive use: Pohl and Hilbig (Pohl, 2006; Hilbig & Pohl, 2008) reported that the recognition heuristic fits the data better when the heuristic would lead to a correct inference than when it would lead to an incorrect one. For instance, in Hilbig and Pohl's (2008) experiment, 76% chose the recognized city in pairs when it was incorrect and 82% when it was correct. The authors interpret this slight difference in means as indicative of additional knowledge being relied on in some cases, but what this knowledge is remains unclear. It could be related to criterion knowledge, an issue that Pachur and Hertwig (2006), and Hilbig et al. (2009) have taken different sides on. What is needed is a model that can predict when this effect occurs. A clarification of these two observations will hopefully contribute to better theories about the evaluation process.

Recognition Plus Additional Knowledge

Pohl (2006) reported that the recognition heuristic predicted inferences better on $R^{+}U$ pairs (comparison between a recognized object about which a person has additional knowledge [R^{+}] and an unrecognized object [U]) than on RU pairs (R = mere recognition). The question is how to explain this difference. Pohl (2006) concluded from this result that some people use a compensatory strategy, but without specifying and testing any such a strategy. Yet this is not the only interpretation. Another is that the difference follows from systematic variations in the strength of the recognition signal and the recognition validity (Marewski et al., 2010; Marewski & Schooler, 2011).

Recognition and Preference Formation

Although we formulated the recognition heuristic as a model for inferences, it can also serve as a model for preferences. Consider consumer choice, in which the classical model of brand preference is a formalization of Fishbein's (1967) work on beliefs and attitudes:

$$A_b = \sum_{i=1}^{N} W_i B_{ib} \tag{2}$$

where A_b = the attitude toward brand *b*, W_i = the weight of the *i*th product attribute, B_{ib} = the consumer's belief about brand *b* where attribute *i* is concerned, and N = the number of attributes deemed important for choosing a brand.

The resemblance to the weighted linear models studied in judgment and decision-making research is clear. With weights and beliefs that are typically elicited from the decision maker, such models do a good job in fitting consumers' brand choices for orange juice, lipstick, and the like (Bass & Talarzyk, 1972). However, *what* people choose is different from *how* people choose, as those studying decision processes have noticed. We illustrate this here with *noncompensatory screening* and *halo effects*.

Before choosing products, consumers often reduce a large number of possible alternatives to a smaller set, which they inspect more closely. Such "consideration sets" turn out to be excellent predictors of what is ultimately chosen (Shocker, Ben-Akiva, Boccara, & Negungadi, 1991; Hauser, 1978). Although these considerations sets can in theory be created by compensatory multiattribute procedures that integrate all available information (Roberts & Lattin, 1991), studies suggest that products are filtered into a consideration set by means of noncompensatory heuristics (Gilbride & Allenby, 2004; Laroche, Kim, & Matsui, 2003; Payne, 1976; Bettman & Park, 1980). The generalization of the recognition heuristic to the domain of preferences and multialternative choice enables its use as a building block in consideration set formation (Marewski et al., 2010). Recognition-based consideration sets facilitate decisions when the initial choice set is large. Sometimes recognition itself is the desirable attribute, as when students choose universities.

Further deviations from the classical linear model of brand preference have been suggested by the presence of halo effects, that is, the tendency for people who favor a brand to evaluate it positively on all attributes and those who dislike it to do the opposite (Beckwith & Lehmann, 1975). Such behavior suggests that the expressed beliefs about attributes may themselves be inferences, as opposed to the result of recall from memory. Beliefs about the attributes of unrecognized brands cannot be stored in memory and must be constructed on the fly. Extending beyond the original domain of the recognition heuristic, one exciting possibility is that the effect of recognition on attribute beliefs can be even stronger than that of direct experience. Experimental studies on food choice indicate not only that people buy the products they recognize but that brand recognition often dominates other cues to a degree that can change the perception of the product. For instance, in a blind taste test, most people preferred a jar of high-quality peanut butter over two alternative jars with low-quality peanut butter. Yet when one familiar and two unfamiliar brand labels were randomly assigned to the jars, preferences changed. When the high-quality product was in a jar with an unknown brand name, it was preferred only 20% of the time, whereas the low-quality product in the jar with the familiar brand name was chosen 73% of the time. When the exact same peanut butter was put into three jars, one showing a familiar brand name and two showing unfamiliar brand names, the (faux) familiar product won the taste test 75% of the time (Hoyer & Brown, 1990; see also Macdonald & Sharp, 2000). One way to interpret this result is that, for the majority of consumers, brand name recognition dominates the taste cues present in the blind test. But there is an interesting alternative to this noncompensatory processing hypothesis. The taste cues themselves might be changed by name recognition—people "taste" the brand name. Such a process could be modeled in a similar way as the change of perceived cue values in the RAFT model of hindsight bias (Hoffrage, Hertwig, & Gigerenzer, 2000). This interpretation—like the halo effect—suggests a model in which recognition imputes or changes attribute values themselves. Such a process is likely to occur when cue values are direct subjective experiences such as tastes, which are neither presented as propositions nor retrieved from memory. It would provide an alternative hypothesis for the processing of brand name recognition, one that is not based on the distinction between noncompensatory and compensatory processes.

Do Humans and Other Animals Share Common Heuristics?

Behavioral biologists have documented in detail the rules of thumb (their term for heuristics) that animals use for choosing food sites, nest sites, or mates. For instance, Stevens and King (2012) discuss how animals use simple heuristics for recognizing kin to facilitate cooperation, and Shaffer, Krauchunas, Eddy, and McBeath (2004) report that dogs, hoverflies, teleost fish, sailors, and baseball players rely on the same heuristics for intercepting prey, avoiding collisions, and catching balls. Biology offers numerous, specific examples, but no systematic theory of heuristics, such as in terms of rules for search, stopping, and decision (Hutchinson & Gigerenzer, 2005). If we can find signs of the same rules across species, this might provide converging evidence for specific models of heuristics. The recognition heuristic seems to be a good candidate. For instance, rats and mice prefer foods they recognize from having tasted or from having smelled on the breath of fellow rats, a tendency known as *neophobia*. They may also rely on recognition to infer which of several foods made them sick. In one experiment, Norway rats were fed two foods. Both were relatively novel, but one was familiar from the breath of a fellow rat. After these rats were given a nauseant, they subsequently avoided the food they did not recognize from the neighbor's breath (Galef, 1987). As in the experiments with humans, one can test whether recognition is overruled by a powerful cue. Consider a similar situation, where one food is recognized from the breath of a fellow rat, but now the fellow rat is also (experimentally made to appear) sick at the time its breath is smelled. Surprisingly, observer rats still chose the recognized food from the breath of the sick neighbor (Galef, McQuoid, & Whiskin, 1990). As in humans, accordance rates were not 100%, but around 80%. Thus, recognition appears to overrule the sickness cue that advises against selecting the recognized food.

The question of which heuristics humans and other animals share has been recently discussed in a target article in *Behavioural Processes* (Hutchinson & Gigerenzer, 2005) and commentaries (e.g., Cross & Jackson, 2005; Shettleworth, 2005), although we know of no systematic research on the topic. Some comparative research has focused on common biases, but few researchers have tested models of common heuristics. The recent dialogue with psychologists studying the adaptive toolbox has led biologists to revisit the fundamental question of how to model behavior. Models of heuristics are not simply approximations to optimizing models; rather, they allow scientists to study behavior in uncertain, complex worlds as opposed to the certain, small worlds required for the ideal of optimization.

Conclusion

The recognition heuristic is a simple model that can be put to many purposes: describing and predicting inferences and preferences, and forecasting such diverse events as the outcomes of sporting events and elections. The research on the recognition heuristic has promoted the use of testable models of heuristics (instead of vague labels), and of simple models in which each parameter can be directly measured rather than fitted. With such precise models, one can easily observe when a heuristic makes correct predictions and when it fails. But the emerging science of heuristics also caused unease in some research communities, breaking with cherished ideals such as general-purpose models of cognition, the assumption of a general accuracy–effort trade-off, and the conviction that heuristics are always inferior to complex strategies. This may well explain why every critique we know of the recognition heuristic claims that minds add and weigh cues; none has proposed and tested a different, perhaps simpler model. As the last decade has shown, however, there is clear evidence that this simple model consistently predicts the judgments of a substantial proportion of individuals, even in the presence of contradicting cues. Moreover, research now suggests that people may use heuristics in an adaptive way, as witnessed in the substantial correlation between recognition validities and accordance rates. We thank all fellow researchers and critics for carving out the details of the adaptive toolbox, and thus contributing, in the words of Herbert Simon (1999), "to this revolution in cognitive science, striking a great blow for sanity in the approach to human rationality."

Appendix Table 8.2: An Overview of Experimental Studies on the Recognition Heuristic (RH) Reporting Mean Correct Predictions (Accordance Rates). Three plusses in Columns 4–6 mean that the domain was one for which the recognition heuristic was proposed as a model: α = substantial recognition validity; Mem = Inferences from memory (as opposed to inferences from givens); Nat = natural recognition (as opposed to experimentally induced). Studies that satisfy these three conditions are listed first; others follow. (RU) = comparison between a recognized object (R) and an unrecognized object; (R^+U) = comparison between a recognized object about which a person has additional knowledge (R^+) and an unrecognized object.

Article/Exp.	Reference Class	Criterion	Task in the Domain of RH? α	Mem	Nat	Alternative Model?	Individual Analysis?	Mean Correct Predictions of Judgments by RH (and Recognition Validity α)
Frosch et al., 2007	Richest individuals in UK	Richest	+	+	+	no	no	(r = .73): 79%
		Poorest	+	+	+	no	no	(r = .73): 83%
Marewski et al., 2009	Political parties	Election result	+	+	+	yes	yes	87%–89% For experimentally induced recognition, RH predicts the forecasts of 49 voters, fluency heuristic of 12, and both equally of 4 voters.
Marewski et al., 2010/1	Political candidates	Election result	+	+	+	yes	Most voters always followed RH, independent of whether there was a conflicting cue	(α = .80): 87% 1 negative cue: 73% RH predicted better than compensatory models tested.
Marewski et al., 2010/2	Political parties	Election result	+	+	+	yes	no	(α = .92): 89% (R^+U) (α = .68): 62% (RU) RH predicted better than compensatory models tested.

Marewski et al., 2010/3	Largest cities	Population	+	+	+	yes	Most voters always followed RH, independent of whether there was additional knowledge (R^+)	($\alpha = .81$): 96% (R^+U) ($\alpha = .74$): 86% (RU) RH predicted better than compensatory models tested.
Marewski et al., 2010/4	Political parties	Election result	+	+	+	yes	For 25 out of 26 participants, RH predicted better than 2 compensatory models	RH predicted better than compensatory models tested.
Pachur & Biele, 2007	2004 European Soccer Championship	Winner	+	+	+	yes	no	($\alpha = .71$): 91% RH predicted better than 4 alternative models.
Pachur et al., 2008/1	British cities	Population	+	+	+	no	More negative than positive cues: 60% always (16 out of 16 trials) followed RH	($\alpha = .70$): 96% No effect of up to 3 negative/positive cues on RH accordance
Pachur et al., 2008/2	British cities	Population	+	+	+	no	3 negative cues: 46% always (18 out of 18 times) followed RH	($\alpha = .72$): 94% 3 negative cues: 85%
Pachur et al., 2008/3	British cities	Population	+	+	+	no	3 negative cues: 48% always (10 out of 10 times) followed RH	($\alpha = .71$): 96% 3 negative cues: 93%
Pachur et al., 2009	Infectious diseases; U.S. cities	Frequency	+	+	+	no	no	Young adults: ($\alpha = .90$): 95% ($\alpha = .62$): 64%
		Population	+	+	+	no	no	Elderly: ($\alpha = .92$): 96% ($\alpha = .60$): 71%

(*continued*)

Table 8.2: Continued

Article/Exp.	Reference Class	Criterion	Task in the Domain of RH? α	Mem	Nat	Alternative Model?	Individual Analysis?	Mean Correct Predictions of Judgments by RH (and Recognition Validity α)
Pohl, 2006/2	Largest Swiss cities mixed with ski resorts	Population	+	+	+	no	no	(α = .72): 75%
Pohl, 2006/3	Italian cities	Population	+	+	+	no	no	(α = .82): 89%
	Belgian cities		+	+	+	no	no	(α = .89): 88%
Richter & Späth, 2006/3	Largest U.S. cities	Population	+	+	+	no	no Reanalysis (see Figure 8.1): Majority consistently followed RH	(rank corr. = −.56) 1 positive cue: 98% no cue: 95% 1 negative cue: 82%
Scheibehenne & Bröder, 2007	Wimbledon Gentlemen Singles 2005	Winner	+	+	+	no	no	Amateurs (α = .71): 89% Laypeople (α = .69): 79%
Serwe & Frings, 2006	Wimbledon Gentlemen Singles 2003	Winner	+	+	+	no	no	Amateurs (α = .73): 93% Laypeople (α = .67): 88%
Snook & Cullen, 2006	NHL hockey players	Career point	+	+	+	no	no	(α = .87): 96%
Volz et al., 2006	Cities from 7 countries	Population	+[1]	+	+	no	no	(α = .63): 84%
Pohl, 2006/4	Mountains	Height	–	+	+	no	no	(α = .49): 89%
	Rivers Islands	Length	+	+	+	no	no	(α = .74): 94%
		Area	+	+	+	no	no	(α = .85): 81%
Oppenheimer, 2003/1	Undefined (real and fictional cities)	Population	–	+/–	+/–	no	no	α cannot be determined Adherence: < chance

Oppenheimer, 2003/2	Undefined (real and fictional cities)	Population	–	+/–	+/–	no	no	α cannot be determined Adherence: < chance
Richter & Späth, 2006/2	Largest airlines in the world	Safety	–[2]	+	+	no	no	α unknown 3 positive cues: 98% 2 positive/1 negative cue: 88% 2 negative/1 positive cue: 81% 3 negative cues: 67%
Bröder & Eichler, 2006	Unknown small towns	Population	–	+	–	no	no	α cannot be determined; reported is validity of repetition (v). (v = .80): 75% (v = .65): 67%
Newell & Shanks, 2004/1	Undefined (fictional company names)	Investment decision	–	–	–	no	yes	α cannot be determined; reported is validity of repetition (v). (v = .80): 88% (v = .65): 62%
Newell & Shanks, 2004/2	Undefined (fictional company names)	Investment decision	–	–	–	no	yes	α cannot be determined; reported is validity of repetition (v). (v = .60): 66%

[1] Low recognition validities (α) introduced to study decisions against RH.

[2] Recognition validity not reported; reported was a partial $r = -.29$ between recognition and number of fatalities when year established was controlled for.

Part IV

Intuitions about Sports and Gender

Chapter 9

The Hot Hand Exists in Volleyball and Is Used for Allocation Decisions

Many people who participate in, watch, or report on sports believe that a player has a higher chance of making a shot after two or three successful shots (hits) than after two or three unsuccessful shots (misses). One might hear a reporter say that "Lincy is a streaky shooter," for instance, or a fan claim that "the Lakers have a run." Such convictions based on the sequential performance of a player or team have been termed the *hot hand belief* (Gilovich, Vallone, & Tversky, 1985).

In this article, we pose two questions. First, does the hot hand belief reflect reality? Second, how is the belief used for decision making? The first question seems to have already been answered in the literature, but a closer look shows that the answer is not clear-cut. One complicating factor is that the "hotness" of a player may not be observable because the other team increases its defense against the player, as, for instance, in basketball. Thus, finding no streaks can be the result of increased defense rather than merely the absence of the hot hand. In the following, we concentrate on a sport that limits this confounding factor: volleyball, where each team remains on a different side of the court. The second question is new, and in our view the more important one. We argue that players use the belief about streaks to alter their allocation behavior. In most team sports, one key strategy is to allocate the ball to the player with the strongest chance of scoring. It is argued that the belief in the hot hand can enhance the chance of scoring and thus may even be adaptive (Burns, 2004). If the hot hand belief is adaptive, then playmakers acting on it will increase the team's chance of winning, for instance, by allocating the ball more often to the player with the higher base rate or a current streak.

This is the first study on the hot hand in volleyball, and we show that (i) coaches, playmakers, and fans are sensitive to streaks, which reinforces the

Originally published as Raab, M., Gula, B., & Gigerenzer, G. (2012). The hot hand exists in volleyball and is used for allocation decisions. *Journal of Experimental Psychology: Applied, 18*, 81–94.

existence of the hot hand belief; (ii) the hot hand exists for half of the players in our study; and (iii) this belief is used to guide playmakers' allocation decisions.

State of the Art

There is a huge body of literature on the hot hand phenomenon, virtually all of which has addressed the first question concerning its existence but not the second concerning its behavioral use. Belief in the hot hand is widespread, according to the first major experimental investigation by Gilovich and colleagues (1985) as well as subsequent studies in specific sport disciplines (baseball: Crothers, 1998; basketball: McCallum, 1993; Wolff, 1998; tennis: Klaasen & Magnus, 2001; golf: Clark, 2003; Gilden & Wilson, 1995b; horseshoe pitching: G. Smith, 2003; bowling: Dorsey-Palmateer & Smith, 2004), sport science (Gould, Tammen, Murphy, & May, 1991; Hales, 1999), perception (Gilden & Wilson, 1995a), psychology (Adams, 1995; Oskarsson, Van Boven, McClelland, & Hastie, 2009), and economics (Camerer, 1989). Although the unanimous conclusion is that the hot hand belief does exist, some, following Gilovich et al.'s original study, argue that "hotness" itself does not exist, whereas others maintain that it does (e.g., Larkey, Smith, & Kadane, 1989). According to the review by Bar-Eli, Avugos, and Raab (2006), the score between these two camps in sports is nearly tied at 14:13.

Gilovich et al. (1985) initiated the debate about the hot hand belief with a set of studies showing that in basketball, individual sequences of hits and misses in field goals are nondependent (Gilovich et al., Study 2) yet that fans nevertheless believe in it (Gilovich et al., Study 1). Furthermore, they showed that independence of shots exists in free shots of professional basketball players (Gilovich et al., Study 3) and in a controlled free-shooting experiment of intercollegiate basketball teams (Gilovich et al., Study 4).

In one of Gilovich et al.'s surveys, 91 of 100 basketball fans believed that a hypothetical player has "a better chance of making a shot after having just made his last two or three shots than he does after having just missed his last two or three shots" (p. 297). The average response was that the chance of another hit was 61% if the player had made the last shots, compared with 42% if he had missed. This difference is the empirical observation that has been labeled the hot hand belief. Further research supported the definition of a streak based on at least three hits (e.g., Carlson & Shu, 2007; Gula & Köppen, 2009), which we will apply in our investigation. Belief in the hot hand was originally attributed to a "misperception of random sequences," assuming that some comparison between an observed sequence of hits and misses and a subjective notion of randomness is performed. Little is known about this process and how to measure streaks.

Comparing the conditional probability of hits after misses to that of hits after hits is one measure for the hot hand. Two others are runs and autocorrelations. A run is defined in the hot hand literature as three or more hits in a sequence (Carlson & Shu, 2007). Imagine a player with a base rate of .5 (e.g., five of 10 shots are hits) who plays 16 rounds of four shots. We expect only one of 16 rounds to have four hits in a row ($1/2^4 = 1/16$). If a player exhibits fewer runs than expected by chance, the conclusion is that this player has a hot hand (Hales, 1999). A third way to test the existence of a hot hand is autocorrelation (Bar-Eli et al., 2006). Imagine a player who makes hits (H) and misses (M) in the sequence MMHHHHHHMM with a base rate of .60 (six hits of 10). Autocorrelation counts the correlations between successive events; only in the third and ninth position in the given example does the following event (H and M, respectively) differ from the preceding event. The lag 1 autocorrelation is a correlation between the original sequence and the sequence moved by one position. If systematic autocorrelations exist, there is evidence of a hot hand.

Cognitive Fallacy or Adaptive Behavior?

Early researchers termed the hot hand belief a fallacy[1] because successive shots were found to be independent (Gilovich et al., 1985) and thus players would be no more likely to hit after several hits than after several misses. From this, the conclusion was drawn that if people rely on wrong beliefs, it will be costly for, say, a team (Gilovich et al., 1985) or an individual placing bets (Camerer, 1989). This assumption is highly plausible, but is it correct?

Let us consider the alternative conclusion: Can false beliefs ever lead to advantageous behavior? For example, consider "as-if" theories in science, the practice of drawing the right conclusion from the wrong assumptions about underlying processes. A number of theories have postulated that cognitive limitations (Hertwig & Todd, 2003; Kareev, 2000), "improper models" (Dawes, 1979), forgetting (Schooler & Hertwig, 2005), and other apparent shortcomings can in fact improve behavior. The importance of judging cognitive processes by adaptive criteria, such as success and speed, is the focus of research on ecological rationality (Gigerenzer, Todd, & the ABC Research Group, 1999). The logical question of whether a belief matches reality should not be confused with the ecological question of how useful the belief is to achieve some goal (Burns, 2004; Gula & Raab, 2004; Hahn & Warren, 2009).

To answer this ecological question, consider the case of two players, one with a base rate of .60 (player A) and the other with a base rate of .40

1. Note that the gambler's fallacy, in contrast, predicts that a streak of events of the same class with stationary probabilities (e.g., "reds" in roulette) will more likely be followed by an event from a different class ("black"; Laplace, 1814/1951). It has been argued that this paradox can be explained by people's beliefs regarding the process that generates the specific sequence at hand (Ayton & Fischer, 2004; Burns & Corpus, 2004; Caruso & Epley, 2004; Oskarsson et al., 2009). The hot hand belief is driven by people's assumption that the generating process is controlled by will (e.g., by athletes), whereas the gambler's fallacy assumes that the generating process is stochastic (e.g., in roulette).

(player B). A playmaker could use many allocation strategies, such as probability matching (Gaissmaier & Schooler, 2008; Gallistel, 1993). Probability matching would mean that of 100 balls, 60 are allocated to player A and 40 to player B, resulting in an expected value of (60 × .6) + (40 × .4) = 52 successful hits out of 100 attempts. The hot hand belief is adaptive if players' performance changes in a systematic way such that momentary increases in performance relative to a player's average performance allow more balls to be allocated to him than to another player and thereby increase the chance of winning. Moreover, the hot hand belief can also be used for an allocation strategy if base rates are not known, such as in pick-up games.

Burns (2001, 2004) and Burns and Corpus (2004) suggested that believing in the hot hand may contribute to an adaptive behavioral strategy in basketball because it leads playmakers to pass the ball to a player with a higher scoring average in a game. Burns (2001) used a simulation of ball allocations to two virtual players and showed that behavior based on the hot hand belief resulted in higher average scores for the team than when the belief was ignored. Allocating the ball to the hot player would result in a small but important advantage of about one point in every seven or eight games. Furthermore, Burns argued that the greater the variability in the base rate of a player's scoring performance, the greater the advantage of the hot hand belief. Burns's mathematical model assumes that playmakers cannot detect base rates directly; their belief in the hot hand provides another, indirect source of information.

To determine whether the hot hand exists in volleyball and is used adaptively, we conducted two sets of studies. In Study 1A we asked whether athletes believe in the hot hand and in Study 1B tested its existence in volleyball. In Study 2A we investigated how sensitive coaches can detect streaks, and in Study 2B we tested whether playmakers' allocation behavior is influenced by the belief. Volleyball is played six against six players: three players in the front row, who are allowed to spike over the net, and three players backcourt. One team serves the ball to the other side of the net. The next serve is given to the team that won the last rally (rotation rule). This rule requires players to rotate one position further when they serve after the opposing team's serve, meaning that players who play in front rotate to the back. After each set, teams change sides of the court. The team that wins three sets wins the game. The final score of the sets is 25 points (15 points in the event of a final fifth set) when one team leads by two points; otherwise, the game is continued until one team leads by two points.

Study 1A: Do Athletes Believe in the Hot Hand in Volleyball?

Previous research argued that even if people believe in the hot hand in basketball, players do not necessarily possess a hot hand (e.g., Gilovich et al., 1985). The opposite result, that is, evidence for streakiness in players' performance, does not guarantee that a hot hand *belief* exists either. Therefore, we tested how strong the belief in the hot hand is in volleyball compared with in other sports and gambling. Previous research indicated that belief in the hot hand is context specific, such that within the same person the belief varies between sport and nonsport situations (Caruso & Epley, 2004; Tsyska, Zielonka, Dacey, & Sawicki, 2007).

Method

Participants

Ninety-four sport science students (78 male, 16 female; mean age 24.2, SD = 2.57, mean sport experience, M = 12.8 years, SD = 5.6) at the German Sport University Cologne participated in this study for partial course credit and provided written informed consent.

Materials

We developed a questionnaire based on previous studies of the hot hand belief in volleyball (Gula & Köppen, 2009) that asked six questions. First, we listed all sports that were previously investigated in a review of the hot hand in sports (Bar-Eli et al., 2006) and added sports from recent studies not included in the review. The order of sports was randomized (archery, basketball, baseball, billiards, bowling, darts, golf, hockey, horseshoe pitching, soccer, table tennis, team handball, tennis, volleyball). Second, we piloted questions about roulette, in which the inversed hot hand belief, known as the *gambler's fallacy*, is found (Sundali & Croson, 2006). Finally, we asked participants directly whether they believed in the hot hand in volleyball.

Procedure

Before receiving the questionnaire, participants were provided with a context-free definition of the hot hand: "Hot hand is defined as the higher probability in, for instance, sports to score again after two or more hits compared with two or three misses." They were then asked five questions: (i) Rank the following sports in the order in which you believe that the hot hand phenomenon is most present. (ii) Imagine that you are playing roulette. What color will occur after a sequence of two or three reds? (a) The probability of red is higher than black; (b) the probability of black is higher than red; (c) both are equally probable. (iii) Do you believe that a player should bet on red after a series of two or three reds? Yes/No. (iv) Do you believe in the hot hand in volleyball? Yes/No. (v) Do you believe that playmakers should play to a player that is hot? Yes/No. In a final open question, they were asked to relate their individual sport experience by naming all sports they performed or had once performed at club level.

Results and Discussion

Eighty-six of the 94 athletes believed in the hot hand in volleyball (about 91%). The average rank order of the 10 sports in which the hot hand was believed to occur was computed for all 94 athletes and was as follows (from most to least likely): basketball, volleyball, darts, billiards, bowling, baseball, golf, tennis, horseshoe pitching, and table tennis. We checked whether participants' individual experience in specific sports might influence belief for a particular sport, but this was not the case. The hot hand belief was distributed over the top 10 sports, independent of the sport in which an athlete had expertise. The majority of participants (about 81%) selected (b) that "the probability of black is higher than red" after two or three reds in roulette.

This study provides the first evidence that athletes believe in the hot hand in volleyball, which supports independent evidence from a study in our lab (Gula & Köppen, 2009). However, does this belief reflect reality?

Study 1B: Is There a Hot Hand in Volleyball?

In Study 1B we tested whether a hot hand in fact exists in volleyball. We began with a brief discussion of how to detect a hot hand, a fiercely debated topic. How can the stability and existence of the hot hand belief be defined? Gilovich et al. (1985) stated that fans are right to believe in the hot hand if players' performance is either nonstationary or dependent. Based on athletes' retrospective reports of experienced "hotness," Hales (1999) maintained that the belief is more appropriately tested by tests of stationarity than by tests of dependence. Both types of tests have the drawback that they rely on perfect runs of either hits or misses. In contrast, Larkey et al. (1989) argued that people may believe that a performance sequence reflects hotness, even if the sequence contains misses, which studies such as those of Gilovich et al. (1985) simply failed to detect. Thus, alternative tests should be developed that include imperfect sequences. However, for the study, we followed the "classical" analytical approach, assuming that if unusual variability in performance can be found, alternative tests are not necessary.

Method

We analyzed the offensive performance (sequences of successive spikes) of male players in Germany's first-division volleyball league. Gilovich et al. (1985) discussed and rejected the hypothesis that the failure to detect the hot hand is a result of the opposing team's intensified defensive strategies against the hot player. In volleyball, however, allocation to a player cannot be hindered directly because the teams are separated by a net.[2] And before the serve, the setter (playmaker) is not able to use many temporal or spatial cues from the opposing team for deciding to whom to pass the ball. Moreover, at first-division level, only one playmaker passes the ball to the final attacker. In these respects, volleyball provides a more controlled environment than basketball for the study of "streakiness" in players' performance data.

Inclusion Criteria

We evaluated the extent to which performance deviated from average within a short time frame (runs within a game) or long time frame (across games). The database we used (TopScorer) consists of 37,000 successful hits (spikes) and misses in the order of their occurrence in each game for more than 100 male players of 226 games in the German first-division volleyball league. In our analyses, we refer to *miss* and *hit* based on the TopScorer definitions, where a miss is an error committed when the attacker smashes the ball into the net or outside the playing field or steps over the attack line. A hit is a point for the attacking team by hitting the ball to the opponents' field such that the ball cannot be successfully defended.[3] First-division volleyball teams, consisting of professionals or semiprofessionals, are the highest performance league in Germany, where there are no collegiate teams as, for instance, in the United States. For the present analysis, we used players from the best teams that entered the play-offs. Play-offs alone were used for calculating base rates and hotness, where the pressure to win is so high that only best performance matters, meaning that effective players will stay on the field and noneffective players will be substituted more quickly than in regular games. To check for individual streaks, one needs long sequences of hits and misses; therefore, only players who were constantly playing—that is, who attacked more than 40 balls in all play-off games—were analyzed, which resulted in a total of 26 top players (see Table 9.2 for the distribution of individual players' attacks between 44 and 319 for all play-off games).

From these sequences, we tested the existence of random or nonrandom sequences using conditional probabilities and a runs test, exactly as in the Gilovich et al. (1985) study. In addition, we analyzed whether higher base rates result in significant lag 1 autocorrelations.

2. It is important to note that the playmaker's ball allocation cannot be directly hindered as it can in open field sports such as basketball, team handball, or soccer, in which man-marking is a typical strategy to reduce allocation to one specific player. In volleyball, the defense can adjust to opponent players' behavior by organizing the block at the net or by positioning the backcourt defense players.

3. In professional volleyball, one definition of a player's base rate is the overall probability of scoring, in which errors also include those spike attempts that were successfully blocked. Because the TopScorer database did not contain this information, the present analysis uses a base rate definition of a conditional probability of scoring given that the player was not blocked.

Results

Probability of Scoring after Hits and Misses

For the set of 26 offensive players we found base rates (p_{hit}) ranging from .59 to .98 (unweighted $M = .9$; $SD = .08$) over the play-offs period (see Table 9.1). These high base rates in volleyball as compared with those in basketball led to a small number of sequences of misses. One reason is that we selected our sample from play-offs reflecting only the best players of the league. Another reason is that the definition of a miss in the database is restricted to spikes that do not go over the net or that land outside the field, meaning that neutral

Table 9.1: Analyses of Conditional Probabilities for the Performance of 26 Top Players in the German Male First-Division Volleyball League

Player	p_{hit}	Total	$p_{hit}/1_{miss}$	Total	$p_{hit}/1_{hit}$	Total	$p_{hit}/2_{hit}$	Total	$p_{hit}/3_{hit}$	Total
1	.91	123	.60	10	.94	112	.94	105	.94	99
2	.98	176		3	.98	172	.99	168	.99	165
3	.95	186		9	.96	176	.96	168	.96	161
4	.94	319	.94	17	.94	301	.96	284	.96	272
5	.93	196	.93	14	.93	181	.93	168	.94	155
6	.93	308	.67	21	.95	286	.96	271	.97	259
7	.95	108		4	.95	103	.95	98	.95	92
8	.86	63		9	.87	53	.87	46	.90	40
9	.85	101	.53	15	.91	85	.91	76	.90	68
10	.95	143		7	.96	135	.96	128	.96	122
11	.89	108	.91	11	.89	96	.88	85	.92	75
12	.81	100	.63	19	.85	80	.91	67	.92	61
13	.89	161	.49	17	.88	143	.90	125	.92	112
14	.59	44	.12	17	.88	26	.87	23	.90	20
15	.95	75		4	.97	70	.97	67	.97	64
16	.73	79	.62	21	.77	57	.79	43	.82	33
17	.92	205	.58	10	.95	194	.95	183	.95	173
18	.85	149	.83	23	.85	125	.87	105	.88	90
19	.93	119		8	.95	110	.96	104	.96	99
20	.98	118		2	.98	114	.98	111	.98	108
21	.89	80		8	.92	71	.91	64	.93	57
22	.89	87	.80	10	.89	76	.91	67	.92	60
23	.97	197		6	.97	190	.97	184	.98	178
24	.92	198	.75	16	.93	181	.94	168	.96	157
25	.84	118	.58	19	.89	98	.91	86	.90	77
26	.92	117		9	.93	107	.94	98	.95	91

Note: Sequences of hits and misses as probabilities after one miss ($p_{Hit}/1_{Miss}$) or one ($p_{Hit}/1_{Hit}$), two ($p_{Hit}/2_{Hit}$), or three ($p_{Hit}/3_{Hit}$) hits. Data are from season 1999/2000. Total refers to the number of attacks; $p_{hit}/1_{miss}$ is calculated only for totals ≥10.

attacks in which the opposing team defends are not counted. Therefore, we could only test whether the probability of a hit is different after one miss or hit, following Gilovich et al.'s (1985) analyses.

Distributions of the conditional probabilities were analyzed across the 15 players for whom the total number of hits following a miss was equal or larger than 10 (column 5, Table 9.1). The unweighted mean probability of hitting conditioned on one miss was .67 (SD = .21) and .90 (SD = .05) conditioned on one hit (one-sided $t(14) = 4.29$; $p = .001$; Cohen's $d = 1.1$).

Table 9.2: Analyses of Runs and Autocorrelation for the Performance of 26 Top Players in the German Male First-Division Volleyball League

Player	Base Rate (p_{hit})	Number of Attacks	Hits	Misses	Observed Runs	Expected Runs	Z	Autocorrelation
1	**.91***	**123**	**112**	**11**	**14**	**21.0**	**−3.98**	.31
2	.98	176	173	3	7	6.9	.25	−.02
3	**.95**	**186**	**177**	**9**	**15**	**18.1**	**−2.56**	.18
4	.94	319	301	18	34	35.0	−.52	.00
5	.93	196	182	14	26	27.0	−.55	.01
6	**.93**	**308**	**287**	**21**	**29**	**40.1**	**−5.05**	.28
7	.95*	108	103	5	10	10.5	−.62	−.04
8	.86*	63	54	9	15	16.4	−.75	.09
9	**.85***	**101**	**86**	**15**	**17**	**26.5**	**−3.81**	.37
10	.95*	143	136	7	13	14.3	−1.22	.10
11	.89	108	96	12	22	22.3	−.17	−.02
12	**.81**	**100**	**81**	**19**	**25**	**31.8**	**−2.23**	.22
13	.89	161	144	17	35	31.4	1.52	−.12
14	**.59**	**44**	**26**	**18**	**6**	**22.3**	**−4.98**	.74
15	**.95**	**75**	**71**	**4**	**6**	**8.6**	**−3.14**	.22
16	.73*	79	58	21	27	31.8	−1.41	.15
17	.95*	205	195	10	21	20.0	.75	−.05
18	.85*	149	126	23	39	39.9	−.29	.02
19	**.93***	**119**	**111**	**8**	**11**	**15.9**	**−3.71**	.33
20	.98	118	115	3	5	6.8	.22	−.02
21	.89*	80	72	8	15	15.4	−1.14	.13
22	**.89**	**87**	**77**	**10**	**17**	**18.7**	**−2.06**	.10
23	**.97**	**197**	**191**	**6**	**11**	**12.6**	**−2.06**	.14
24	**.92***	**198**	**182**	**16**	**25**	**30.4**	**−2.63**	.18
25	**.84***	**118**	**99**	**19**	**23**	**32.9**	**−3.41**	.31
26	.92*	117	108	9	17	17.6	−.41	.04

Note: Bolded numbers represent players with Z-values of ≥1.96. Asterisk (*) in base rate column (p_{hit}) indicates significant variation between sets and games. Data are from season 1999/2000. Z refers to the standardized test statistic, that is, the difference between the observed and the expected number of runs divided by the standard deviation of the expected number of runs. Autocorrelation scores reflect lag 1 autocorrelation.

Weighting the probabilities by the ratio of the respective individual to the column total leads to similar results but with a reduced effect size (one-sided $t(14) = 1.93$; $p = .037$; Cohen's $d = .5$). Whether these results provide evidence for cold or for hot streaks was explored by testing both conditional probabilities of a hit against the overall base rate. Effect sizes were higher for the difference between the overall base rate and the probability of a hit after a miss, $t(14) = 4.78$; $p = .001$; Cohen's $d = 1.2$; weighted by relative total: $t(14) = 1.97$; $p = .035$; Cohen's $d = 0.5$, than between the base rate and the probability of a hit after a hit, $t(14) = 1.98$; $p = .034$; Cohen's $d = 0.5$; weighted by relative total: $t(14) = 1.53$; $p = .074$; Cohen's $d = 0.4$. This suggests that the streakiness resulted more from "coldness" than from hotness. As can be seen from Table 9.1, similar tests of the conditional probabilities of a hit after two or more misses are not feasible because the corresponding totals are too low. Instead, we assessed streakiness by testing whether the players' sequences contained fewer runs than expected.

Runs Test

We conducted a Wald-Wolfowitz runs test for all 26 players and used the same criterion as in Gilovich et al. (1985) for classifying a player as hot, which in our data results in a player's Z-value above 1.96. For this test, each sequence of consecutive hits or misses counts as a run. Thus, the more consecutive hits (or misses) a sequence contains, the fewer runs can be observed. In Table 9.2, the observed number of a player's runs is compared with the expected number of runs according to his base rate. Twelve of the 26 players had fewer runs than expected (bolded in Table 9.2); only one or two of the 12 would be accounted for by chance. The number of volleyball players with fewer runs than expected is substantially higher than among the basketball players studied by Gilovich et al. (1985) and Koehler and Conley (2003), where only one or two players, respectively, were found to have a significantly lower or higher number of runs than expected. The analysis of runs leads to the same conclusion as in the previous analysis of conditional probabilities: In volleyball, a substantial number of players exhibit streakiness.

Autocorrelations

In a third test of the hot hand, we computed the autocorrelations as described above and found higher autocorrelations between successive hits in volleyball, compared with Gilovich et al.'s (1985) basketball data. The autocorrelation values (lag 1) for each player are shown in Table 9.2. We used lag 1 values as indicators for a positive correlation as these measure the direct influence of the previous hit/miss to the next trial. Autocorrelations around zero (±.1) are found for 14 of 26 of the top players; the remaining 12 players show significant autocorrelations between .1 and .7, indicating that successive attacks are dependent. Twelve players could be termed streaky according to both autocorrelation and the runs test. A similar result was not observed in the Gilovich et al. (1985) data, possibly because base rates were lower (around .5) and less variable. However, the only player with a base rate higher than .6 in their basketball study also received a significant but negative autocorrelation.

Variability of Players' Individual Base Rates between Sets and Games

A key reason for playmakers to rely on the hot hand belief to make better allocations would be a systematic variability of players' performance between sets (up to five sets are played in a game of volleyball) and between games. For instance, local base rates[4] in individual games within the regular season ranged from .45 to .70 for player A (no. 1 in Table 9.1) and from .54 to .86 for player B (no. 18 in Table 9.1). These players' variability was used as a guideline for a realistic manipulation of within-game sequences in the subsequent experiments (Studies 2A and 2B). When looking at the within-game variability from set to set, player A showed a range of local base rates from .20 to .87 and player B from .33 to .90. We found this large variability in more than half of the players, both within a game (i.e., between the maximum of five sets) and between the maximum of 15 games (see Table 9.2 base rate column for significant variation indicated by an asterisk). This variability between sets and games is consistent with the findings in the previous sections that the hot hand exists in the data set and provides the precondition for an adaptive reliance on the hot hand.

Correlations between Number of Allocations and Players' Performance

We correlated the number of playmakers' allocations to players and players' performance that may be indicative of the relation between both. First we correlated the observed number of runs and the number of allocations of a playmaker (Table 9.2, Total number of attacks). We found a correlation of .47 ($p = .008$), which is about the same as a correlation of .49 ($p = .005$) between the base rate (Table 9.2, p_{hit}) and the number of allocations. Thus, both base rate and runs relate to allocation behavior. If we control for base rate in a partial correlation between observed runs and number of allocations, we find a correlation of .61 ($p = .001$). Assuming that fewer observed runs than expected represent streakiness, we computed the degree of streakiness as the difference between the expected and observed number of runs. For the 12 streaky players in Table 9.2, we found a partial correlation of .66 ($p = .026$) between the degree of streakiness and the number of allocations (controlling for base rate). These results indicate that allocations are associated with

4. We use the term *local* base rate when referring to the relative performance within a game, set, or half-set.

runs and streakiness (hot hands) and thus playmakers' allocations cannot be explained solely by players' base rates.

Discussion

In our analysis of 26 top players, we found that (i) all three measures—conditional probabilities, runs, and autocorrelation—provided evidence of streakiness in about half of the players' hit-and-miss patterns and (ii) players' individual base rates varied between sets and games. This result differs from the stable patterns found in basketball by Gilovich et al. (1985) and Koehler and Conley (2003). One possible structural explanation for the difference between the studies in volleyball and basketball lies in the nature of each sport. If there is a hot hand in basketball, it may not be detected in the hit-and-miss data, because opponent players can react by using counter-strategies against the hot player. This is less possible in volleyball, where a net separates opposing teams. However, even when defense was absent in basketball, that is, when players had free throws (Gilovich et al., 1985) or in three-point shootout contests (Koehler & Conley, 2003), hot hand was not found either. Whether the hot hand belief is restricted to basketball or can be generalized to other sports or other domains (see Gilovich, 1993; Gilovich et al., 1985; Tversky & Gilovich, 1989) is still under debate, but a recent review concluded that the hot hand should be more present in sports or situations within a sport in which defense can be eliminated or limited (Bar-Eli et al., 2006).

Study 1 provides the first evidence of streakiness in offensive performance in volleyball. There are two possible limitations to this study. First, the autocorrelations were calculated throughout the full sequence of hits and misses; therefore, the last hit or miss in play-off game 1 was compared with the first attack in play-off game 2. However, because there are only three to five play-off games in which an individual player can participate, this effect does not influence the results. Alternatively, running separate autocorrelations for each player for each game would potentially inflate alpha errors. A second limit is that defense pressure can change between games even with the same opponent in play-offs and therefore could inflate the streakiness in the data. Although we chose play-offs of the top teams to limit this specific effect, the variability within a game and between games suggests that there is variability that could be perceived by coaches or fans, and could also alter playmakers' allocation strategies.

The current data suggest that in volleyball there is both streakiness ("hot hand") for half of the players and no streakiness for the other half. Thus, the hot hand belief reflects reality for half of the players but not for everyone. It is therefore necessary to show that coaches and playmakers can actually detect the subtle but systematic changes in players' performances beyond evidence of correlations. Study 2 addressed how sensitive coaches can perceive changes in performance (Study 2A) and how playmakers use the hot hand belief in their allocation decisions (Study 2B).

Study 2A: Are Coaches Able to Detect Players' Changes of Performance over Time?

In Study 2A we evaluated how sensitive coaches are in detecting changes of performance and whether they used these to instruct playmakers how to allocate the ball. Sensitivity was measured for local changes of the average base rate of a player. A local base rate needs to be specified with respect to a reference class; here we consider one natural unit, the set in a volleyball game (e.g., six of 10 hits in the first set is better than two of 12 in the second). Although there is evidence that people are sensitive to changes in base rates (e.g., Oskarsson et al., 2009; Koehler, 1996b for overviews) nothing is known in the context of sports. For instance, a general finding is that within-subject variation in the base rate draws attention to the base rate (Gigerenzer, Hell, & Blank, 1988; Mueser, Cowan, & Mueser, 1999), but it is not known how accurately changes in the base rate of an individual player can be detected. Furthermore, Castaneda and Rodrigo (1998) showed that people are more sensitive to perceptual information than to information presented semantically (as is used in most experiments) when estimating the base rates of events. Because Castaneda and Rodrigo used *stable* nonspecific visual information, but in a real game coaches react to *variable* sport-specific stimuli, it is unclear how their result applies to sports. We therefore conducted an experiment that evaluated the sensitivity to picking up base rates of hits in dynamic and visually presented volleyball attacks.

Method

Participants

Sixteen coaches from Berlin with a B- or C-level coaching license in volleyball and a mean age of 37.5 years ($SD = 10.3$) participated in the study. In Germany, B-level coaches are allowed to coach up to the third highest and C-level coaches up to the second highest professional league level. They received 10 euros for participating. All coaches gave written informed consent beforehand.

Apparatus

The stimuli were video files showing the attacks of the two teams in a volleyball game. These were projected onto a large screen (1.5 × 2 m) to ensure a realistic perceptual setup. Video presentations are typically used by coaches and athletes to evaluate performance after a game (Lyons, 2003; Paiement,

Baudin, & Boucher, 1993). In the videos, we limited the attacks to sequences of two volleyball players, arranged so that player A always attacked from the left side of the display and player B always attacked from the right. A computer program randomly displayed either the left or the right player, who then performed either a hit (defined as a spike into the opponent's field) or a miss (a spike that sent the ball either outside the field or into the net).

Task

The task was to watch a game, which consisted of four sets of 44 attacks (176 attacks in total). Coaches were asked to watch the distribution of the setters' passes to two attacking players (as they would in a real game). At the end of each set they were asked to write down instructions for the setters for allocating the ball to the two players as well as to recall the observed players' performances.

Coaches were given written instructions that they were required to analyze a playmaker's allocation strategy on a video. They did not know how the two attackers would perform beforehand. Before the experiment began, they viewed two scenes with each player (one hit, one miss) to familiarize themselves with the task. Subsequently, they were asked to (i) watch attacks by players A and B in a random pattern of sequences of 11 attacks by each player, (ii) after each sequence of 22 attacks, write down the distribution of balls to player A and to player B for the next 10 allocation decisions, and (iii) recall the performances of the two observed players (how many out of the last 11 attacks did player A—or B—hit?).

At the end of the experiment we distributed a questionnaire about the coaches' hot hand belief that contained the same items as in Gilovich et al. (1985) but adjusted to volleyball. We asked: (i) Does a player have a better chance of making a successful strike after having made two or three previous spikes than he does after having missed the previous ones? (ii) Consider a hypothetical player who has an average spike hit rate of 50%. Please estimate (from 0% to 100%) the player's average hit percentage after just having successfully spiked before ____ and after just having missed the last spike ____. (iii) Do you believe that playmakers play to a player that is hot? Yes/No.

Material

The sequences of players were distributed as ecologically validly as possible by matching sequences of successful hits and misses found for two players in the TopScorer database (Study 1B). The Institute for Applied Training Sciences (IAT) in Leipzig, Germany, provided us with audiovisual (AV) files of these players' successful hits and misses from both sides of the court, making it possible to manipulate the sets (after each set, teams change court sides) accordingly. The base rates used, the sequence of successful hits and misses, and the number of hits an average player attempted before the next side-out attacker rotated from the back row to the front in order to hit near the net (based on the rotation rule) were organized to reflect real-life game situations.

The AV files of the two players were edited in the following way. One team served and a player passed the ball to the playmaker, who then set the ball to the left outside position, where a player either hit or missed. This player who hit was either player A (when shown on the left) or player B (when shown on the right).

Four sets of stimuli were prepared such that each set ended when a team reached 25 points, as in a real game. Each set consisted of 44 attacks (22 by each player). After half of the set (22 attacks, 11 for each player), a time-out was displayed. The base rates for the players in each half-set (22 points) were either the same (twice) or different (six times). The difference between the players' number of hits was varied from a one-point difference to a five-point difference to see how easily the base rates and differences between players could be picked up and used when defining allocation strategies for the playmakers. To avoid any potential ambiguities (e.g., a ball that was very close to an offside line), the score was displayed after each point.

Results

92.3% of the coaches believed that a player has a better chance of making a successful strike after having made two or three previous spikes than he does after having missed the previous ones, indicating a general belief in the hot hand. Coaches' estimation of a player's hitting probability just after having made a strike was 55.5%; their average estimation of a player's strikes just after having missed a strike was 46.8%. This difference is smaller than Gilovich's results for basketball (61% to 42%). 84.6% of the coaches believed that playmakers play the ball in accordance with this belief, indicating that coaches consider more factors for allocation than merely the hit rate of a player.

On average, the 16 coaches wrongly estimated the number of hits in the preceding half-set by only 1.2 balls out of 22. License class, age, and gender did not influence the coaches' level of performance. Player A was evaluated as a slightly better hitter (by about 3 points), even though the mean base rates were the same over the game. The correlation between the estimated base rate of a player (how many hits of the last 11) and the allocation strategy (how many balls to player A and to player B) was $r = .55$, $p = .002$, over the entire game. This result indicates that the coaches were sensitive to changes in base rate and corrected their allocation strategies according to their base rate judgments.

Discussion

The results of the experiment indicate that coaches are sensitive to base rate changes and use them as cues for allocation decisions. This conclusion is supported by coaches' allocation instructions that correlated with the perceived base rate changes within the two players. The observed number of hits influenced the coaches' instructions for allocating the next 10 attacks. All volleyball coaches' base rate sensitivity was highly accurate, with an average error of about 1 hit or miss for each player out of 22 attacks. In general, coaches believe in the hot hand and also say that their playmakers use the hot hand belief for their allocations. From the coaches' allocation instructions, it is evident that they would not play all balls to the player with the momentarily higher performance because of potential side effects, such as a specific player's fatigue.

The changes in coaches' allocation instructions show that they go beyond using the overall base rate of players and take short-term performances into account, which in the present study would result in equal distribution to both players. In the next step, we tested whether players actually allocate balls in an adaptive way depending on the perceived "hotness" of players and base rate. For Study 2B, we therefore used the same paradigm as in Study 2A and asked players to allocate balls to two players who have either the same or a different base rate and who vary in their degree of being hot.

Study 2B: How Do Playmakers Use the Belief for Allocation Decisions?

Method

Participants

Twenty-one persons (11 male, 10 female; age: $M = 23.5$ years, $SD = 2.9$) with volleyball experience took part in the experiment. Volleyball experience was defined as participating in a major course in volleyball that requires playing at adult club level for at least one year. All participants were students in physical education or sport science. Participants played on average at regional level (from third lowest level to third highest national level) and thus below the professional first and second league. None of them had taken part in Study 2A. They received 10 euros for their participation and could win another 20 euros based on their performance. All gave written informed consent before participating in the experiment.

Apparatus and Material

The task was again presented as a digital video test on a computer, projected at a presentation size of 1.5 × 2 m at a distance of 3 m from the participant. The software (C++, Czienskowski, 2002) displayed AV files and collected data from the participants. In this study the participants' task was to allocate the ball to one of two players before every attack, define their allocation strategy in advance, and estimate the two players' base rates. Note that in the present context the term *base rate* is used to refer to the relative frequency of hits of a player per half-set of the game.

The same AV files as in Study 2A were used, but the order of presentation was varied. The manipulated base rate changes were—according to our analysis of the TopScorer database—a typical distribution of players' real base rate changes.

Four conditions counterbalanced in order were designed (see Table 9.3): (i) base rates are equal for both players and both show a pseudorandom distribution of hits and misses (no hot hand); (ii) base rates are equal for both players but one player is hot; (iii) base rates are different and both players demonstrate a pseudorandom distribution of hits and misses; and (iv) base rates are different and one player is hot. We varied the base rates in both the "same base rate" and "different base rate" conditions to model real base rate changes between sets (see Table 9.3). A pseudorandom distribution was defined as no more than two hits or two misses in a row for each player. In the hot hand condition, a sequence of three hits occurred at least twice. The number of attacks (hits and misses) was arranged based on the distributions found in the database of real competitions. We restricted the hot hand sequences so that no more than four spikes in a row were made by one player.

To determine experimentally whether belief in the hot hand influences behavior, we constructed series of hits and misses that were either pseudorandom (no more than two consecutive hits or misses) or "nearly hot" for each player and half-set (three to four consecutive hits). When constructing the pseudorandom condition we used a positive Z-value between 0 and 1 as a criterion, reflecting slightly more or exactly the same number of actual runs as expected given the players' base rates. In the hot condition we deliberately introduced runs ("nearly hot") that would not satisfy the statistical definition of a hot hand streak ($Z > 1.96$, see Study 1B) yet were long enough for participants to associate them with an increased probability of a hit. We assumed that if participants estimated the probability of a hit to be higher after two or three hits than after two or three misses (the hot hand question in Study 1A), then runs of length three and four should be perceived as streaky. The goal was to examine the effect of perceived hotness, as opposed to actual streaks, on allocation decisions.

In comparison, in the original study of the hot hand phenomenon (Gilovich et al., 1985), the nine players of the Philadelphia 76ers had a mean number of runs of about 215 and an expected number of runs of 210, which results in a small negative Z-value of -0.56. Daryl Dawkins had the highest negative Z-value of -3.09, representing an expected number of runs of about 190 and an observed number of runs of 220.

We controlled for the influence of a specific player to ensure that both players were hot equally often and had the same base rate over the entire game. In half-sets 1 and 7 (Condition 1) base rates were the same for both players, and their hits and misses were distributed pseudorandomly. Neither cue (differences in base rates or streaks) in this control condition provided any information that might be exploited for an allocation strategy other than random. In half-sets 2 and 8 (Condition 2) the base rates were also the same

Table 9.3: Tests Consisted of Four Sets (Eight Half-Sets With 22 Trials/Attacks Each) in Which Base Rates of Player A and Player B Were the Same (1, 2, 7, 8) or Different (3–6), and Either One Player Was Hot ("Hot: Yes") and the Other Player had a Pseudorandom Sequence ("Hot: No") or Both Had Pseudorandom Sequences

Half-set		1	2	3	4	5	6	7	8
Player A	BR	.50	.64	.40	.73	.45	.67	.50	.64
	Hot	No	*Yes*	No	No	No	*Yes*	No	No
Allocation mean, (*SD*)		12.4 (3.5)	12.3 (3.1)	10.2 (3.1)	10.0 (2.0)	12.1 (3.2)	13.3 (3.7)	12.4 (3.6)	9.5 (3.8)
Estimation mean, (*SD*)		52.38 (13.4)	59.74 (13.01)	40.69 (11.36)	52.81 (15.64)	60.17 (17.59)	61.03 (16.55)	48.05 (13.21)	50.21 (8.43)
Player B	BR	.50	.64	.58	.55	.73	.40	.50	.64
	Hot	No	No	No	*Yes*	No	No	No	*Yes*
Allocation mean, (*SD*)		9.6 (3.5)	9.7 (3.1)	11.8 (3.1)	12.0 (2.0)	11.9 (3.1)	8.7 (3.6)	9.6 (3.6)	12.5 (3.8)
Estimation mean, (*SD*)		47.6 (11.46)	43.72 (13.05)	58.44 (9.77)	56.71 (13.45)	41.99 (15.34)	43.28 (15.72)	45.88 (10.93)	64.5 (10.32)
Condition		1	2	3	4	3	4	1	2

Note: BR = base rate; Hot = hot hand; Allocation = mean number of allocations to the respective player (*SD*); Estimation = estimation of local bases rates. Conditions are explained in the text.

but the sequence of spikes and misses was manipulated systematically to be hot (3 to 4 hits in a row) for one player and pseudorandom for the other.

Because there is no difference in performance between the players with equal base rates, only the hot hand can serve as an allocation cue in these conditions. This environmental structure resembles a typical "pickup" game situation, where base rates cannot be used as an allocation cue (given that they are not known) and streaks may instead serve as cues to infer the underlying level of performance and influence allocation. In the other half-sets, the players' base rates were varied.

In half-sets 3 and 5 (Condition 3) the sequences of successful spikes and misses were pseudorandom for both players; hence, any consistent allocation behavior of participants can be attributed only to the differences in the players' base rates. Note that Conditions 2 and 3 were especially designed to determine whether the two kinds of information provided (i.e., sequence structure and base rates) independently influence allocation behavior.

In Condition 4 the effect of both cues on allocation behavior was investigated. First, the sequence in half-set 6 was manipulated to be hot for the player with the higher base rate. Then the sequence in half-set 4 was manipulated to be hot for the player with the lower base rate.

Procedure

Participants were individually tested in a session that took about 60 min. At the beginning of the session, the experimenter presented a written introduction to the experiment and explained the performance-based payment system. Participants could then warm up for the test by observing 10 pilot video clips and pressing the button corresponding to the player to whom they wanted to allocate the next ball. Four sets of a volleyball game were presented. Each set consisted of 44 attacks (balls in play until the final score, with one team winning 25 points). After the first 22 attacks (the time-out) participants were asked to type in a displayed text box on the screen how they wanted to continue their ball allocations to the players. Specifically, they were asked how many of the next 10 balls they wanted to distribute to player A and how many they wanted to distribute to player B.

After participants made their allocations for the following half-set, a new query appeared on the display, asking them to estimate for player A and player B how many of their 11 attacks in the preceding half-set were successful, that is, were hits. The final 22 attacks for the first set were presented when participants finished their base rate estimates. After the first set, the questions about their allocation strategy and base rate estimates were repeated. This procedure was repeated for all sets within the experiment. After the final set, participants were asked to fill in a questionnaire containing items on personal data and a section on their decision strategy, in which they could rank possible factors (base rate and structure of sequence) that influenced their decisions.

In addition, the questionnaire included items regarding participants' belief in the hot hand. For the German participants, we translated the term *hot hand* literally, saying that someone is hot ("ist heiß"). In particular, the questionnaire explicitly asked about the use of the structure of the sequence and of the base rates, and to whom they would allocate the ball if, for instance, toward the end of the game the player with the lower base rate hit two or three balls in a row.

Results

Hot Hand Belief Questionnaire

As in Study 2A, two questions were asked regarding belief in the hot hand. These were followed by three additional questions concerning the information that had been manipulated in the experiment. First, we asked if it was important to allocate the ball to the player who had just hit two or three times. Nineteen of 21 participants (about 90%) answered in the affirmative. Second, we asked if a player who had just hit the last two or three times had a better chance of hitting the next ball than a player who had just missed the last two or three balls. Fifteen of 21 participants (about 71%) said yes. These results support the rationale behind the constructed sequences that runs of three and four hits would be perceived as hot by the majority of participants. When asked the third question, to rank the importance of different factors that influenced their decisions, participants ranked the players' hot hand as most important, followed by the "cold hand" (number of misses in a row), and the overall base rate in the set. Participants also stated that they used the hot hand or cold hand of the players to allocate their passes in seven of 10 decisions. When asked explicitly about a situation where one player has the higher base rate and the other has just made two or three hits but has a lower base rate, 62% of the participants claimed that they would allocate the ball to the latter player.

Allocation Behavior When Base Rate and Hot Hand Information Are Present

We found that participants (setters in the experiment) were sensitive to both the hot hand and the players' base rates, as indicated by their real allocation behavior. The hot hand cue was stronger than base rates in determining participants' allocation. That is, most participants relied on the hot hand consistently (see Figures 9.1, 9.2, and 9.3).

Condition 1. In this control condition, the players' base rates were equal and the sequences for both players were pseudorandom. The allocation to player A or player B was close to equal, with a nonsignificant tendency to pass more often to player A (see Table 9.3). Similarly, participants' estimates of the number of hits did not differ from the actual base rates in the videos.

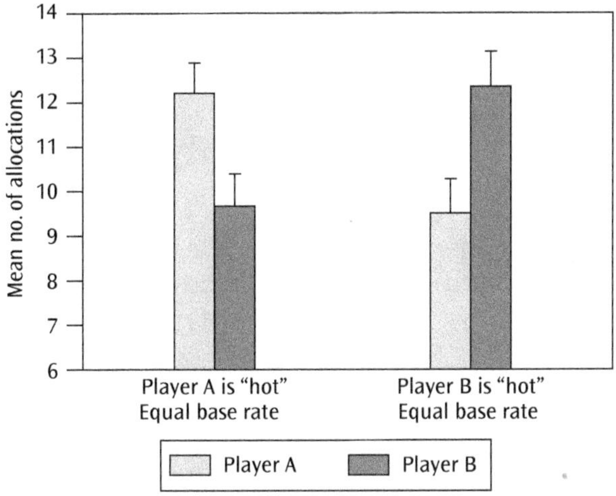

Figure 9.1: Mean number of allocations of the ball to player A and player B when base rates were equal but a hot hand existed for player A (left) or player B (right). Error bars represent standard errors.

Condition 2. We then considered the critical case in which the base rates of hits for the two players were the same but one of the players was hot. If participants used base rates alone, allocations to both players should be equal. Yet as shown in Figure 9.1, participants allocated balls more often to the hot player, even though the base rates were the same. This result holds regardless of whether player A or B was hot (*t* test player hot/not hot, $\eta^2 = .24$; $p = .02$).

As in Condition 1, the base rate estimates for the player with a hot hand were almost perfect. However, for the players with pseudorandom sequences, base rates were underestimated by about two more misses when player A was hot (*t* test real vs. estimated base rate, $\eta^2 = .34$; $p = .01$) and similarly, when player B was hot ($\eta^2 = .63$; $p = .001$). This is a surprising finding, where a hot player appears to lead to a devaluation of other players (see General Discussion).

Condition 3. In this condition, one player had a higher base rate but neither player was hot. The question was whether participants allocate balls in a way that is sensitive to base rate differences when there is no hot hand. Figure 9.2 shows that allocations increased in both situations from about 10 to 12 (*t* test player with higher vs. lower base rate, $\eta^2 = .16$; $p = .07$).

Condition 4. Here, participants were shown videos of players who had different base rates, in which one player was hot. The hot player had either the higher or the lower base rate. When player B had a hot hand but player A had the higher base rate (left side of Figure 9.3), more allocations were made to the hot player. The same result was found when player A had a hot hand and player B had the higher base rate (*t* test hot hand vs. higher base rate, $\eta^2 = .42$; $p = .01$). When the player with the hot hand also had

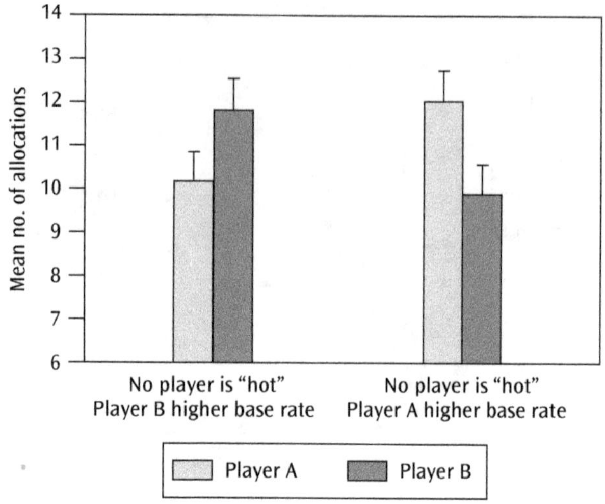

Figure 9.2: Mean number of allocations of the ball to player A and to player B when no hot hand existed for the players but the base rate for player B (left) or player A (right) was higher. Error bars represent standard errors.

the higher base rate (right side of Figure 9.3), the allocation to this player was higher (*t* test hot hand and higher base rate vs. no hot hand and lower base rate, $\eta^2 = .15$, $p = .03$). In this situation, more allocations were made than in Conditions 2 and 3, where only hot hand or base rate was available (ANOVA, $\eta^2 = .35$; $p = .01$).

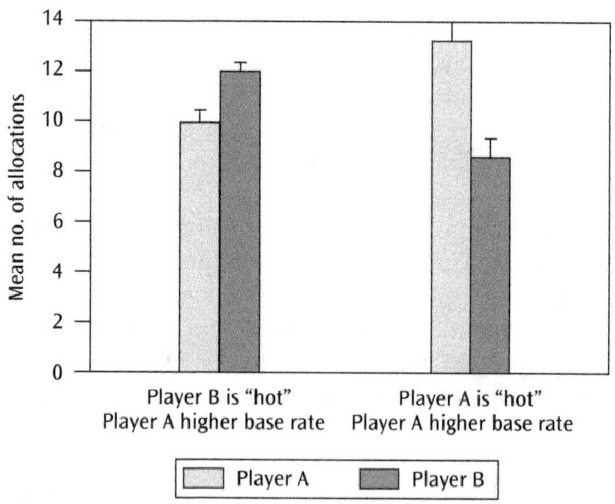

Figure 9.3: Mean number of allocations of the ball to player A and to player B when a hot hand and the higher base rate were present in the same player (right) or not (left). Error bars represent standard errors.

An analysis of the base rate estimates showed the same surprising result as in Condition 2 but extended it to the case where base rates differed. If the player was hot and had the higher base rate, the participants estimated his base rate almost perfectly but underestimated the base rate of the other player by about two hits (*t* test estimated base rate for player A vs. player B, $\eta^2 = .41$; $p = .001$).

Do Players React in an Adaptive Way to Local Base Rate Changes?

Because each participant was exposed to the four conditions, all of them were confronted with changing base rates, enabling us to test whether they could allocate balls in an adaptive way when a player's base rate suddenly changed. After viewing 22 attacks of player A and B in half-set 1, they were asked for estimates of both players' base rates (they were never asked whether a player was hot) and then asked how they would allocate the balls for the next half-set. Yet unknown to them at that point, the base rate in the next half-set changed for both players from .50 to .64. Whereas player A and player B had the same base rate in the first two half-sets, in half-set 3 player B had a higher base rate than player A. If participants reacted to this change in an adaptive way, the announced allocation should not be identical with the actual allocation in the following half-set but change in the direction of the new base rate. We analyzed the number of adaptive changes compared with nonadaptive behavior. An adaptive change is present, for instance, if an intended allocation prefers player A to B, but in the following half-set, where player B has a higher base rate, participants abandon their intended allocation to player A and allocate more balls to player B. An adaptive change is also present if intended allocation is equal to both players because their base rates in the present set are equal, but in the next half-set, where player A has a higher base rate, participants allocate more balls to player A. A nonadaptive allocation strategy is present if participants remain loyal to their intended allocation, such as allocating more balls to player A in the next half-set, even if player B now has a higher base rate. Because there are eight half-sets, we can observe seven changes of base rates for player A/ player B. We found that, averaged over participants, an adaptation in the direction of the new base rates was present in five out of these seven cases. The correct changes from the intended allocations in response to the new base rates were on average one or two balls in the correct direction, reflecting base rate changes between half-sets of about .1 or .2. In the two cases in which intended allocation was not in the direction of the actual base rates, all half-sets had one player with a "nearly hot hand." Thus, the intended future allocations always followed the base rates significantly (all *ps* $<.05$), except in the two cases where the base rate conflicted with being hot. For these cases, where the hot hand dominated, using the hot hand belief can be labeled nonadaptive.

Discussion

Study 2B showed that participants were able to detect a hot hand and used this information for their allocation decisions. Moreover, the hot player was allocated more balls even if his base rate was equal to or lower than that of the other player. Whereas the hot hand dominated allocation decisions, comparison of the conditions showed that base rates alter the allocation frequencies. An unexpected result was that if one player was hot, the other player's base rate was systematically underestimated. This effect could be explained by increased attention to the hot player and more cognitive resources allocated to monitoring his performance. The trials results suggest that strategies based on few preceding events may perform better than those based on complete information about attackers' performance.

General Discussion

The previous studies addressed three questions: Do athletes and coaches believe in a hot hand in volleyball? Does the hot hand exist in volleyball? And is the hot hand belief used to inform allocation decisions? The results indicate that the answer is "yes" for all three questions. In our samples, 91% of 94 athletes (Study 1A), 90% of 21 athletes (Study 2B), and 92% of 16 coaches (Study 2A) believe in the hot hand. Depending on the criterion used, 53% of the 26 German top players show significant streaks, and 46% show significant autocorrelations (Study 1B). The key evidence that suggests an adaptive use of the hot hand belief is that playmakers allocate the ball more often to players with streaks and that this leads to better performance than when allocating the ball to the player with the higher average base rate (Study 2B).

In what follows, we discuss several open questions resulting from the present and previous research.

How Is the Hot Hand Belief Used in Real Allocation Decisions?

Study 2B demonstrated that the hot hand of a player can directly influence allocation strategy. As shown in Study 2A, coaches use information on streaks to determine behavioral strategies, that is, allocation, and they believe that other professionals do the same. Study 1B showed that their belief in the hot hand is partly correct.

Moreover, information on streaks is not used merely because players' base rates or changes in their base rates cannot be directly detected (Burns, 2004); rather, the hot hand of players is used as information *in addition to* their base rates. Study 1B was the first to show correlations between number of playmakers' allocations and observed runs as well as correlations between allocations and base rates. As of yet we are not able to differentiate the information that influences allocation.

It is important to distinguish between hot hand defined as a performance difference within a player (the classical definition) and as a performance difference between players. The latter is the important information for allocation in the real game. This interpretation is supported by the results of Study 2B, in which the estimation of the two players' base rates was influenced nonequally such that the base rates were underestimated if the player was not hot. To explain this striking and potentially costly result, further investigation is necessary (Gilovich et al., 1985). For instance, it could be assumed that a scout watching a game focuses mainly on absolute performance of an individual player, whereas the playmaker of a team as tested in this set of studies may rely on relative performance for making decisions.

A model for allocations in sports that goes beyond the models using streaks and base rate explored in Study 2B should incorporate further cues that players might use to allocate a ball. In the present research, we considered only the influence of players' base rates and hot hand on playmakers' decisions. Drawing on the studies conducted so far, we can begin to extract process details of these decisions. However, the possible mechanisms that a playmaker uses to allocate balls in volleyball and potentially in other sports are still far from being understood, and one step toward answering this question is taken by Köppen and Raab (2012). Another issue highlighted by Study 2B is that streaks influence the allocations more strongly when base rate differences between players are small. Finally, because the current manuscript focuses on positive streaks, it is not yet shown whether beliefs about the cold hand in sports result in fewer allocations to the player in question and in earlier substitution of that player in the game (but see Köppen & Raab, 2012).

When Is the Use of the Belief in the Hot Hand Adaptive?

Our key argument is that streaks within a small time unit of a set of a game exist and when these are picked up by the playmaker, better performance can result. By better we mean better than a random allocation strategy or a base rate strategy. The results of Study 2B indicate that the belief in the hot hand is used when base rates change within a player across sets and between players within a set.

However, a hot-hand-only allocation strategy that ignores base rates does not appear foolproof. It can decrease performance if the hot player's base rate is lower than those of the other players. For instance, Figure 9.1 suggests that the effect of the hot hand allocation strategy is about two to three balls (of 22) more for the hot player. The strategy will lead to nonadaptive allocations if the base rate differences between the hot and the not-hot player compensate for this effect. This can only happen if the hot player has the lower base rate and the base rate difference results in three or more balls. In Study 2B, Condition 4, we created such a situation, and playmakers continued to apply the hot hand strategy (if a hot player exists, allocate to him) even though it led to worse outcomes. In this condition, the base rates differed from .55 to .73, that is, by 18 percentage points.

It remains unclear how often the combination of a hot hand with a substantially lower base rate occurs within a real volleyball team. In our data we have exactly one player (player 14, Table 9.2) who has a base rate of .59 (we used .55 in Study 2B, Condition 4) and a hot hand indicated by a high autocorrelation and *Z*-score. Let us consider this player in a team of six volleyball players on the court, of whom one is a playmaker and one is a defense player who does not attack, resulting in a choice to allocate the ball to one of four players. If the playmaker chose to allocate the ball to the hot player with the lower base rate systematically more often than reflected by his low base rate, this allocation would be ineffective. However, this can only occur if the other players are not hot and the hot player is on the court. Note that players with a lower base rate do not play often (e.g., player 14 played the lowest number, 44 attacks from a total of 3,804, compared with the highest number, 319 attacks, by player 4). Furthermore, in our database only one such player exists in one team of the final four teams of the play-offs, meaning that this situation is rare.

Based on autocorrelations and analyses of runs tests we know that players (at least in volleyball) have high base rates and that nonrandom sequences exist. We also know that coaches and playmakers are able to detect base rates and changes in base rates of players accurately in the simple and fairly realistic setting of our experiment. However, we do not know what playmakers more likely believe: (a) that a player is hot if he overshoots a specific level of base rates or (b) that a player is hot if a difference between players' base rates under monitoring meets a threshold. These issues deserve further experimental investigation to understand the important relation between the gradual increase or decrease of the hot hand belief during changes in situations, base rates, and individual behavior. Burns (2004) cited unpublished data that suggest that people are more likely to follow streaks (in our experiment, allocate balls to a hot hand player) in situations in which they perceive that options may differ in their probabilities of success. In turn, these are exactly the conditions under which following streaks is most adaptive, implying that people are sensitive in some way to the implications of the hot hand phenomenon.

The behavioral results of Study 2B suggest that both cues (hot hand and base rate) are independently used for allocation strategy. These results extend the work of Burns (2004), who excluded equal base rates of the two players in his simulation. For instance, Study 2B showed that even if players had equal base rates, participants used the hot hand cue and passed the ball more often to the player with streaks. Participants were clearly following streaks. Following streaks when both players' base rates are equal is not maladaptive per se. If the defense cannot detect or exploit the allocation behavior, following streaks has at the very least a neutral impact on performance.

What Do We Learn from Analyzing Streaks?

In this article we showed that the hot hand as well as the belief in it exists in volleyball. One may ask why this result is of any importance. Yet every day millions of people watch, participate in, or report on sports events. Research on the hot hand belief and hot hand behavior has the potential to connect sports fans and coaches with psychological research and to apply experimental studies to a topic that has great meaning to many across the world. Moreover, beliefs about short-term sequential dependencies are also formed outside sports. Many if not most natural events (e.g., the daily weather) and human behaviors (e.g., parents' interaction with their children) are characterized by naturally occurring sequential dependencies.

The potential reasons for streak detection can be phylogenetic, as shown by Wilke and Barrett's evolutionary approach (2009), where the authors concluded that people have an inborn competence for detecting and using streaks based on a cognitive adaptation to the environment. Streak detection may be a result of experience, in which momentum and contrarian strategies are used based on experience or as a result of human learning, as shown by Carlson and Shu (2007; see Oskarsson et al., 2009 for an overview). The motivation and ability to detect streaks can help us beyond the domain of ball allocations, from making careful scientific observations to avoiding blunt superstition.

Chapter 10

Stereotypes about Men's and Women's Intuitions

A Study of Two Nations

In 1904, G. Stanley Hall, president of Clark University, and founder and first president of the American Psychological Association, explained how the mind of a woman differs from that of a man:

> She works by intuition and feeling; fear, anger, pity, love, and most of the emotions have a wider range and greater intensity. If she abandons her natural naiveté and takes up the burden of guiding and accounting for her life by consciousness, she is likely to lose more than she gains, according to the old saw that she who deliberates is lost. (p. 561)

Hall's statement stood in a long intellectual European tradition, from Kant's distinction between men's mastery of abstract principles and women's grasp of concrete detail to Darwin's opposition of male energy and genius to female compassion and intuition (Schiebinger, 1989). This article examines the existence and nature of stereotypes about men's and women's intuitions one century after Hall's pronouncement.

We investigate the existence, strength, and contents of stereotypes about intuitions of men and women in two large European countries that differ with respect to the traditional roles women have in their society, Germany and Spain. Although extensive cross-cultural comparisons of gender stereotypes exist (e.g., Williams & Best, 1990), none dealt with intuition. In contrast to previous studies that relied on convenience samples, we obtained representative national samples from the general population of both countries. Germany and Spain have populations of about 80 and 46 million inhabitants, which amounts to a third of the population in the European Union. The two countries differ in social values and in the relative socioeconomic status of men and women. Germans tend to hold

Originally published as Gigerenzer, G., Galesic, M., & Garcia-Retamero, R. (2014). Stereotypes about men's and women's intuitions: A study of two nations. *Journal of Cross-Cultural Psychology*, *45*, 62–81. This chapter has been slightly updated.

postmaterialist values such as individual improvement, personal freedom, citizen input in government decisions, the ideal of a society based on humanism, and maintaining a clean and healthy environment to a greater extent than Spaniards (Inglehart, Basánez, Díez-Medrano, Halmann, & Luijkx, 2004). Gender egalitarianism is more valued in German than in Spanish society, and German women are accorded a higher status in their society (House, Hanges, Javidan, Dorfman, & Gupta, 2004). More women are employed in Germany than in Spain, although they more often work part-time (Eurostat, 2006). Per-capita income earned by women as a percentage of that earned by men is higher in Germany than in Spain (57% vs. 44%; Eurostat, 2006). Germany has a higher female literacy rate than Spain and a larger proportion of educated women (Eurostat, 2006). While in both countries women spend more time than men on domestic tasks, from food preparation to childcare, German men take on a larger share of household tasks. Whereas German women carry 67% of the daily workload in the household (4 hr 41 min for German women vs. 2 hr 21 min for German men), Spanish women carry 75% (4 hr 55 min for Spanish women vs. 1 hr 37 min for Spanish men; Eurostat, 2006). In sum, these differences in values and socioeconomic indicators suggest a larger gender gap in Spain than in Germany.

These statistics may reflect the history of the two countries. The women's movement was more active in Germany than in Spain, helping German women to gain the right to vote much earlier (1919 vs. 1931). In the 1930s, both countries' existing democratic systems were overthrown by dictatorships, the Nazi regime and the Franco regime. Although these differed in important respects, both supported a traditional division of labor between men and women, expecting women to take care of the household and children while men worked (Morant, 2006; Rosenfeld, Trappe, & Gornick, 2004). In Spain, this regime lasted 30 years longer than in Germany, into the mid 1970s. It upheld the Roman Catholic tradition, prohibiting divorce, abortion, and the sale of contraceptives. Even into the 21st century, divorce statistics indicate that marriage is a more stable institution in Spain than in Germany, where the number of divorces per 100 new marriages is more than twice as high (57 vs. 23; Lanzieri, 2006).

Social Role Theory (Eagly, 1987; Eagly, Wood, & Diekman, 2000; Wood & Eagly, 2002) assumes that gender roles lead to the particular gender stereotypes in a society. If the historical gender roles described above are relevant, Social Role Theory would predict that Spaniards should have more conservative stereotypes of gender differences than Germans. This is what we found in past work on gender stereotypes of leadership (Garcia-Retamero & López-Zafra, 2006a, 2006b; Garcia-Retamero, Müller, & López-Zafra, 2011). Yet, as we will see, current gender roles have changed in some domains of life in both countries and more rapidly in Spain. That could reduce or even reverse the differences in gender stereotypes between the two countries.

Gender-Specific Intuitions

We define an intuition as a judgment that has three characteristics: (a) it appears quickly in consciousness, (b) the underlying reasons for it are not in awareness, but (c) it nevertheless guides behavior (Gigerenzer, 2007). Thus, unlike in a conscious process of deliberation, a person cannot explain the reasons for an intuition.

We define a stereotype as an individual's or a group's set of beliefs about the characteristics of a target group (Judd & Park, 1993). Stereotypes need not be negative; they can be positive as well as accurate, inaccurate, or not testable (Diekman & Eagly, 2000). The target group can differ from the group holding the belief (e.g., male beliefs about women's intuitions) or be the same (e.g., female beliefs about women's intuitions). Because this article discusses both genders' stereotypes about both genders' intuitions, we will avoid potential confusion by referring to the target groups as *men and women* and the subject group, that is, the carrier of the stereotypes, as *males and females*.

What is known about men's and women's intuitions from earlier research? On one hand, in the large body of literature on gender stereotypes, we could not find any studies on gender stereotypes about men's and women's intuitions. On the other hand, in the literature on intuitive judgments, gender is hardly ever an issue (e.g., Epstein, Pacini, Denes-Raj, & Heier, 1996; Evans, 2008; Gigerenzer, 2007; Kruglanski & Gigerenzer, 2011). Thus, there is little connection between research on intuitive judgments and research on gender roles.

Domains of Intuition

In this article, we distinguish between domain-general and domain-specific stereotypes about men's and women's intuitions. A domain-general stereotype is of the kind "women are intuitive" or "women have better intuitions than men." That is, it has the form "group *A* has attribute *X*." A domain-specific stereotype, in contrast, is more differentiated and has the form "group *A* has attribute *X* in domain *Y*." The belief that women have better intuitions about good leadership but men have better intuitions about the right business partner would be an example of domain-specific stereotypes.

We chose nine domains relevant for intuition that can be categorized into three classes: intuitions about personal affairs (choosing the right romantic partner, understanding intentions of women, and understanding intentions of men), intuitions about professional social tasks (good leadership, choosing the right business partner, and political decision making), and intuitions about professional individual tasks (scientific discoveries, reactions to dangerous situations, and investment in stocks). The three classes differ in the degree to which they challenge traditional gender roles. Believing that women have better intuitions in the first set of domains reinforces the stereotype expressed by Stanley Hall and poses no threat to male dominance in professional affairs. However, believing that women have equal or better intuitions than men in the second and third sets of domains is at odds with the traditional division of labor. These nine domains are not exhaustive but represent a broad variety of tasks. Note that we are concerned with the intuitive competences in each domain, not with deliberate or analytical competences such as in scientific methods or financial calculations.

Research Questions

In this study, we asked our participants whether one of the genders has better intuitions than the other in nine different domains. The participants could vote for either men or women, or say that there is no difference between the genders. We define the *strength of a stereotype in a sample in domain i* (s_i) as the difference between the percentage s_{wi} of people who believe that women have better intuitions in domain *i* and the percentage s_{mi} of those who believe that men have better intuitions:

$$s_i = s_{wi} - s_{mi},$$

with $-100 \leq s_i \leq 100$, and $i = 1, \ldots, 9$. A plus sign means that the stereotype favors women, a minus sign that it favors men. For instance, if all participants hold the belief that there is no difference between the genders, then $s_i = 0$; if everyone believes that women (men) have the better intuitions, then $s_i = 100$ (-100); if 10% believe that there is no difference, but 60% believe that women and 30% that men have better intuitions, then $s_i = 30$. This value of s_i measures both the intensity and the direction of the stereotype. Note that the s_i measure does not capture whether a stereotype is empirically correct.

With these distinctions in mind, we formulate four research questions.

Research Question 1: Do stereotypes about gender-specific intuition exist?

In societies with a general sense of gender equality in intuitive judgment, the answer should be negative. If that is the case, then $s_i = 0$ for all domains. Note that $s_i = 0$ does not mean that every member of the sample believes that there is no difference. If the balance between those who favor men and those who favor women is equal, s_i would also equal zero. Given that gender roles differ between men and women, Social Role Theory predicts that the answer to this research question is "yes." If that is the case, then $s_i > 0$ for all or some domains.

Research Question 2: If stereotypes about gender-specific intuition exist, are these domain general or domain specific?

An example of a domain-general stereotype is Stanley Hall's claim that women's minds generally work by intuition, without differentiating the object of intuition. Similarly, some contemporary theories assume an intuitive and a rational system but do not explicitly distinguish domains and thus implicitly assume domain-general processes (Evans, 2008; Kahneman, 2011). However, to the degree that gender roles differ between domains (as the Social Role

Theory predicts), we would expect domain-specific stereotypes about intuition. For instance, given that men still play the leading role in business, politics, and science (Tables 10.1–10.3), they should be judged to have better intuition than women in these domains. Domains in which women traditionally play a more active role, such as interpersonal communication, should be those in which women are judged to have better intuitions. A domain-general stereotype is defined as one for which (a) all $s_i \neq 0$ and (b) s_i should be the same for all i domains. If stereotypes are, in contrast, domain specific, we should observe that (a) there exists at least one $s_i \neq 0$ and (b) s_i varies systematically across domains. Note that this definition of "domain general" refers to whether stereotypes about intuition generalize across specific domains; it does not refer to what answer people would give if they were asked whether men or women have better intuition without a domain being specified.

Research Question 3: Do stereotypes about gender-specific intuition differ between Germany and Spain?

To the degree that socioeconomic, political, and historical differences between the two countries affect men and women's judgments today, one would expect more conservative stereotypes of gender differences in Spain than in Germany. Therefore, the hypothesis is that across domains, s_i is larger in Spain than in Germany. However, despite the history of greater conservatism in Spain, recent changes suggest that Spain might have overtaken Germany in terms of gender equality in some domains. For example, Spanish women now work more often full-time than German women (Table 10.1), there is a larger percentage of female politicians in Spain than in Germany at almost all political levels (Table 10.2), and women comprise a larger percentage of scientists in Spain than in Germany (Table 10.3). Therefore, in these professional domains, the opposite might be true, namely, s_i is smaller in Spain than in Germany.

Research Question 4: Do stereotypes about gender-specific intuition exhibit in-group preferences?

In-group preference refers to the phenomenon that members of a group esteem or favor members of their own group more highly than they do those of other groups (Swim, 1994). If more males than females think that men have better intuitions, and more females than males think that women have

Table 10.1: Employment Statistics for Germany and Spain

	Year	Germany	Spain	Source
Employment rate				
Women	2005	59	51	Eurostat (2006)
Men	2005	71	75	
Part-time employment (% of all employed)				
Women	2005	44	25	Eurostat (2006)
Men	2005	8	5	

Table 10.2: Political Statistics for Germany and Spain

	Year	Germany	Spain	Source
Percentage of women in national parliaments				
Lower/single houses	2009	33	37	Inter-Parliamentary Union (2010)
Upper houses/senate	2009	22	31	
Percentage of women in executive power				
Ministers	2008	38	53	Council of Europe (2010)
Deputy/junior ministers	2008	30	37	
Heads of regional government	2008	0	5	
Members of regional government	2008	22	40	
Mayors	2008	8	15	
Municipality councilors	2008	24	31	

better intuitions, this amounts to a case of in-group preference. The key alternative to in-group preference is that males and females share the same average s_i. We define in-group preference among females by,

$$s_i^f > s_i,$$

where s_i^f is the strength of the stereotype for the subgroup of females. Note that this condition implies $s_i^m < s_i$, where s_i^m is the strength of the stereotype

Table 10.3: Science Statistics for Germany and Spain

	Year	Germany	Spain
Percentage of female researchers in the higher education sector, by discipline			
Natural sciences	2006	24	39
Engineering and technology	2006	16	35
Medical sciences	2006	41	40
Agricultural sciences	2006	42	39
Social sciences	2006	30	39
Humanities	2006	42	40
Percentage of female researchers in the government sector, by discipline			
Natural sciences	2006	28	42
Engineering and technology	2006	20	39
Medical sciences	2006	44	50
Agricultural sciences	2006	36	49
Social sciences	2006	41	45
Humanities	2006	46	47

Note: ISCO = International Standard Classification of Occupations. Researchers are persons employed in science and technology as "professionals" (ISCO-2) or "technicians and associate professionals" (ISCO-3).

Source: European Commission (2009).

for the subgroup of males. The corresponding definition will be used for in-group preferences of males. If this condition holds, we can measure the degree g_i of in-group preference in domain i as:

$$g_i = \left|s_i^f - s_i^m\right|.$$

If stereotypes about intuition exist and these exhibit in-group preference, then we should find that (a) there exists at least one $s_i \neq 0$ and (b) $g_i > 0$.[1] Note that in-group preference can occur for domain-general as well as domain-specific stereotypes.

Research Question 5: Are stereotypes about gender-specific intuition weaker for younger people?

Because of the societal changes after 1945 in Germany and the 1970s in Spain, age appears likely to influence the strength of stereotypes. Given that many women have joined the workforce in recent decades, younger people might show less intensive stereotypes than older ones do. This expectation can be formalized in the following hypothesis:

> The average absolute value of s across all domains i is smaller for younger age groups.

Method

Sample

To ascertain representative samples of the general public, we obtained large nationwide quota samples of 1,016 adults in Germany in December 2006, and 1,002 adults in Spain in May 2009. The data were collected by the international survey company GfK Group, based in Nuremberg, Germany, and Valencia, Spain. The samples were selected according to quotas designed to make the sample representative for the populations of the two countries in terms of four variables: age, gender, region, and size of settlement. Table 10.4 shows the characteristics of the two samples, as well as error margins (95% confidence intervals [CI]) for estimates based on different subgroups of the samples, assuming simple random sampling. When using 95% CI, our sample size of approximately 1,000 participants per country provides a power of .99 to detect a small effect size (corresponding to Cohen's h = .2), and a power of over .995 to detect a medium effect size (corresponding to Cohen's h = .5). When using 99% CI, the power is .97 and over .995, respectively

1. The logic of this measure is as follows. Consider first the case of $s_i > 0$, that is, the stereotype favors women. To the degree that in-group preference exists, the strength s_i^f of females' stereotypes is higher than s_i while the strength s_i^m of males' stereotypes is lower, resulting in a positive g_i. If $s_i < 0$, that is, the stereotype favors men, then in-group preference results again in a higher (less negative or more positive) value for females and a lower (more negative or less positive) value for males, resulting also in a positive g_i.

Table 10.4: Sample Structure and Error Margins for Estimates Based on Different Subgroups, for Germany and Spain

	Germany		Approximate 95% CI When Percentage is		Spain		Approximate 95% CI When Percentage is	
	n	%	10%	50%	*n*	%	10%	50%
Total	1,016	100	±1.8	±3.1	1,002	100	±1.9	±3.1
Gender								
Male	494	48.6	±2.6	±4.4	491	49.0	±2.7	±4.4
Female	522	51.4	±2.6	±4.3	511	51.0	±2.6	±4.3
Age								
18–35	265	26.1	±3.6	±6.0	322	32.1	±3.3	±5.5
36–50	291	28.6	±3.4	±5.7	304	30.3	±3.4	±5.6
51+	460	45.3	±2.7	±4.6	376	37.5	±3.0	±5.1
Religious practice per month								
0 times	700	68.9	±2.2	±3.7	699	69.8	±2.2	±3.7
1–2 times	175	17.2	±4.4	±7.4	176	17.6	±4.4	±7.4
3+ times	140	13.8	±5.0	±8.3	127	12.7	±5.2	±8.7
Need for certainty								
Low	587	57.8	±2.4	±4.0	435	43.4	±2.8	±4.7
High	429	42.2	±2.8	±4.7	567	56.6	±2.5	±4.1

Note: Confidence intervals differ depending on the size of percentage (they are largest when the percentage is around 50%).

(Cohen, 1988). CIs are more informative and robust than significance levels (Cumming, 2008).

Procedure and Materials

Participants were interviewed individually at their homes. We used face-to-face rather than more impersonal survey modes to increase the quality of the data. Face-to-face contact enabled the interviewer to establish rapport with participants, respond to queries, and explain the meaning of the questions when needed. Participants were asked the same question for each of nine domains, "Who has better intuitions about (name of domain)?" The response alternatives were "men," "women," and "no difference." We also collected data about participants' gender, age, and religious practice (measured as frequency of attending religious services in the last month).

All materials were developed in German, translated into Spanish by a proficient translator, and back-translated for control into German by another person with equivalent language skills. The Ethics Committee of the Max Planck Institute for Human Development and of the University of Granada approved the methodology.

Results

Do Stereotypes about Gender-Specific Intuition Exist? If So, Are These Domain General or Domain Specific?

These two questions need to be considered together. The average *s*-measure across the nine domains, for instance, might be close to zero, apparently confirming the absence of a stereotype. But this could also result from opposing stereotypes in different domains, some of which favor men and others women. Averaged across domains, the size of the gender stereotype was close to zero in both countries, $s = -4.0$ in Germany and $s = 3.9$ in Spain, and both CIs included 0 (95% CI is [−23.6, 31.6] in Germany and [−27.4, 19.6] in Spain).

However, as Figure 10.1 shows, there are large *s*-values in specific domains favoring either men or women. For instance, the size of the stereotype concerning intuitions about the right romantic partner is as large as $s = 53$ in Spain, meaning that the difference between the percentage of Spaniards who trust the intuition of women (62%) and those who trust that of men (9%) is 53 percentage points. Consistently in both countries, women are judged to have the better intuitions for each domain within the class "personal affairs." The average strength of the stereotype in personal domains is $s = 37$ for Germany and 44 for Spain. In contrast, for most of the other domains men are considered to have better intuitions: The average strength of stereotypes across the six professional domains is $s = -25$ in Germany but only −16 in Spain. In professional social tasks, the average stereotype is in favor of men, with $s = -11$ for Germany and −13 for Spain. In professional individual tasks, the stereotype is in the same direction but stronger: $s = -38$ for Germany and −19 for Spain.

Across all domains, 33% of the responses in Germany and 38% in Spain were "no difference." That is, in two thirds of the cases, respondents believed that there are differences between the genders. Thus, the answer to the first question is "yes": Stereotypes about gender-specific intuitions do exist. This analysis also provides the answer to the second question: Stereotypes exist but are domain specific rather than domain general.

Do Stereotypes about Gender-Specific Intuition Differ between Germany and Spain?

Gender-specific intuitions are somewhat stronger in Germany than in Spain, with average absolute values of s_i of 30 and 25, respectively. However, in neither of the two countries are women or men judged to have better intuition in general. Rather, citizens make sharp distinctions between the domains to which an intuition applies. This result means that we must continue our analysis domain by domain. We begin with the three domains that can be classified as personal affairs.

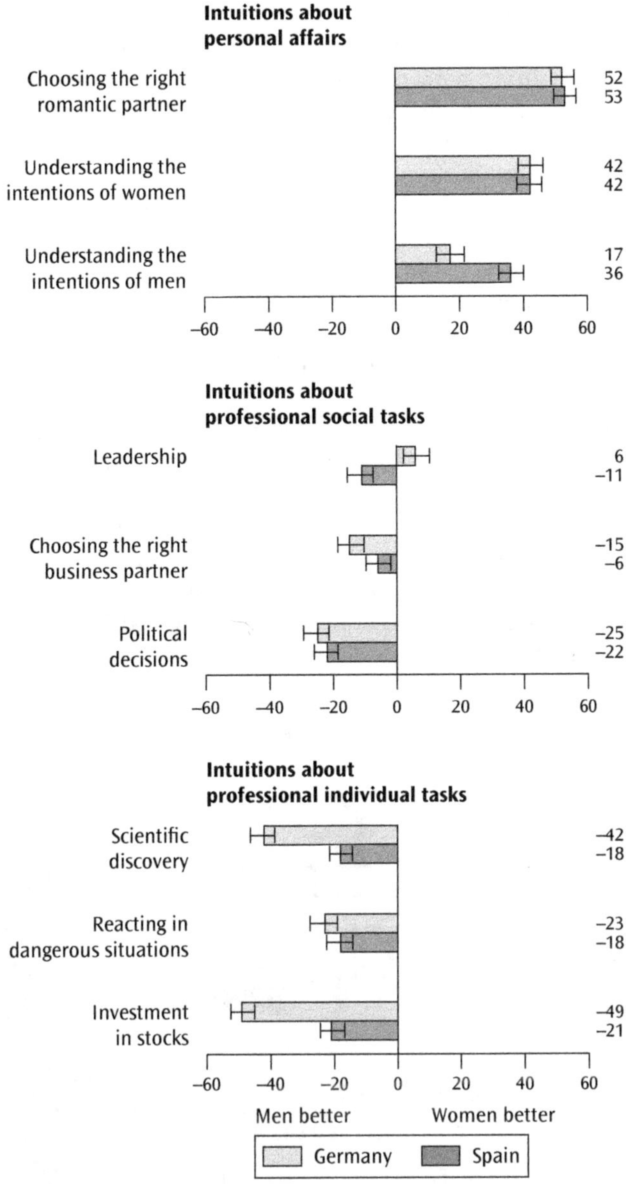

Figure 10.1: Intensity of stereotypes about intuitions of men and women in Germany and Spain. Note: Bars show, for each domain i, the difference s_i between the percentage of participants who believe that women have better intuitions and the percentage of those who believe that men have better intuitions in that domain. For instance, 63% of Germans believe that women have better intuitions about the right romantic partner, while 11% believe that men have better intuitions, resulting in $s_i = 52$. Error bars show 95% confidence intervals.

Choosing the Right Romantic Partner. Figure 10.1 shows a strong stereotype in favor of women's superior intuition about the right romantic partner. Sixty-three percent of Germans believe that women's intuitions are better, 11% men's, and only 26% think that there is no difference. The numbers are almost exactly the same in Spain. Figure 10.2 shows that this belief is shared by both genders in both countries: Not only females but also males believe that women have better intuitions about choosing their romantic partners. However, there are also traces of gender in-group preference: More males (14% in Germany and 10% in Spain) believe in men's superior intuitions than females do (8% and 7%, respectively). Similarly, more females (66% and 64%) believe in women's superior intuitions than males do (60% and 59%). This leads to values of $g = 12$ in Germany and $g = 8$ in Spain.

Understanding the Intentions of Women. Figure 10.1 shows that the stereotype about who best understands the intuitions of women is nearly as strong ($s = 42$) and in the same direction as that about choosing the right romantic partner. As Figure 10.2 shows, the majority of males and females believe that women have better intuitions about the intentions of women. Nevertheless, more females than males believe it, and substantially more males than females believe that men have better intuitions. The strength of this in-group preference is $g = 24$ in Germany and 16 in Spain.

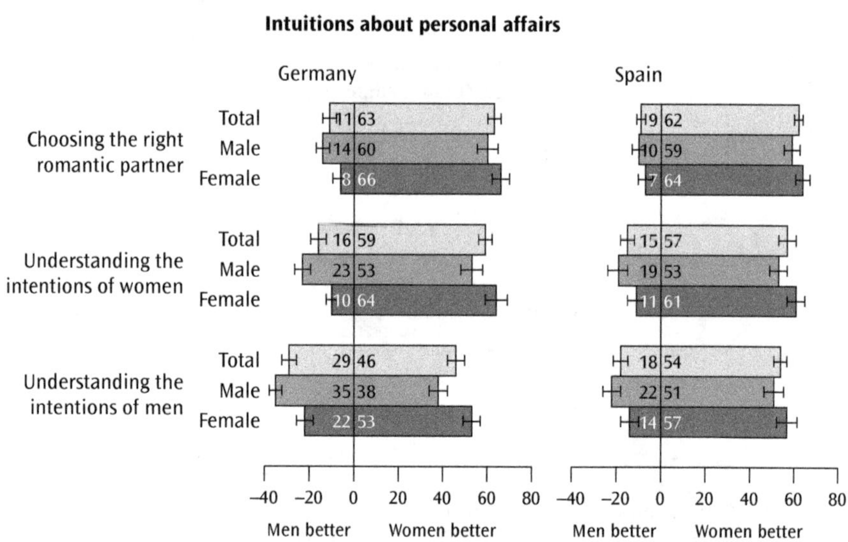

Figure 10.2: Stereotypes of males and females related to intuitions about personal affairs, by country. Note: The figure shows the percentages of participants answering "men" or "women"; the percentage answering "no difference" is the difference to 100%. For instance, 14% of German males believe that men have better intuitions for choosing the right romantic partner, compared with 8% of German females. Error bars show 95% confidence intervals.

Understanding the Intentions of Men. As Figure 10.1 shows, the prevailing view is that women, not men, have better intuitions about men's intentions. This belief is held by both genders (Figure 10.2). Both countries show in-group preference of size g = 28 for Germany and 14 for Spain. As in the previous domains, g is larger for Germans. Unlike in the previous domains, the strength of the stereotype differs between the two countries, and in the direction predicted: The belief that women's intuitions are superior is stronger in Spain than in Germany (Figure 10.1). Figure 10.2 reveals that this difference is largely due to the beliefs of German males, who are the only subgroup where the strength of the stereotype is close to zero.

The next six domains concern judgment in professional contexts, as opposed to intuitions in personal affairs.

Leadership. Leadership is the domain in which gender stereotypes are the weakest (Figure 10.1), followed by choosing the right business partner—a similarly unexpected finding (see below). More Spaniards believe that men have better intuitions about leadership than women do (s_i = −11), whereas the opposite holds for Germans (s_i = 6). As Figure 10.3 shows, underlying the low average s-values is a disagreement between males and females that was absent in the three domains concerning personal affairs. German males believe that men have better intuitions about good leadership, whereas German females believe that women have the better intuitions. A similar pattern of disagreement holds in Spain. This disagreement between genders is reflected in in-group preference values of g = 28 in Germany and 25 in Spain.

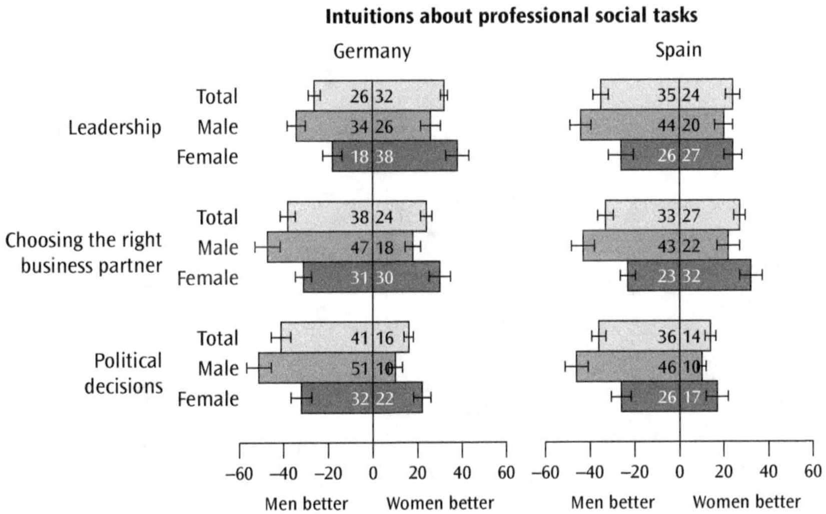

Figure 10.3: Stereotypes of males and females related to intuitions about professional social tasks, by country. Note: Error bars show 95% confidence intervals.

Choosing the Right Business Partner. The stereotypes in both countries prefer men, but their strength is relatively weak ($s = -15$ in Germany and −6 in Spain; see Figure 10.1). However, similar to the domain of leadership and in contrast to the three domains of personal affairs, when it comes to business-related intuitions, males and females disagree about who has the better intuitions. On average, males believe that men are better at choosing the right business partner, whereas females believe that women are better (Figure 10.3). This disagreement is reflected in a good deal of in-group preference, with $g = 28$ for Germany and 30 for Spain.

Political Decisions. Men's intuitions about politics are perceived to be better than women's in both countries (Figure 10.1). On average, males and females favor male intuitions in political affairs, and the strength of the stereotype is about the same in both countries, albeit with a higher tendency among Germans (Figure 10.3). Stereotypes about intuitions in political decision making show a strong degree of in-group preference, with $g = 31$ for Germany and 27 for Spain.

The final set of three questions concerns competences that differ from personal affairs and professional social decisions in that they are primarily individual professional tasks. This is not to say that social competences do not play a role in these domains, but they are not central.

Scientific Discovery. As Figure 10.1 shows, the stereotype that men have better intuitions for scientific discovery than women exists in both countries, but is much stronger in Germany than in Spain ($s = -42$ vs. −18). Both genders share this stereotype, with a certain degree of in-group preference, with $g = 28$ for Germany and 14 for Spain (Figure 10.4).

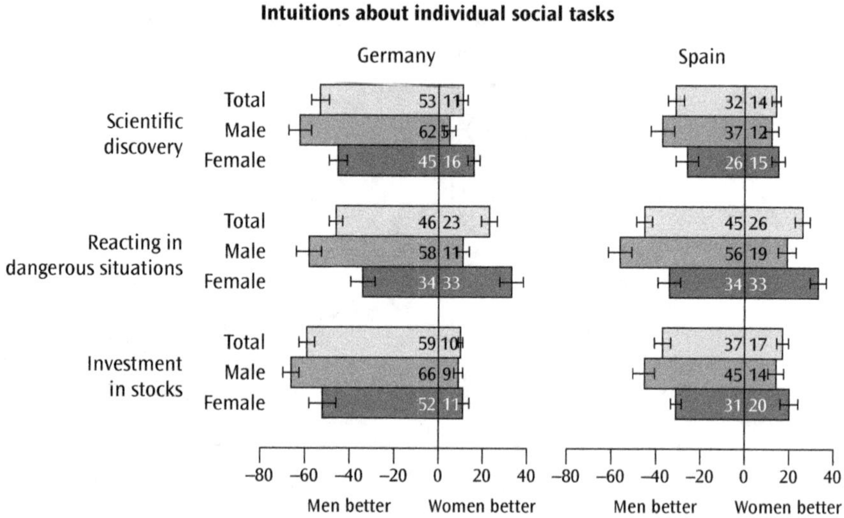

Figure 10.4: Stereotypes of males and females related to intuitions about professional individual tasks, by country. Note: Error bars show 95% confidence intervals.

Reacting in Dangerous Situations. The stereotypes in the two countries about better intuitions in reacting in dangerous situations favor men (Figure 10.1). Yet a closer look shows that this stereotype is based entirely on males' beliefs alone (Figure 10.4). Among the females in both countries, the strength of the stereotype is virtually zero (s = 1 in both countries), whereas it is one of the strongest among males (s = 47 in Germany and 37 in Spain). A substantial amount of in-group preference exists, g = 46 in Germany and 36 in Spain. There is no evidence supporting the historically motivated hypothesis that stereotypes for this domain are stronger in Spain. On the contrary, these stereotypes are less evident among Spanish males.

Investment in Stocks. The final domain we analyze is investing in stocks, where intuition plays as large a role as analytical investment methods do. The majority of both Germans and Spaniards believe that men have better intuitions about investing in stocks (see Figures 10.4). As for the domain of scientific discovery, the stereotype is shared by both genders, but there is again evidence for in-group preference, with g = 16 for Germans and 20 for Spaniards.

Contrary to the historically based hypothesis about national differences, Spaniards show more egalitarian views than Germans. These results reflect the pattern of beliefs for intuitions about choosing the right business partner and about scientific discovery.

Do Stereotypes about Gender-Specific Intuition Exhibit In-Group Preferences?

Although males and females agree on the direction of most stereotypes, they show a preference for members of their own gender. For all nine domains, males favor intuitions of men more strongly than females do, and females favor intuitions of women more strongly than males do. Figure 10.5 shows the values of g_i—the absolute difference between the strength of stereotype s_i for females and males—for each domain. Overall, in-group preferences are higher for professional tasks (average g is 30 in Germany and 25 in Spain) than for personal affairs (average g is 21 in Germany and 12 in Spain). In both countries, g is highest for intuitions about reacting in dangerous situations (46 in Germany and 36 in Spain) and lowest for intuitions about the right romantic partner (12 in Germany and 8 in Spain). For all professional tasks, in-group preferences are stronger for men's intuitions than for women's. In other words, males have stronger in-group preferences than females in these domains. All in all, in-group preferences exist, and their size is substantial.

Are Stereotypes about Gender-Specific Intuition Weaker for Younger People?

Contrary to the hypothesis that younger age groups show weaker stereotypes, the average s-values of the three German age groups are 31 (95% CI = [23.4, 38.6]) for the youngest; 29, CI = [21.5, 36.5] for the middle; and 32,

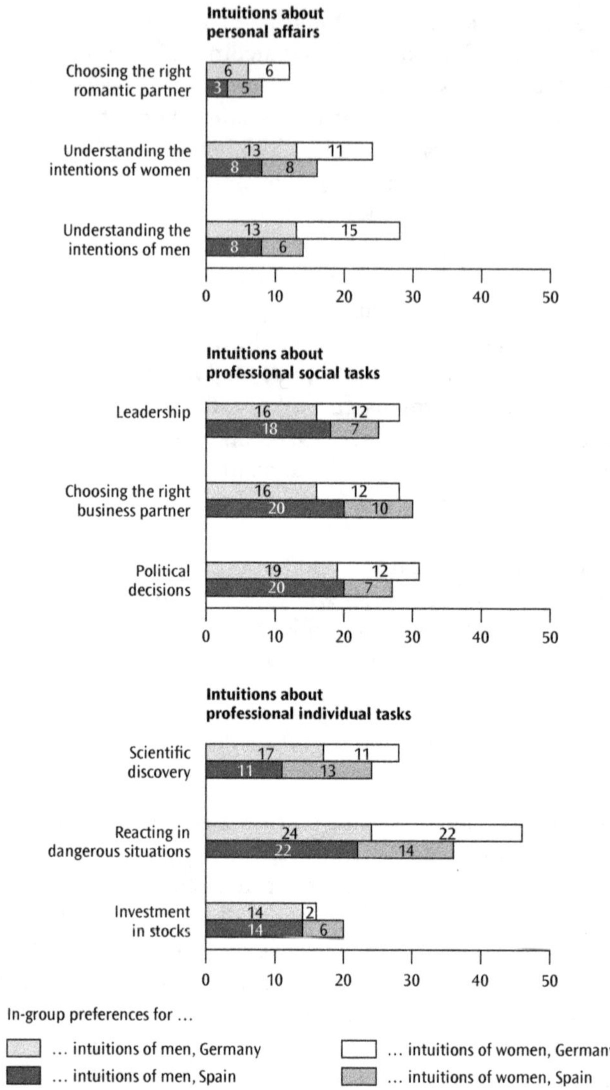

Figure 10.5: Preferences for intuitions of own gender. Note: The top bar for each domain shows the strength of in-group preferences among Germans, the second bar that among Spaniards. Consider Germans' beliefs about understanding the intentions of women. Twenty-three percent of German males believe that men have better intuitions, compared with only 10% of German females (Figure 10.2). The difference of 13 percentage points is the shaded, left part of the bar. Fifty-three percent of German males but 64% of German females believe that women have better intuitions. The difference of 11 percentage points is the right part of the bar. The sum of both parts is the strength of in-group preference, $g_i = g_{mi} + g_{fi}$. Overall, in-group preferences are higher for professional tasks than for personal affairs. For all professional domains, in-group preferences are stronger for men's intuitions than for women's intuitions.

CI = [26.2, 37.8] for the oldest age group, thus showing no trend. Similarly, the average *s*-values for the three age groups in Spain are 24, CI = [17.4, 30.6]; 25, CI = [18.2, 31.8]; and 28, CI = [21.7, 34.3], for the youngest to oldest participants. Among the 18 tests on age differences (nine domains × two countries), only two show a more conservative stereotype with age: The belief that men have better intuitions about political decisions and scientific discovery is strongest in the oldest age group (51+) in Spain. Thus, counter to the hypothesis that the absolute values of *s* increase with age, the direction and the strengths of gender-specific stereotypes remains constant across generations.

Finally, using multinomial logistic regression analyses, we tested whether the differences in stereotype strength between countries as shown in Figures 10.1 depended on age and religious practice and found that these differences remained stable.

Discussion

Research on gender stereotypes has relied mostly on convenience samples of students or highly educated people. To the best of our knowledge, this study is the first that obtained large nationwide samples, representative for the general population of Germany and Spain in terms of age, gender, region, and size of settlement. These samples allow us to describe the entire population of a country and analyze differences between countries. Nevertheless, this study also has its limits. We turn to these, and then to the major results.

Limitations

First, this study does not consider whether stereotypes are the cause or the consequence of gender roles or skills. Second, it does not assess the reliability of participants' responses, for instance, by repeating questions at different time points. However, we used face-to-face interviews rather than telephone calls to create personal interaction and thus reduced the likelihood of random and false answers (De Leeuw & van der Zouwen, 1988). A third limitation is provided by the *s*-measure we used throughout this study, which does not distinguish between "no difference" judgments and an equal proportion of responses that favor men and women. For instance, an *s*-value of 0 (no stereotype) can result from 50% favoring men and 50% favoring women, or from 100% answering "no difference." Nevertheless, the "no difference" proportions can be determined for all domains from Figures 10.2 to 10.4. Finally, this study is cross-sectional rather than longitudinal, and thus the finding that the signs and sizes of gender-specific stereotypes were constant over age groups does not allow for the conclusion that they also remain constant across a lifetime.

Stereotypes about Gender-Specific Intuitions Exist and Are Domain Specific

We found that in both countries, stereotypes about gender-specific intuitions exist. Yet, these are not domain general. Between the nine domains, the strength of the stereotypes, as measured by *s*, ranges from $s = -49$ to +53 (Figure 10.1). In the three domains labeled "personal affairs," women are consistently believed to have better intuitions. In the three domains labeled "professional social tasks," a different picture emerges. Here, the average stereotype is in favor of men, although this is mainly due to the views of males, not females. For the third group, "professional individual tasks," the stereotype in favor of men is even stronger: Men are consistently believed to have the better intuitions. We next discuss the possible origins of these domain differences.

Personal Affairs. Taken together, the results shown in Figure 10.2 are surprising: On average, males trust women's intuitions more, no matter whether they concern choosing the right romantic partner, women's intentions, or their own (men's) intentions. Is there empirical evidence to support a general superiority of women's intuitions about others' intentions? Although we are not aware of an experimental study that has answered this question, research has addressed a related topic. It is commonly reported that women are better than men at recognizing emotional expressions. Meta-analyses of gender differences in decoding emotional signals revealed small- to medium-sized effects of about $d = .40$ for nonverbal cues (Hall, 1978), and .13 to .32 for facial cues (McClure, 2000). Female advantage in processing emotion seems to be relatively stable from infancy to adolescence (McClure, 2000). This indirect evidence suggests that women could indeed be better in judging the intentions of other people—both men and women. Thus, the stereotype expressed by the majority of our participants may not be entirely without reason.

As mentioned above, the literature reports that German society upholds gender equality to a greater extent than Spanish society does and that until the mid 1970s, the role of women in Spanish society was largely confined to home and children. Consistent with the hypothesis stated above, this would lead to stronger stereotypes about gender-specific intuitions for personal affairs in Spain than in Germany. However, this is not the case. The number of participants who believe that women or men have better intuitions for choosing the right romantic partner and understanding intentions of women is essentially the same in the two countries (Figure 10.2). A difference emerges only for understanding intentions of men, with more Germans than Spaniards favoring intuitions of men rather than women.

Professional Social Tasks. For a long time, leadership has been considered to be a male domain (Chemers, 2001). Across different countries, characteristics associated with leadership roles—such as power, competition, and authority—are ascribed more frequently to men than to women (Garcia-Retamero & López-Zafra, 2008; Schein, Mueller, Lituchy, & Liu, 1996; Sczesny, Bosak, Neff, & Schyns, 2004). These studies suggest a strong stereotype that men have better intuitions for leadership than women. However, that is not the case (Figure 10.1). The general absence of a strong gender-specific stereotype about intuitions for leadership in both countries is supported by empirical evidence for the equal qualification of men and women as good leaders. For instance, studies show that the financial performance of companies employing relatively more female managers is, after controlling for other relevant variables, higher than or at least equal to the performance of companies with fewer women in managerial positions (Krishnan & Park, 2005; Singh & Vinnicombe, 2005). The absence of a strong stereotype may also reflect the fact that males and females practice leadership in different organizational contexts. Men are rated as better leaders in male-dominated environments such as the military, but women are rated better in environments such as educational, governmental, and social service organizations (Eagly, Karau, & Makhijani, 1995). Consistent with our hypothesis about country differences and the results of previous research on the differences between the two countries in gender stereotypes related to leadership (Garcia-Retamero & López-Zafra, 2006a, 2011), we find that more Spaniards believe that men have better intuitions about leadership than women do ($s_i = -11$), whereas the opposite holds for Germans ($s_i = 6$). The fact that these differences are small is in line with the available statistics: While the percentage of women in all managerial positions is higher in Spain than in Germany (32% vs. 26%; Eurostat, 2006), in the biggest companies, the reverse is the case (4% vs. 12%; Eurostat, 2008b).

Unlike for intuitions about good leadership, concerning the right business partner Spaniards do not show a stronger stereotype favoring men than the Germans do, contradicting the general hypothesis based on the two countries' histories. A possible reason is the faster development of gender equality in Spain. As mentioned before, although more women work in Germany than in Spain, German women are much more often employed only part-time (Table 10.1). This may be linked to the lower use of childcare in Germany. In Spain, 38% of children up to 3 years of age are enrolled in formal childcare, while in Germany, only 19% of children are enrolled (Eurostat, 2008a). In addition, there is a greater tradition of grandparents and other family members helping young parents with childcare in Spain than in Germany. Having children under 12 decreases the employment rate of Spanish women aged 25 to 49 by 11 percentage points compared with women without children; in Germany, this difference is 16 percentage points (Eurostat, 2008a). For men, the difference goes in the opposite direction: Having children increases their employment rate by 7 and 8 percentage points in Germany and Spain, respectively (Eurostat, 2008a). In sum, fewer Spanish women than German women work, but they more often work full-time.

The direction and the strength of the stereotypes about political intuitions are practically the same in Germany and Spain. Although women's political rights were suppressed for a longer period in Spain than in Germany, this historical trajectory appears to be compensated for by the fact that there is a larger percentage of female politicians in Spain today than in Germany at almost all political levels (Table 10.2). Although Germany currently has a female chancellor, the majority of politicians in Germany and Spain are still men.

Professional Individual Tasks. Gender differences in prominence and interest in science are well documented. Of 558 Nobel prizes awarded in physics, chemistry, and medicine from 1901 to 2010, only 16 were awarded to women. A longitudinal study showed that males were twice as likely as females to favor investigative interests, whereas women were three to four times more likely to favor artistic and social interests (Achter, Lubinski, & Benbow, 1996). Gender stereotypes appear to play a substantial role in these and similar differences. When children reach the age of 11 to 13 years, their parents believe that science is less interesting and more difficult for girls than for boys (Tenenbaum & Leaper, 2003), and mothers underestimate daughters' and overestimate sons' mathematics abilities (Frome & Eccles, 1998). Moreover, science had been closed to women well into the middle of the 20th century, when women were finally able to enter academic careers and take a public part in the scientific enterprise. Based on the historical differences between the two countries described earlier, a stronger stereotype regarding intuitions in science could be expected in Spain than in Germany. As in the fields of business and politics, however, the percentage of women working in almost all areas of science in present-day Spain is similar to or higher than in Germany (Table 10.3). This discrepancy is particularly high in the areas of natural sciences and engineering and technology. Indeed, the majority of Spaniards believe that men and women have equally good intuitions for scientific discoveries, whereas only one third of Germans think the same (Figure 10.4). Similarly, 62% of German males but only 37% of Spanish males believe that men have better intuitions. These country-specific stereotypes are mirrored among females: 45% of German females agree with the majority of German males and believe in male superiority in scientific intuition, compared with only 26% of Spanish females. This negative self-perception among German females might decrease their interest in science—and thus contribute to the actual difference between professional scientists in the two countries.

As for science, strong stereotypes about the superiority of intuitions of men exist for reacting in dangerous situations and investments in stocks. In popular Hollywood culture, those who master and survive dangerous situations are typically men, be it Tarzan, James Bond, or Spiderman. In real life, an indicator of strong male bias when it comes to working in dangerous situations is the low percentage of women in active military service: Only 6% of active military troops in Germany and 19% in Spain are women. In fact, only since 2001 have women in Germany had entry to military branches involving the use of arms. Accordingly, the stereotypes about better intuitions in reacting in dangerous situations favor men (Figure 10.1).

As billionaire George Soros (2003) explained his success in stock investments, in addition to his theory, "I relied on my instincts and intuition" (p. 35). These intuitions are believed to be superior in men (Figure 10.1). Investing in stocks is indeed historically a male domain; for instance, more than two thirds of investment bankers in the United States are men (U.S. Equal Employment Opportunity Commission, 2003). Men are reported to have more investment-related knowledge than women do (e.g., Goldsmith & Goldsmith, 1997; Goldsmith, Goldsmith, & Heaney, 1997). This does not imply that men also have more success in stock picking. Ortmann, Gigerenzer, Borges, and Goldstein (2008) found that public stock portfolios based on women's recognition of stocks made more money than those based on men's recognition. In sum, although men have more knowledge about stocks on average, there is no firm evidence that men have superior intuitions about investing in stocks.

Stereotypes Are Stable across Age Groups

The domain-specific stereotypes about women's and men's intuitions are stable across age groups. This is an unexpected result. We found no difference in the direction of the stereotype (the sign of *s*) and in its size in Germany, and little difference in Spain. This result suggests that present-day gender roles influence gender-specific stereotypes more than historical trajectories.

Males and Females Share Stereotypes But Show Strong In-Group Preferences

Our study was not designed to explain these in-group preferences, but there are possible motivational hypotheses, such as self-enhancement (Kruger & Dunning, 1999; but see Krueger & Mueller, 2002). However, what looks like a bias may be a by-product of a rational strategy (Funder, 1987). One rational, ecological explanation of in-group preferences is based on an argument by Fiedler (1996) and Gigerenzer, Fiedler, and Olsson (2012). It begins with the observation that in most societies, males encounter more males and females more females. To simplify, let us code people's intuitions as either good or bad and assume that the ratio of good and bad intuitions is the same for both genders. The sample sizes of observations, however, differ between genders. For instance, assume that the ratio of people with good and bad intuitions about business partners is 9 to 4, and the same among men and women. Assume further that a particular male has twice as large a sample of men than of women:

Men: 18 good and 8 bad intuitions.
Women: 9 good and 4 bad intuitions.

If this male now tested whether there is a difference in the number of good and bad intuitions among men and women, a binomial test would find significantly more good than bad intuitions among men (p = .037) but not among women (p = .13). A female would reach the opposite conclusion, given that she has a larger sample of women than of men. This simple model explains in-group preferences in the absence of real differences between genders. To obtain this surprising result, it is not necessary to assume that people calculate the exact probabilities; an approximate "statistical sense" would be sufficient (Gigerenzer & Murray, 1987).

The same ecological model also implies increasing in-group preferences with increasing sample size differences. To the degree that sample size differences are more extreme in professional domains than in personal domains, for instance, because of professions in which most employees are of the same gender (e.g., pilots, nurses), the in-group preferences will be larger in professional domains. This consequence can explain the observed differences between domains in Figure 10.5. Finally, the observation that males have stronger in-group preferences than females do in professional domains also follows from this model. The necessary condition is that more men are employed than women, which is the case in both Germany and Spain. For instance, if a firm employs 12 men and 3 women, and everyone has a preference toward socializing with one's own gender and wants to have at least 4 friends, then for males, the ratio of men to women could be as high as 4:0 but that for females at most 2:2 (not counting oneself as a friend). This implies higher in-group preferences among males than females, as observed in Figure 10.5 for all professional domains.

Stereotypes about Gender-Specific Intuition Are Stronger in Germany Than in Spain

Across domains, stereotypes about gender-specific intuitions are stronger in Germany than in Spain. Similarly, more Spaniards than Germans see no difference between intuitions of men and women: 38% versus 33% on average across domains.

These results are surprising given that Spaniards have a recent history of a conservative regime that strongly supported traditional male and female roles. However, the current environment appears to have a stronger influence on gender stereotypes. Spain has presently surpassed Germany in the number of women in leadership positions, full-time employment (Table 10.1), politics (Table 10.2), and science (Table 10.3). Women in Spain are also more present in active military service. Although women are far from being equally represented in professional life, more Spanish women participate in professional life than German women do. Consistent with this state of affairs, the average strength of stereotypes across the six professional domains is larger in Germany than in Spain.

Social Role Theory (Eagly, 1987) argues that beliefs about gender are derived from observations of actual gender roles and thus reflect the division of labor and gender hierarchy of a society. In this view, stereotypes are a consequence rather than a cause of societal gender roles. Although the present study was not designed to examine causal relationships between stereotypes and gender roles, our results show that the current environment rather than history affects gender stereotypes. Specifically, our findings suggest that people attribute better intuitions to the gender that has more experience in a particular domain. This is in line with studies showing that experts in a certain task are able to find good solutions using simple rules and only a few relevant cues, whereas novices engage in detailed consideration of different options (Beilock, Carr, MacMahon, & Starkes, 2002; Garcia-Retamero & Dhami, 2009; Johnson & Raab, 2003; Shanteau, 1992). Finally, the present study shows that widespread stereotypes about men's and women's intuitions still exist, even a century after the first president of the American Psychological Association made his infamous statement.

Part V

Theory

Chapter 11

As-If Behavioral Economics

Neoclassical Economics in Disguise?

For a research program that counts improved empirical realism among its primary goals, it is surprising that behavioral economics appears indistinguishable from neoclassical economics in its reliance on "as-if" arguments. "As-if" arguments are frequently put forward in behavioral economics to justify "psychological" models that add new parameters to *fit* decision outcome data rather than specifying more realistic or empirically supported psychological processes that genuinely *explain* these data. Another striking similarity is that both behavioral and neoclassical research programs refer to a common set of axiomatic norms without subjecting them to empirical investigation. Notably missing is investigation of whether people who deviate from axiomatic rationality face economically significant losses. Despite producing prolific documentation of deviations from neoclassical norms, behavioral economics has produced almost no evidence that deviations are correlated with lower earnings, lower happiness, impaired health, inaccurate beliefs, or shorter lives. We argue for an alternative nonaxiomatic approach to normative analysis focused on veridical descriptions of decision process and a matching principle—between behavioral strategies and the environments in which they are used—referred to as *ecological rationality*. To make behavioral economics, or psychology and economics, a more rigorously empirical science will require less effort spent extending "as-if" utility theory to account for biases and deviations, and substantially more careful observation of successful decision makers in their respective domains.

Introduction

Behavioral economics frequently justifies its insights and modeling approaches with the promise, or aspiration, of improved empirical realism

Originally published as Berg, N., & Gigerenzer, G. (2010). As-if behavioral economics: Neoclassical economics in disguise? *History of Economic Ideas, 18*, 133–165. This chapter has been slightly updated.

(Rabin 1998, 2002; Thaler 1991; Camerer 1999, 2003). Doing economics with "more realistic assumptions" is perhaps *the* guiding theme of behavioral economists, as behavioral economists undertake economic analysis without one or more of the unbounded rationality assumptions. These assumptions, which count among the defining elements of the neoclassical, or rational choice, model, are: unbounded self-interest, unbounded willpower, and unbounded computational capacity.

Insofar as the goal of replacing these idealized assumptions with more realistic ones accurately summarizes the behavioral economics program, we can attempt to evaluate its success by assessing the extent to which empirical realism has been achieved. Measures of empirical realism naturally focus on the correspondence between models on the one hand, and the real-world phenomena they seek to illuminate on the other. This includes both theoretical models and empirical descriptions. Of course, models by definition are abstractions that suppress detail in order to focus on relevant features of the phenomenon being described. Nevertheless, given its claims of improved realism, one is entitled to ask how much psychological realism has been brought into economics by behavioral economists.

We report below our finding of much greater similarity between behavioral and neoclassical economics' methodological foundations than has been reported by others. It appears to us that many of those debating behavioral versus neoclassical approaches, or vice versa, tend to dramatize differences. The focus in this paper is on barriers that are common to both neoclassical and behavioral research programs as a result of their very partial commitments to empirical realism, indicated most clearly by a shared reliance on Friedman's as-if doctrine.

We want to clearly reveal our own optimism about what can be gained by increasing the empirical content of economics and its turn toward psychology. We are enthusiastic proponents of moving beyond the singularity of the rational choice model toward a tool kit approach to modeling behavior, with multiple empirically grounded descriptions of the processes that give rise to economic behavior and a detailed mapping from contextual variables into decision processes used in those contexts (Gigerenzer & Selten, 2001).[1]

Together with many behavioral economists, we are also proponents of borrowing openly from the methods, theories, and empirical results that neighboring sciences—including, and perhaps, especially, psychology—have to offer, with the overarching aim of adding more substantive empirical

1. Singular definitions of what it means to behave rationally are ubiquitous in the behavioral economics literature. One particularly straightforward articulation of this oddly neoclassical tenet appearing as a maintained assumption in behavioral economics is Laibson (2002, p. 22), who writes: "There is basically only one way to be rational."

This statement comes from a presentation to the Summer Institute of Behavioral Economics organized by the influential Behavioral Economics Roundtable under the auspices of the Russell Sage Foundation (see http://www.russellsage.org/programs/other/behavioral/, and Heukelom [2007], on the extent of its influence).

content. As the behavioral economics program has risen into a respectable practice within the economics mainstream, this paper describes limitations, as we see them, in its methodology that prevent its predictions and insights from reaching as far as they might. These limitations result primarily from restrictions on what counts as an interesting question (i.e., fitting data measuring outcomes, but not veridical descriptions of decision processes leading to those outcomes); timidity with respect to challenging neoclassical definitions of normative rationality; and confusion about fit versus prediction in evaluating a model's ability to explain data. We turn now to three examples.

As-If Behavioral Economics: Three Examples

Loss-Aversion and the Long-Lived Bernoulli Repair Program

Kahneman and Tversky's 1979 prospect theory provides a clear example of as-if behavioral economics—a model widely cited as one of the field's greatest successes in "explaining" many of the empirical failures of expected utility theory, but based on a problem-solving *process* that very few would argue is realistic. We detail why prospect theory achieves little realism as a decision-making process below. Paradoxically, the question of prospect theory's realism rarely surfaces in behavioral economics, in large part because the as-if doctrine, based on Friedman (1953) and inherited from neoclassical economics, survives as a methodological mainstay in behavioral economics even as it asserts the claim of improved empirical realism.[2]

According to prospect theory, an individual chooses among two or more lotteries according to the following procedure. First, transform the probabilities of all outcomes associated with a particular lottery using a nonlinear probability–transformation function. Then transform the outcomes associated with that lottery (i.e., all elements of its support). Third, multiply the transformed probabilities and corresponding transformed lottery outcomes, and sum these products to arrive at the subjective value associated with this particular lottery. Repeat these steps for all remaining lotteries in the choice set. Finally, choose the lottery with the largest subjective value, computed according to the method above. How should one assess the empirical realism achieved by this modeling strategy relative to its predecessor, expected

2. Starmer (2005) provides an original and illuminating methodological analysis that ties as-if theory, which appeared in Friedman and Savage a few years before Friedman's famous 1953 essay, to potential empirical tests that no one has yet conducted. Starmer shows that both Friedman and Savage defended expected utility theory on the basis of the as-if defense. Paradoxically, however, both of them wind up relying on a tacit model of mental process to justify the proposition that mental processes should be ignored in economics. Starmer writes: "This 'as if' strategy entails that theories *not* be judged in terms of whether they are defensible models of mental processes. So to invoke a model of mental process as a defence of the theory would . . . not seem . . . consistent." (p. 286)

utility theory? Both prospect theory and expected utility theory suffer from the shortcoming of assuming that risky choice always emerges from a process of weighting and averaging (i.e., integration) of all relevant pieces of information. Both theories posit, with little supporting evidence (Starmer, 2005) and considerable opposing evidence (e.g., Brandstätter, Gigerenzer & Hertwig, 2006; Leland, 1994; Payne & Braunstein, 1978; Rubinstein, 1988; Russo & Dosher, 1983), that the subjective desirability of lotteries depends on all the information required to describe the lottery's distribution, in addition to auxiliary functions and parameters that pin down how probabilities and outcomes are transformed. This is not even to mention the deeper problem that in many, if not most, interesting choice problems (e.g., buying a house, choosing a career, or deciding whom to marry), the decision maker knows only a tiny subset of the objectively feasible action set (Hayek, 1945), the list of outcomes associated with lotteries, or the probabilities of the known outcomes (Knight, 1921). These assumptions in both expected utility theory and prospect theory—of transforming, multiplying and adding, as well as exhaustive knowledge of actions and outcomes (i.e., event spaces associated with each action)—are equally defensible, or indefensible, since they play nearly identical roles in both theories.

The similarities between prospect theory and expected utility theory should come as no surprise. Gigerenzer (2008, p. 90) and Güth (1995, 2008) have described the historical progression—from expected value maximization (as a standard of rationality) to expected utility theory and then on to prospect theory—as a "repair program" aimed at resuscitating the mathematical operation of weighted integration, based on the definition of mathematical expectation, as a theory of mind. Expected-value maximization was once regarded as a proper standard of rationality. It was then confronted by the St. Petersburg Paradox, however, and Daniel Bernoulli began the repair program by transforming the outcomes associated with lotteries using a logarithmic utility of money function (or utility of *change* in money—see Jorland [1987], on interpreting Bernoulli's units in the expected utility function). This modification survived and grew as expected utility theory took root in 20th-century neoclassical economics. Then came Allais' Paradox, which damaged expected utility theory's ability to explain observed behavior, and a new repair appeared in the form of prospect theory, which introduced more transformations with additional parameters, to square the basic operation of probability-weighted averaging with observed choices over lotteries.

Instead of asking how real people—both successful and unsuccessful—choose among gambles, the repair program focused on transformations of payoffs (which produced expected utility theory) and, later, transformations of probabilities (which produced prospect theory) to fit, rather than predict, data. The goal of the repair program appeared, in some ways, to be more statistical than intellectual: adding parameters and transformations to ensure that a weighting-and-adding objective function, used incorrectly as a model of mind, could fit observed choice data. We return to the distinction between fit versus prediction below. The repair program is based largely on tinkering with the mathematical form of the mathematical expectation operator and cannot be described as a sustained empirical effort to uncover the process by which people actually choose gambles.

Fehr's Social Preference Program

The insight that people care about others' payoffs, or that social norms influence decisions, represents a welcome expansion of the economic analysis of behavior, which we applaud and do not dispute.[3] Fehr and Schmidt (1999), and numerous others, have attempted to demonstrate empirically that people generally are other-regarding. Other-regarding preferences imply that, among a set of allocations in which one's own payoff is exactly the same, people may still have strict rankings over those allocations because they care about the payoffs of others. Fehr and Schmidt's empirical demonstrations begin with a modification of the utility function and addition of at least two new free parameters. Instead of maximizing a "neoclassical" utility function that depends only on own payoffs, Fehr and Schmidt assume that people maximize a "behavioral" or other-regarding utility function. This other-regarding utility function, in addition to a standard neoclassical term registering psychic satisfaction with own payoffs, includes two arguments that are nonstandard in the previous neoclassical literature: positive deviations and negative deviations of own payoffs, each weighted with its own parameter.

As a psychological model, Fehr and Schmidt are essentially arguing that, although it is not realistic to assume individuals maximize a utility function depending on own payoffs alone, we can add psychological realism by assuming that individuals maximize a more complicated utility function. This social preferences utility function ranks allocations by weighting and summing to produce a utility score for each allocation, and choice is by definition the allocation with the highest score. The decision process that maximization of a social preferences utility function implies begins, just like any neoclassical model, with exhaustive search through the decision maker's choice space. It assigns benefits and costs to each element in that space based on a weighted sum of the intrinsic benefits of own payoffs, the psychic benefits of being ahead of others, and the psychic costs of falling behind others. Finally, the decision maker chooses the feasible action with the largest utility score based on weighted summation. If the weights on the "social preferences" terms in the utility function registering psychic

3. Binmore and Shaked (2010) argue that the tools of both classical and neoclassical economics can easily take social factors into account and, therefore, the inclusion of social factors in economic analysis should not automatically be classified as a behavioral methodology. But although Binmore and Shaked are correct, in principle, that utility theory does not preclude other people's consumption from entering the utility function, they fail to acknowledge the key role of the no-externalities assumption (i.e., no channels other than price for individuals to affect each other) in the Welfare Theorems and for normative economics in general.

satisfaction from deviations between own and other payoffs are estimated to be different than zero, then Fehr and Schmidt ask us to conclude that they have produced evidence confirming their social preference model.

This approach almost surely fails at bringing improved psychological insight about the manner in which social variables systematically influence choice in real-world settings. Think of a setting in which social variables are likely to loom large, and ask yourself whether it sounds reasonable that people deal with these settings by computing the benefits of being ahead of others, the costs of falling behind the others, and the intrinsic benefits of own payoffs—and then, after weighting and adding these three values for each element in the choice set, choosing the best. This is not a process model but an as-if model. Could anyone defend this process on the basis of psychological realism? In addition, the content of the mathematical model is barely more than a circular explanation: When participants in the ultimatum game share equally or reject positive offers, this implies non-zero weights on the "social preferences" terms in the utility function, and the behavior is then attributed to "social preferences."

A related concern is the lack of attempts to replicate parameter estimates. Binmore and Shaked (2010) raise this point in a critique of Fehr and Schmidt (1999)—and of experimental economics more generally. Binmore and Shaked point out that, if Fehr and Schmidt's model is to be taken seriously as an innovation in empirical description, then a single parameterized version of it should make out-of-sample predictions and be tested on multiple data sets—without adjusting parameters to each new data set. According to Binmore and Shaked, Fehr and Schmidt use very different (i.e., inconsistent) parameter estimates in different data sets. To appreciate the point, one should recall the large number of free parameters in the Fehr and Schmidt model when subjects are allowed to all have different parameters weighting the three terms in the utility function. This huge number of degrees of freedom allows the model to trivially fit many sets of data well without necessarily achieving any substantive improvements in out-of-sample prediction over neoclassical models or competing behavioral theories. Binmore and Shaked (2010) write:

> [T]he scientific gold standard is prediction. It is perfectly acceptable to propose a theory that fits existing experimental data and then to use the data to calibrate the parameters of the model. But, before using the theory in applied work, the vital next step is to state the proposed domain of application of the theory and to make specific predictions that can be tested with data that was used neither in formulating the theory nor in calibrating its parameters. (p. 89)

This may seem so basic as to not be worth repeating. Yet the distinction between fit and prediction, which has been made repeatedly by others (Roberts & Pashler, 2000), seems to be largely ignored in much of the behavioral economics literature. Behavioral models frequently add new parameters to a neoclassical model, which necessarily increases R-squared. Then this increased R-squared is used as empirical support for the behavioral models without subjecting them to out-of-sample prediction tests.

Hyperbolic Discounting and Time Inconsistency

Laibson's (1997) model of impulsiveness consists, in essence, of adding a parameter to the neoclassical model of maximizing an exponentially weighted sum of instantaneous utilities, in order to choose an optimal sequence of quantities of consumption. Laibson's new parameter reduces the weight of all terms in the weighted sum of utilities except for the term representing utility of current consumption. This, in effect, puts more weight on the present by reducing weight on all future acts of consumption.

Thus, the psychological process involved has hardly changed at all relative to the neoclassical model from which the behavioral modification was derived. The decision maker is assumed to make an exhaustive search of all feasible consumption sequences, compute the weighted sum of utility terms for each of these sequences, and choose the one with highest weighted utility score. The parameters of this model are then estimated. To the extent that the estimated value of the parameter that reduces weight on the future deviates from the value that recovers the neoclassical version of the model with perfectly exponential weighting, Laibson asks us to interpret this as empirical confirmation—both of his model, and of a psychological bias to overweight the present over the future.

Another example is O'Donoghue and Rabin (2006), who suggest that willpower problems can be dealt with by taxing potato chips and subsidizing carrots, to induce people to overcome their biased minds and eat healthier diets. This formulation, again, assumes a virtually neoclassical decision process based on constrained optimization in which behavior is finely attuned to price and financial incentives, in contrast to more substantive empirical accounts of actual decision processes at work in food choice (Wansink, 2006).

Neoclassical + New Parameters with Psychological Names = Behavioral Economics?

A Widely Practiced Approach to Behavioral Economics: "More Stuff" in the Utility Function

In a frequently cited review article in the *Journal of Economic Literature,* Rabin (1998) argues that "greater psychological realism will improve mainstream economics." He then goes on to describe the improvement to economics that psychology has to offer, not as a more accurate empirical description of the decision processes used by firms and consumers, and not as a broad search for new explanations of behavior. Rather, Rabin states that the motivation for behavioral economists to borrow from psychology is to produce a more detailed specification of the utility function: "psychological research can teach us about the true form of the function U(x)." Thus, rather than questioning the rationality axioms of completeness, transitivity, and other technical requirements for utility function representations of preferences to exist—and ignoring the more substantive and primitive behavioral question of how humans actually choose and decide—Rabin lays out a behavioral economic research program narrowly circumscribed to fit within the basic framework of Pareto, Hicks, and Samuelson, historical connections that we return to below. According to Rabin, the full scope of what can be accomplished by opening up economics to psychology is the discovery of new inputs in the utility function.

Behavioral Utility Functions: Still Unrealistic as Descriptions of Decision Process

Leading models in the rise of behavioral economics rely on Friedman's as-if doctrine by putting forward more *unrealistic* processes—that is, describing behavior as the process of solving a constrained optimization problem that is more complex—than the simpler neoclassical model they were meant to improve upon. Many theoretical models in behavioral economics consist of slight generalizations of otherwise familiar neoclassical models, with new parameters in the objective function or constraint set that represent psychological phenomena or at least have psychological labels.

To its credit, this approach has the potential advantage of facilitating clean statistical tests of rational choice models by nesting them within a larger, more general model class so that the rational choice model can be tested simply by checking parameter restrictions. But because the addition of new parameters in behavioral models is almost always motivated in terms of improving the realism of the model—making its descriptions more closely tied to observational data—one can justifiably ask how much additional psychological realism is won from this kind of modeling via modification of neoclassical models. The key point is that the resulting behavioral model hangs on to the central assumption in neoclassical economics concerning behavioral process—namely, that all observed actions are the result of a process of constrained optimization. As others have pointed out, this methodology, which seeks to add behavioral elements as extensions of neoclassical models, paradoxically leads to optimization problems that are more complex to solve (Winter, 1964, p. 252, quoted in Cohen & Dickens 2002; Sargent, 1993; Gigerenzer & Selten, 2001).[4]

4. Lipman (1999) argues that it is okay if the model representing boundedly rational agents who cannot solve problem P is the solution to a more complex problem P'. Lipman's argument is that the solution to this more complex problem is the modeler's "representation" and should not be interpreted as a claim that the decision maker actually solves the harder problem P'. But this strikes us as an indirect invocation of Friedman's as-if doctrine.

Aside from this paradox of increasing complexity found in many bounded rationality models, there is the separate question of whether any empirical evidence actually supports the modified versions of the models in question. If we do not believe that people are solving complex optimization problems—and there is no evidence documenting that the psychological processes of interest are well described by such models—then we are left only with as-if arguments to support them.

Commensurability

A more specific methodological point on which contemporary behavioral and neoclassical economists typically agree is the use of standard functional forms when specifying utility functions, which impose the assumption—almost surely wrong—of universal commensurability between all inputs in the utility function. In standard utility theory, where the vector $(x_1, \ldots, x_j, \ldots, x_k, \ldots, x_N)$ represents quantities of goods with the jth and kth element represented by x_j and x_k, respectively, commensurability can be defined as follows. For any pair of goods represented by the indexes j and k, $j \neq k$, and for any reduction r in the jth good, $0 < r < x_j$, there exists a quantity of compensation in units of the kth good, $c > 0$, such that the consumer is at least as well off as she was with the original commodity bundle:

$$U\left(x_1,\ldots,x_j - r,\ldots,x_k + c,\ldots x_N\right) \geq U\left(x_1,\ldots,x_j,\ldots,x_k,\ldots x_N\right).$$

This is sometimes referred to as the Archimedean principle. Geometrically, commensurability implies that all indifference curves asymptote to the x-axis and y-axis. Economically, commensurability implies that when we shop for products represented as bundles of features (e.g., houses represented as vectors of attributes, such as square footage, price, number of bathrooms, quality of nearby schools, etc.), then no undominated items can be discarded from the consideration set. Instead of shoppers imposing hard-and-fast requirements (e.g., do not consider houses with less than 2000 square feet), commensurable utility functions imply that smaller houses must remain in the consideration set. If the price is low enough, or the number of bathrooms is large enough, or the quality of schools is high enough, then a house of any size could provide the "optimal" bundle of features.

Edgeworth included commensurability among the fundamental axioms of choice. Psychologists since Maslow have pointed out, however, that people's preferences typically exhibit distinctly lexicographic structure. Moreover, the structures of environments that elicit compensatory and noncompensatory strategies are relatively well known. An early review of process tracing studies concluded that there is clear evidence for noncompensatory heuristics, whereas evidence for weighting and adding strategies is restricted to tasks with small numbers of alternatives and attributes (Ford et al., 1989).

Recently, researchers in psychology and marketing have produced new evidence of lexicographic strategies that prove very useful in high-dimensional environments for quickly shrinking choice sets down to a manageable set of alternatives. The reduction of size in the consideration sets proceeds by allowing a few choice attributes to completely overrule others among the list of features associated with each element in the choice set. This obviates the need for pairwise tradeoffs among the many pairs of choices and enables choice to proceed in a reasonable amount of time (Yee, Dahan, Hauser, & Orlin, 2007). In a choice set with N undominated elements where each element is a vector of K features, complete ranking (needed to find the optimum) requires consideration of $KN(N-1)/2$ pairwise tradeoffs, which is the number of features of any alternative multiplied by a quadratic in the number of elements that represents the number of unordered pairs in the choice set.

Although interesting game-theoretic treatments of lexicographic games have appeared (Binmore & Samuelson, 1992; Blume, Brandenburger, & Dekel, 1991), behavioral and neoclassical economists routinely seem to forget the absurd implications of universal commensurability, with its unrealistic implication of ruling out lexicographic choice rules. If, for example, x represents a positive quantity of ice cream and y represents time spent with one's grandmother, then as soon as we write down the utility function $U(x, y)$ and endow it with the standard assumptions that imply commensurability, the unavoidable implication is that there exists a quantity of ice cream that can compensate for the loss of nearly all time with one's grandmother. The essential role of social interaction, and time to nurture high-quality social interactions as a primary and unsubstitutable source of happiness, is emphasized by Bruni and Porta's (2007) recent volume on the economics of happiness. The disadvantage of ruling out lexicographic choice and inference also rules out their advantage of time and effort savings, in addition to improved out-of-sample prediction in some settings (Czerlinski, Gigerenzer, & Goldstein, 1999; Gigerenzer & Brighton, 2009).

Fit Versus Prediction

Given that many behavioral economics models feature more free parameters than the neoclassical models they seek to improve upon, an adequate empirical test requires more than a high degree of within-sample fit (i.e., increased R-squared). Arguing in favor of new, highly parameterized models by pointing to what amounts to a higher R-squared (sometimes even only slightly higher) is, however, a widely practiced rhetorical form in behavioral economics (Binmore & Shaked, 2010).

Brandstätter et al. (2006) showed that cumulative prospect theory (which has five adjustable parameters) overfits in each of four data sets. For instance, among 100 pairs of two-outcome gambles (Erev et al., 2002), cumulative prospect theory with a fit-maximizing choice of parameters chooses 99% of the gambles chosen by the majority of experimental subjects. That sounds impressive. But, of course, including more free parameters always improves.

The more challenging test of a theory is in prediction using a single set of fixed parameters. Using the parameter values estimated in the original Tversky and Kahneman (1992) study, cumulative prospect theory could predict only 75% of the majority choices. The priority heuristic (a simple lexicographic heuristic with no adjustable parameters), in contrast, predicts 85% of majority choices. Moreover, when the ratio of expected values is larger than two (so-called "easy problems" where there is wide consensus among most subjects that one gamble dominates the other), cumulative prospect theory does not predict better than expected value or expected utility maximization (Brandstätter, Gigerenzer, & Hertwig, 2008, fig. 1). When the ratio of expected values is smaller, implying less consensus among subjects about the ranking of two gambles, the priority heuristic predicts far better than cumulative prospect theory. Thus, in prediction, cumulative prospect theory does not perform better than models with no free parameters.

Examples of psychological parameters introduced to generalize otherwise standard neoclassical models include Kahneman and Tversky's (1979) prospect theory in which new parameters are needed to pin down the shape of functions that under- or overweight probabilities; Laibson's (1997) model of impulsiveness expressed in terms of new parameters controlling the shape of nonexponential weights in the intertemporal optimization problem referred to as hyperbolic discounting; and Fehr and Schmidt's (1999) psychic weights on differences between own and others' payoffs. There are many other examples, which include overconfidence (with at least three different versions concerning biases in first and/or second moments and own beliefs versus the beliefs of others); biased belief models; "mistake" or tremble probabilities; and social preference utility functions with parameters that measure subjective concern for other people's payoffs.

By virtue of this modeling strategy based on constrained optimization, with virtually all empirical work addressing the fit of outcomes rather than justifying the constrained optimization problem-solving process itself, behavioral economics follows the Friedman as-if doctrine in neoclassical economics focusing solely on outcomes. By adding parameters to increase the *R*-squared of behavioral models' fit, many behavioral economists tacitly (and sometimes explicitly) deny the importance of correct empirical description of the processes that lead to those decision outcomes.

Behavioral and Neoclassical Economics Share a Single Normative Model

Is there such a thing as normative behavioral economics? At first, behavioral economists such as Tversky, Kahneman, Frank and Thaler almost unanimously said no (Berg, 2003).

The Early Normative View: Deviations Are Strictly Descriptive, No Normative Behavioral Economics Needed

Tversky and Kahneman (1986) write:

> The main theme of this article has been that the normative and the descriptive analysis of choice should be viewed as separate enterprises. This conclusion suggests a research agenda. To retain the rational model in its customary descriptive role, the relevant bolstering assumptions must be validated. Where these assumptions fail, it is instructive to trace the implications of the descriptive analysis. (p. S275)

Perhaps it was a reassuring sales pitch when introducing behavioral ideas to neoclassical audiences. But for some reason, early behavioral economists argued that behavioral economics is purely descriptive and does not in any way threaten the normative or prescriptive authority of the neoclassical model. These authors argued that, when one thinks about how he or she ought to behave, we should all agree that the neoclassical axioms ought to be satisfied. This is Savage's explanation for his own "mistaken" choice after succumbing to the Allais Paradox and subsequently revising it "after reflection" to square consistently with expected utility theory (Starmer, 2004). In this unquestioning view toward the normative authority of the neoclassical model, the only work for behavioral economics is descriptive—to document empirical deviations from neoclassical axioms: transitivity violations, expected utility violations, time inconsistency, non-Nash play, non-Bayesian beliefs, etc.

Fourteen years before writing "Libertarian Paternalism," Thaler also explicitly warns not to draw normative inferences from his work (Thaler, 1991, p. 138):

> A demonstration that human choices often violate the axioms of rationality does not necessarily imply any criticism of the axioms of rational choice as a normative idea. Rather, the research is simply intended to show that for descriptive purposes, alternative models are sometimes necessary.

Continuing this discussion of what behavioral economics implies about the use of rationality axioms in normative analysis, Thaler (1991, p. 138) argues that the major contribution of behavioral economics has been the discovery of a collection of "illusions," completely analogous to optical illusions. Thaler interprets these "illusions" as unambiguously incorrect departures from the "rational" or correct way of making decisions. Thaler is explicit in accepting neoclassical axioms of individual preferences (e.g., transitivity, completeness, nonsatiation, monotonicity, and the Savage axioms, which guarantee that preferences over risky payoffs can be represented by an expected utility function) as the proper normative ideal when he writes: "It goes without saying that the existence of an optical illusion that causes us to see one of two equal lines as longer than the other should not reduce the value we place on accurate measurement. On the contrary, illusions demonstrate the need for rulers!"

In his interpretation of optical illusions, Thaler does not seem to realize that, if the human faculty of visual perception mapped two-dimensional images directly onto our retinas and into the brain without filtering, then we would have an objectively inferior grasp on reality. Consider a photograph of railroad tracks extending into the distance, which appear narrower and narrower when projected into two-dimensional space but are filtered in our minds as maintaining constant width in three-dimensional space. Thaler seems to suggest that when we see the train tracks narrowing in their two-dimensional representation, it would be more rational to see them as narrowing rather than synthesizing the third dimension that is not really there in the photo. Without deviating from this direct translation of the information in two-dimensional space, our minds would perceive the tracks as uneven and unsuitable for any train to run on.

To correctly perceive reality, perceptive faculties must add information, make intelligent bets, and consequently get it wrong some of the time. A line that extends into the third dimension has a shorter projection on the retina than a horizontal line of the same length. Our brains correct for this by enlarging the subjective length of the line that extends into the third dimension, which works in the real three-dimensional world, but results in optical illusions when interpreting information on two-dimensional paper. Our brains are intelligent exactly because they make informed guesses, and go beyond the information given. More generally, intelligent systems depend on processes that make useful errors (Gigerenzer, 2008).

Yet, in showing that human decisions contradict the predictions of expected utility theory, there is no analog to the straight lines of objectively equal length. Unlike the simple geometric verification of equal lengths against which incorrect perceptions may be verified, the fact that human decisions do not satisfy the axioms underlying expected utility theory in no way implies an illusion or a mistake. Expected utility theory is, after all, but one model of how to rank risky alternatives. Those who insist that standard neoclassical theory provides a singularly correct basis for normative analysis in spite of systematic departures in the empirical record assert, in effect, that behavioral economics is a purely descriptive field of inquiry (Berg, 2003).

A Second Normative View: Designing Policy to Achieve Conformity with Neoclassical Norms

Fast-forward 10 years, and behavioral economists now can be found regularly offering prescriptive policy advice based on behavioral economics models. The stakes have risen in recent years and months, as financial market crises generate new skepticism about the "rationality of markets." Behavioral economists who decades ago pitched the behavioral approach to the neoclassical mainstream as a purely descriptive enterprise (e.g., Tversky & Kahneman,

1986, Thaler, 1991, Frank, 1991—and nearly everyone else published in top-ranked economics journals), now advocate using behavioral concepts for prescriptive policy purposes (Thaler & Sunstein, 2008; Frank, 2008; Amir et al., 2005). This evolution in boldness about looking for prescriptive implications of behavioral economics does not, unfortunately, imply increased boldness about modifying the neoclassical axiomatic formulations of rationality as the unquestioned gold standard for how humans ought to behave.

One specific example of this view that humans are biased and pathological—based on the biases and heuristics literature's abundant empirical accounts of deviations from neoclassical rationality axioms (but not tied empirically to substantive economic pathology)—is Bernheim and Rangel (2005). They suggest new approaches to regulation and policy-making based on the dominant behavioral economics view of ubiquitous behavioral pathology. Jolls, Sunstein, and Thaler (1998) write of the need to write laws that "de-bias" individual decision making. Rather than resting on direct observation of badly performing decision-making processes embedded in real-world domains, these prescriptive claims follow from psychological parameter estimates fitted, in many cases, to a single sample of data. The estimated parameter that maximizes fit leads to a rejection of the neoclassical model nested within the encompassing behavioral model, and readers are asked to interpret this as direct, *prima facie* evidence of pathological decision making in need of correction through policy intervention.

Predictably Stupid, Smart, or None of the Above

Rabin (2002) says psychology teaches about departures from rationality. Diamond (2008, p. 1859) writes that a major contribution of "behavioral analysis is the identification of circumstances where people are making 'mistakes.'" Beshears, Choi, Laibson, and Madrian (2008) introduce a technique for identifying mistakes, formulated as mismatches in revealed preference versus what they call normative preferences, which refer to preferences that conform to neoclassical axioms. To these writers (and many if not most others in behavioral economics), the neoclassical normative model is unquestioned, and empirical investigation consists primarily of documenting deviations from that normative model, which are automatically interpreted as pathological. In other words, the normative interpretation of deviations as mistakes does not follow from an empirical investigation linking deviations to negative outcomes. The empirical investigation is limited to testing whether behavior conforms to a neoclassical normative ideal.

Bruni and Sugden (2007) point out the similar methodological defense needed to rationalize the common normative interpretations in both neoclassical and behavioral economics:

> The essential idea behind the discovered preference hypothesis is that rational-choice theory is descriptive of the behaviour of economic

> agents who, through experience and deliberation, have learned to act in accordance with their underlying preferences; deviations from that theory are interpreted as short-lived errors. (p. 148)

The discussion of methodological realism with respect to the rational choice framework almost necessarily touches on different visions of what should count as normative. It is a great irony that most voices in behavioral economics, purveyors of a self-described opening up of economic analysis to psychology, hang on to the idea of the singular and universal supremacy of rational choice axioms as the proper normative benchmarks against which virtually all forms of behavior are to be measured. Thus, it is normal rather than exceptional to read behavioral economists championing the descriptive virtues of expanding the economic model to allow for systematic mistakes and biased beliefs and, at the same time, arguing that there is no question as to what a rational actor ought to do.

This odd tension between descriptive openness and normative dogmaticism is interesting, and future historians of behavioral economics will surely investigate further the extent to which this hardening of the standard normative model in the writings of behavioral economists served as compensation for out-and-out skeptics of allowing psychology into economics—perhaps, in order to persuade gatekeepers of mainstream economics to become more accepting of behavioral models when pitched as an exclusively descriptive tool. One reason why the tension is so interesting is that almost no empirical evidence exists documenting that individuals who deviate from economic axioms of internal consistency (e.g., transitive preferences, expected utility axioms, and Bayesian beliefs) actually suffer any economic losses. No studies we are aware of show that deviators from rational choice earn less money, live shorter lives, or are less happy. The evidence, to date, which we describe in a later section, suggests rather the opposite.

Like neoclassical economists, behavioral economists assert that logical deduction rather than inductively derived descriptions of behavioral process are the proper starting point for economic analysis. Behavioral economists allow that real people's beliefs (and nearly everything else the neoclassical model specifies) may deviate from this deductive starting point in practice. But they insist that individuals who deviate from axiomatic rationality should aspire to minimize deviance and conform to the neoclassical ideal as much as possible.

Ecological Rationality

A Definition Based on the Extent of Match between Behavior and Environments

It is no trivial question as to whether substantive rather than axiomatic rationality requires preferences to exist at all. The essentializing concept of a stable preference ordering ignores the role of context and environment as explanatory variables that might condition what it means to make a good decision. In this regard, preferences in economics are analogous to personality traits in psychology. They seek to explain behavior as a function of exclusively inherent and essential contents of the individual rather than investigating systematic interaction of the individual and the choice or decision environment.

In contrast, the normative framework of ecological rationality eschews universal norms that generalize across all contexts, and instead requires decision processes to match well with the environments in which they are used (Gigerenzer, Todd, & the ABC Research Group, 1999). Ecological rationality focuses on the question of which heuristics are adapted to which environments. Vernon Smith's definition of ecological rationality is virtually the same, except that he replaces "heuristics" with "institutions" or "markets."

When heuristics, or decision processes—or action rules—function well in particular classes of environments, then ecological rationality is achieved. When systematic problems arise, the diagnosis does not lay blame exclusively on badly behaved individuals (as in behavioral economics) or external causes in the environment (as in many normative analyses from sociology). Rather, problems are diagnosed in terms of mismatched decision process and environment, which suggests more degrees of freedom (than the universally pathological view based on a normative ideal of omniscience) when prescribing corrective policy and new institutional design.

Better Norms

Given the explicitly stated commitment in behavioral economics to empiricism and broader methodological openness (borrowing from psychology and sociology), it is surprising that behavioral economics would adhere so closely to the normative neoclassical model, because there are real alternatives in terms of positive normative frameworks from fields such as psychology, Austrian economics, social economics, biology, and engineering. In spite of hundreds of papers that purport to document various forms of "irrationality" (e.g., preference reversals, deviations from Nash play in strategic interaction, violations of expected utility theory, time inconsistency, non-Bayesian beliefs), there is almost no evidence that such deviations lead to any economic costs.[5] Thus—separate from the lack of evidence that humans make high-stakes decisions by solving constrained optimization problems—much of the behavioral economics research program is predicated on an important normative hypothesis for which there is, as yet, very little evidence.

Are people with intransitive preferences money-pumped in real life? Do expected utility violators earn less money, live shorter lives, or feel less

5. One recent example is DeMiguel et al. (2009) who find that portfolios that deviate from the normative CAPM model by using a simple 1/N heuristic produce higher expected returns and lower risk, relative to portfolios chosen according to CAPM.

happy? Do non-Bayesians systematically misperceive important frequencies and incur real economic losses as a result?

These questions would seem to be the key stylized facts in need of firm empirical justification in order to motivate the prolific research output in behavioral economics documenting biases and deviations. But instead of empirical motivation, behavioral economics—while justifying itself in terms of more rigorous empiricism—puzzlingly follows the neoclassical tradition laid out by Pareto in justifying its normative positions by vague, introspective appeals to reasonableness, without empirical inquiry (Starmer, 2005).

Our own empirical research tries to answer some of these questions about the economic costs of deviating from neoclassical axioms, with surprising results. Expected utility violators and time-inconsistent decision makers earn more money in experiments (Berg, Eckel, & Johnson, 2009). And the beliefs about PSA testing of non-Bayesians are more accurate than those of perfect Bayesians—that is, better calibrated to objective risk frequencies in the real-world decision-making environment (Berg, Biele, & Gigerenzer, 2013). So far, it appears that people who violate neoclassical coherence, or consistency, axioms are better off as measured by correspondence metrics such as earnings and accuracy of beliefs. Recall that according to rationality norms requiring only internal coherence, one can be perfectly consistent, and yet wrong about everything (Hammond, 1996).

There are a growing number of theoretical models, too, where individuals (Dekel, 1999, Compte & Postlewaite, 2004) and markets (Berg & Lien, 2005) do better with incorrect beliefs. These results pose fundamental questions about the normative status of assumptions regarding probabilistic beliefs and other core assumptions of the rational choice framework. If individuals and aggregates both do better (Berg & Gigerenzer, 2007) when, say, individuals satisfice instead of maximize, then there would seem to be no market discipline or evolutionary pressure (arguments often invoked by defenders of the normative status of rationality axioms) to enforce conformity with rationality axioms, which focus primarily on internal consistency rather than evaluation of outcomes themselves.

In a variety of binary prediction tasks, Gigerenzer, Todd and the ABC Research Group (1999) have shown that simple heuristics that ignore information and make inferences based on lexicographic rather than compensatory (weighting and adding) decision procedures are often more accurate in prediction than regression models that simultaneously weight and consider all available information. Berg and Hoffrage (2008) provide theoretical explanations for why ignoring free information can be adaptive and successful. Starmer (2005) makes a number of relevant points on this issue, and Gilboa, Postlewaite, and Schmeidler (2004) expand on the arguments of Hammond's (1996) regarding the normative insufficiency of internal coherence alone. These authors are highly unusual in expressing doubt about whether Bayesian beliefs, and other normative axioms of internal consistency, should be relied upon as normative principles.

Gaze Heuristic

How do baseball players catch fly balls? Extending Friedman's as-if model of how billiards players select their shots, one might follow the neoclassical as-if modeling approach and assume that baseball players use Newtonian physics. According to this as-if theory of catching a fly ball, players would rely upon variables such as initial position, initial velocity, rotation, and wind speed to calculate the terminal position of the ball and optimal direction in which to run.

There are several observable facts that are inconsistent with this as-if model, however. First, baseball players catching fly balls do not typically run to the landing position of the ball and wait for it there. They frequently run away from the ball first, backing up, before reversing course inward toward the ball, which is not predicted by the as-if theory. Finally, experiments that ask baseball players to point to the landing location of the ball reveal that experts with specialized training in catching balls have a very difficult time pointing to the landing position of the ball. Nevertheless, because they consistently catch fly balls, these players are employing a decision process that gets them to the proper location at the proper time. This process is the gaze heuristic (Gigerenzer & Selten, 2001).

The gaze heuristic is a genuine process model that explains *how* the player puts his or her body in the proper position to catch fly balls. When a fly ball is hit, the player waits until the ball reaches a sufficiently high altitude. The player then fixes this angle between his or her body and the ball and begins running to maintain this angle at a nearly constant measure. To keep the angle fixed as the ball begins to plummet toward earth, one must run to a position that eventually converges to directly under the ball.

Maintaining a fixed angle between the player and the ball gets the body to the right place at the right time. This process of maintaining the angle implies that sometimes players will have to back up before running inward toward home plate. This process also does not depend on any optimally chosen parameters. For example, there is a wide and dense range of angles that the player can choose to maintain and still catch the ball. No "optimal angle" is required.

The benefits of this genuine process model are many. For one, we have a realistic explanation of how balls are caught, answering to the descriptive goal of science. For the normative and prescriptive dimensions, the benefits are perhaps even more noticeable. Suppose we were to use the as-if model to design a policy intervention aimed at inducing better performance catching fly balls. The as-if theory suggests that we should provide more or clearer information about initial position, initial velocity, wind speed, and ball rotation. That could mean, for example, that a computer monitor in the outfield instantly providing this information to outfielders would improve their performance. Should we take this seriously?

In contrast, the gaze heuristic suggests that patience to allow the ball to reach high overhead, good vision to maintain the angle, and fast running speed are among the most important inputs into success at catching fly balls. Thus, process and as-if models make distinct predictions (e.g., running in a pattern that keeps the angle between the player and ball fixed versus running directly toward the ball and waiting for it under the spot where it will land; and being able to point to the landing spot) and lead to distinct policy implications about interventions, or designing new institutions, to aid and improve human performance.

Empirical Realism Sold, Bought and Resold

This section summarizes the historical trajectory of debates about empirical realism in economics in the 20th century that is more stylized than detailed, but nevertheless describes a hypothesis about the status of claims to realism in economics. This summary underscores links between debates about, and within, behavioral economics, and the longstanding influence of Pareto in the shift away from psychology toward the as-if interpretation of models and deemphasis of decision-making process in economics. Dismissing empirical realism as an unneeded element in the methodology of economics, the post-Pareto neoclassical expansion under the guidance of Paul Samuelson might be described as "empirical realism sold." In other words, after Pareto's arguments took root in mainstream English language economics, the field proceeded as if it no longer cared much about empirical realism regarding the processes that give rise to economic decisions.

When behavioral economics arrived upon the scene, its rhetoric very explicitly tied its own program and reason for being to the goal of improved empirical realism. This initial phase of behavioral economics could be referred to as "empirical realism bought," because practitioners of behavioral economics, as it was first trying to reach a broader audience, emphasized emphatically a need for psychology and more empirical verification of the assumptions of economics.

Then, perhaps after discovering that the easiest path toward broader acceptance into the mainstream was to put forward slightly modified neoclassical models based on constrained optimization, the behavioral economics program shed its ambition to empirically describe psychological process, adopting Friedman's as-if doctrine. Thus, the second phase in the historical trajectory of behavioral economics is described here as: "empirical realism resold."

Realism Sold

Bruni and Sugden (2007) point out interesting parallels between proponents of behavioral economics (who argued for testing the assumptions of the rational choice model with observational data against defenders of neoclassical economics arguing in favor of unbounded rationality assumptions) and participants in an earlier methodological debate. The earlier debate took place within neoclassical economics about the role of psychology in economics, in which Vilfredo Pareto played a prominent role. According to Bruni and Sugden, the neoclassical program, already underway as Pareto wrote, took a distinct turn as Hicks and Allen, Samuelson, and Savage, made use of Pareto's arguments against using anything from psychology (e.g., the Fechner-Weber Law used earlier as a foundation for assuming diminishing marginal utility, or the beginnings of experimental psychology as put forth in Wilhelm Wundt's *Grundzüge der physiologischen Psychologie* published in 1874) in economics. Pareto argued in favor of erecting a clear boundary insulating economic assumptions from certain forms of empirical inquiry and, rather than inductive empiricism, he advocated much greater emphasis on logical deduction.

The psychology of Pareto's day was hardly vacuous as some defenders of the Pareto-led shift away from psychology in economics have claimed. And Pareto was enthusiastic about using psychology and sociology to solve applied problems, even as he argued that economics should be wholly distinct and reliant solely on its own empirical regularities. Pareto argued for a deductive methodology very much like the contemporary rational choice model in which all decisions were to be modeled as solutions to constrained optimization problems. To understand how Pareto could use ideas and data from psychology and sociology in some settings but argue unequivocally for eliminating these influences from economics, Bruni and Sugden (2007) explain that the neoclassical economics of Pareto's time, which changed dramatically as a result of his positions, was seen as encompassing complementary psychological and economic branches within a common research paradigm:

> This programme was not, as behavioural economics is today, a self-consciously distinct branch of the discipline: it was a central component of neoclassical economics. Neoclassical economics and experimental psychology were both relatively young enterprises, and the boundary between them was not sharply defined. According to what was then the dominant interpretation, neoclassical theory was based on assumptions about the nature of pleasure and pain. Those assumptions were broadly compatible with what were then recent findings in psychophysics. Neoclassical economists could and did claim that their theory was scientific by virtue of its being grounded in empirically-verified psychological laws. . . .Viewed in historical perspective, behavioural economists are trying to reverse a fundamental shift in economics which took place from the beginning of the twentieth century: the "Paretian turn." This shift, initiated by Vilfredo Pareto and completed in the 1930s and 1940s by John Hicks, Roy Allen and Paul Samuelson, eliminated psychological concepts from economics by basing economic theory on principles of rational choice. (p. 149)

Pareto's deliberate shift away from psychology also entailed a shift away from large categories of empirical source material. In this sense, the so-called Paretian turn in the history of economics can be summarized, perhaps too simply, but not inaccurately, as a divestiture of earnest empirical inquiry into the processes by which firms and consumers make decisions. The question of decision process, in the eyes of Pareto, Hicks, and Samuelson, was a solved problem with a singular answer: choice in economics was defined as the solution to an appropriately specified constrained optimization problem. This relieved economics from investigating further the question of how firms and consumers actually make decisions, and shifted the terms of economic analysis toward the business of discovering parameters in objective functions and constraint sets, whose maximizing action rule (mapping exogenous parameters into actions) seemed to capture the regularities that economists regarded, based on introspection, as natural and self-evident, such as downward-sloping demand curves or diminishing marginal utility.

Pareto argued that, for simplification, economics should assume that subjective beliefs about the economic environment coincide with objective facts. Thus, for Pareto and many who relaunched Pareto's program in the 1930s, the question of how well people's subjective experience of economic phenomena matches the objective structure of the environment is assumed away. There is no question of calibration, or correspondence to the real world. Pareto defended this by limiting the domain of phenomena to which economic theory was to be applied, in sharp contrast to promulgators of the Pareto program who later claimed that deductive logic of rational choice models vastly expanded the range of real-world phenomena to which the theory applies.

Realism Bought

Advocates for behavioral economics who have come to prominence in the last two decades frequently make the case that economics will benefit by more openly embracing the empirical lessons of psychological experiments, economic experiments, and standard econometric data sources filtered through models that allow for behavioral phenomena, such as loss aversion in choice under uncertainty and quasi-hyperbolic discounting in intertemporal choice. This phase in the history of behavioral economics can be described as "empirical realism bought"—bought in the sense of the economics discipline siding with arguments made by contemporaries of Pareto who disagreed with him, arguing in favor of using psychological data and behavioral regularities put forward by psychologists in economics (e.g., Pantaleoni, 1898/1889).

Realism Resold

In the earlier section "As-If Behavioral Economics," we considered three prominent theories, often cited as leading examples of the success of behavioral economics. We argued, however, that these three models are not serious attempts at psychological realism and rather rely on Friedman's as-if defense to justify modeling psychological choice as the solution to an even more elaborate constrained optimization problem. These models exemplify the "realism resold" phase in the historical trajectory of behavioral economics. "Realism resold" describes behavioral economics' retreat from investigating actual decision processes, conforming instead to Friedman's as-if defense of unrealistic models. The unrealistic models now being defended are endowed with additional parameters given psychological labels, resting on the claim that people behave as if they are solving a complicated constrained optimization problem with bounds on self-interest, willpower, or computational capacity explicitly modeled in the objective function or constraint set. This strange new methodological configuration, motivated in terms of improved empirical realism, and defended according to the as-if line of defense, can be described as as-if behavioral economics.

Pareto as Precursor to As-If

To the neoclassicals following Pareto's position, an economics defined by axioms of perfect internal consistency as *the* standard of rationality was to provide essential insights into how consumers' and firms' behavior would change when shifting from one equilibrium to another as a result of a change in a single exogenous parameter. Thus, the methodology was to maintain in all cases—rather than test or investigate—the assumptions of transitive preference orderings, expected utility axioms (after Savage), and beliefs that are internally coherent by satisfying Bayes rule. A number of neoclassical economists acknowledged that predicted changes in behavior generated by shifting from one equilibrium to another in response to an exogenous change, of course, abstracts from many other influences that are potentially important (i.e., those that psychologists and sociologists focus on).

The neoclassicals argued, however, that their predictions, free from psychological or sociological factors, were good enough (ironically, a satisficing argument about the aspirations of their theory), and should be interpreted as predictions about behavior after many repetitions when, it was assumed, behavior would converge to the ideal choice predicted by rational choice theory. Bruni and Sugden (2007) point out problems with this position, some of which Pareto was aware of, and some of which seem to persist in the defenses of rational choice theory offered today.

An interesting contrast emerges when comparing very recent justifications for behavioral economics put forward by leading behavioral economists such as Rabin and Thaler, and these authors' earlier writings in which deeper skepticism was occasionally expressed about the utility function framework. An example is Thaler's writing in the first round of *Journal of Economic Perspectives* "Anomalies" series, where Thaler's conclusions sometimes mention deep doubts that good descriptions of behavior could ever be achieved without deeper methodological revisions in economics. Not surprisingly, the part of the behavioral economics movement that won easiest acceptance was the part that was methodologically closest to neoclassical norms, following the path of constrained optimization models with an additional psychological parameter or two.

It is striking that the behavioral economists who successfully sold psychology to neoclassical economists are among the most hardened and staunch defenders of the normative status of the neoclassical model. Whereas neoclassical economists frequently interpret their models as essentialized approximations, from which deviations are expected to average out in the aggregate, many behavioral economists use the rationality standard of neoclassical economics more literally and rigidly than their neoclassical colleagues.

In contrast to the unpsychological spirit of much writing on psychology in behavioral economics, there are some, such as Conlisk (1996), who appreciate that contemporary psychology's use of the term *heuristics* (i.e., shortcut decision processes not generally derived by solving a constrained optimization problem) often implies a useful shortcut to solving a difficult problem—and not a pathological deviation from axiomatic rationality. Particularly when the cost of information is high, or the optimization problem has many dimensions that make its solution very costly or impossible, a heuristic can provide a valuable procedure for making the decision well. The study of ecological rationality (Chapter 7) has shown that the function of heuristics is not restricted to this shortcut interpretation, also known as the accuracy–effort trade-off. By ignoring information, a heuristic can be more accurate in making predictions in a changing and uncertain world than a strategy that does not condition on all available information—so-called less-is-more effects (Gigerenzer & Brighton, 2009).

The debates between behavioral economics and neoclassical economics echo earlier debates in economics from the first half of the 20th century. An interesting dimension of historical similarity are the debates about decision-making processes, prominent in the psychology literature, but virtually absent in both postwar neoclassical economics and contemporary behavioral economics. These missing debates about decision-making process in economics concern whether constrained optimization is realistic or empirically justified, and whether a more directly empirical account of decision-making process can lead to better descriptive and normative economics. The seemingly opposing subfields of neoclassical and behavioral economics, it seems, rely on a common rhetorical strategy that traces back to the methodological shifts in economics away from psychology around the time of Pareto.

If Economics Becomes an Empirical Science . . .

Critiques of Rationality Assumptions Are Nothing New

Long before the contemporary behavioral economics program came to prominence, the economics discipline saw a good deal of complaining about the strictures of rationality assumptions—especially the ones required to rationalize a utility function representation of a preference ordering, and the self-interested rational actor model—long before Herbert Simon or the current leaders of the behavioral economics program began writing. One recalls Veblen's conspicuous consumption in *The Theory of the Leisure Class* (1899/1994), Keynes's "animal spirits" in the *General Theory* (1936/1974), Galbraith's "Rational and Irrational Consumer Preference" (1938), and Hayek's (1945) critique of the disconnect between maximization of given preferences over known choice sets versus "the economic problem which society faces," which rests on the radical limitations on economic actors' knowledge.

In fact, earlier writers before the rise of general equilibrium theory and subsequent ascendancy of highly rationalist game theory in the 1980s frequently expressed interest in decision processes other than those posited in the rational choice model. One finds deep sympathy in Smith's (1759/1997) writings on sentiments, and in writers going back to antiquity (Bruni & Porta, 2007), for the proposition that economic behavior takes multiple forms depending on social context.[6] In this light, it would seem that the singularity of the rational choice model within neoclassical economists' methodological tool kit in post-war North American economics (together with its strict normative interpretation) is anomalous when compared to longer standing norms allowing for a much wider range of behavioral models in economics.

Proponents of genuine process models would argue that, especially when predicting how a new policy or institution will perform, the range of variation in the data used to fit various models may not give illuminating predictions over the relevant ranges of variables after policy and institutions shift. If the actual process generating economic decisions is better understood, however, then social science has a firmer basis to make important predictions about behavior under new and imagined institutional arrangements. Process models would therefore play a crucial role in furthering both the creativity and predictive accuracy of economists attempting to imagine and design new institutions—where success hangs upon how such institutions might interact with the repertoire of heuristics and behavioral rules widely used in a population.

Naming Problem

In thinking about future histories of behavioral economics, the term *behavioral* itself is already problematic on two counts at least. First, as many have pointed out, it seems ironic that a social science would need to call itself "behavioral"—distinguishing itself from apparently nonbehavioral social sciences? Given the antiempirical flavor of as-if defenses of economic

6. Ashraf, Camerer and Loewenstein's 2005 article, "Adam Smith, Behavioral Economist," pushes this claim to an extreme.

analysis that is explicitly uncurious about the "black box" of mind that generates economic decisions, the behavioral label could have implied a useful critique. However, when one digs into the methodological arguments put forward in behavioral economics, the apparent distinctions appear slight. (The term *behavioral economics* seems to have been coined by the psychologist George Katona, who established the Survey Research Center (SRC), part of the Institute for Social Research (IRS) at the University of Michigan. Amos Tversky obtained his PhD at the University of Michigan under the supervision of Clyde Coombs and Ward Edwards.)

At a recent meeting of the Society for the Advancement of Behavioral Economics, one board member suggested that the group dissolve, arguing that behavioral economics had become mainstream, and therefore no distinction or group to advocate on its behalf was needed. Whether this merging of behavioral economics and mainstream economics represents a change in the mainstream or a capitulation of the motive behind the behavioral program aimed at improved realism is open to debate.

A second aspect of the naming problem inherent in "behavioral economics," which may seem trivial, but underscores links to another research program that has run into serious barriers, is potential confusion with the behaviorist movement. (John Broadus Watson published his treatise on the behaviorist approach to psychology in 1913). Bruni and Sugden (2007) describe the behaviorist movement in psychology as having "denied the scientific status of introspection." This is almost equivalent to the denial by some economists, both behavioral and neoclassical, that actual decision processes of firms and consumers are important—that only outcomes of decision processes are appropriate objects for scientific inquiry. Thus, one important theme of the behaviorist program agrees with the as-if Friedman doctrine, carried forward in contemporary behavioral economics by those who argue that the goal of their models is not to provide a veridical description of the actual decision processes being used by economic agents, but to predict the outcome (a particular action or decision).

The Route Not (Yet?) Taken: Process Models Addressing Expected Utility (EU) Violations, Time Inconsistency, and Other-Regarding Behavior

Economists like Herbert Simon, Reinhard Selten, and Vernon Smith illustrate that there is a positive route not taken in behavioral economics, which is more empirical, more open to alternative normative interpretations of deviations from neoclassical theory, and more descriptive of actual decision processes rather than reliant on extensions of Friedman's as-if methodology. Perhaps counterintuitively, the issue of normative interpretation is critical for these thinkers in gauging how far their descriptive work can move away from neoclassical theory and achieve more data-driven descriptions of how decisions are made. Simon, for example, thought that expected utility theory was both normative and descriptively inadequate. Selten proposes elaborate satisficing explanations of choice under uncertainty. And Vernon Smith holds that if someone consciously violates EU, then this does not imply that he or she made an error. The systematic study of heuristic decision making, based on the work of Simon and Selten, provides a program for the "route not taken" (Chapter 7; Gigerenzer et al., 2011).

Regarding the three examples of as-if behavioral economics given in the second section in this paper, one can point to genuine process models that handle the very same behavioral phenomena without as-if justification. Tversky's elimination by aspects described a process to choose between two alternatives that could be gambles. Unfortunately, Tversky abandoned his attempts to use lexicographic structure to model choice under uncertainty when he joined Kahneman and turned to the repair program. The priority heuristic, mentioned earlier, is another process model, and it predicts the experimental data better than as-if cumulative prospect theory. Moreover, the priority heuristic logically implies (rather than merely fits) the major decision "anomalies," including the fourfold patterns of risk attitudes (Katsikopoulos & Gigerenzer, 2008).

Regarding time inconsistency, Rubinstein (2003) put forward a process model for temporal discounting that provides an attractive alternative to the as-if hyperbolic discounting story. The ecological rationality of various forms of time inconsistency was documented by Leland (2002), Rosati et al. (2007), and Heilbronner et al. (2008), who showed that characteristics of the decision maker's environment can explain some differences in discount rates. For example, if one lives among lots of greedy companions rather than alone, this tends to make one less patient.

With respect to other-regarding behavior, Henrich et al. (2001) tried but could not find *Homo economicus* in 15 small-scale societies in remote locations. They found that offers and rejections in the ultimatum game are related to the extent to which these societies' production technologies required workers to cooperate (e.g., hunting in groups) or fend for themselves (e.g., gathering food alone). Carpenter and Seki (2006) report a similar finding about two groups of Japanese fishermen and women. They find that groups who pool the payoffs from all boats' daily catches play the ultimatum game much more cooperatively than groups that reward the members of each boat more individualistically based on the value of each boat's own daily catch.

Empirical Realism: Past to Present

Bruni and Sugden (2007), in their discussion of Hicks and other founders of contemporary neoclassical economics (vis-à-vis neoclassical economics before Pareto's influence came to dominate), write:

> If economics is to be a separate science, based on laws whose truth is to be treated as axiomatic, we have to be very confident in those laws. Otherwise, we are in danger of creating a complex structure of internally consistent theory which has no correspondence with reality. (p. 160)

This correspondence with reality is the essence of the empirical approach to economics. How else do we get to be "very confident" in the laws of economics?

The origins of behavioral economics are many, without clear boundaries or singularly defining moments. And yet, even a cursory look at articles published in economics today versus, say, 1980, reveals a far-reaching, distinctly behavioral shift.[7] A striking element in the arguments of those who have successfully brought behavioral economics to mainstream economics audiences is the close similarity to Friedman's as-if defense.

In prospect theory, behavioral economics has added parameters rather than psychological realism to model choice under uncertainty. In modeling other-regarding behavior, utility functions have been supplemented with parameters weighting decision makers' concern for receiving more, or less, than the group average. Time inconsistency observed in experiments has prompted a large empirical effort to pin down parameters in objective functions that hang onto the assumption of maximization of a time-separable utility function, but with nonexponential weighting schemes that have taken on psychological labels that purport to measure problems with willpower. Described as a new empirical enterprise to learn the true preferences of real people, the dominant method in behavioral economics can be better described as filtering observed action through otherwise neoclassical constrained optimization problems with new arguments and parameters in the utility function.

We have tried to investigate to what extent behavioral economists' attempts to filter data through more complexly parameterized constrained optimization problems succeeds in achieving improved empirical realism and, in so doing, distinguishing behavioral from neoclassical economics. The primary finding is that of widespread similarity in the neoclassical and behavioral research programs. This suggests common limitations in their ultimate historical trajectories and scientific achievements. To become more genuinely helpful in improving the predictive accuracy and descriptive realism of economic models, more attention to decision process will be required, together with bolder normative investigation using a broader set of prescriptive criteria.

7. One can cite many concrete events as markers of the emergence of behavioral economics, or psychology and economics, onto a broader stage with wide, mainstream appeal. One might imagine that such a list would surely include Herbert Simon's Nobel Prize in 1978. But that was a time at which very little behavioral work appeared in the flagship general-interest journals of the economics profession. A concise and of course incomplete timeline would include: Richard Thaler's "Anomalies" series, which ran in the *Journal of Economic Perspectives* starting in 1987; hiring patterns at elite business schools and economics departments in the 1990s; frequent popular press accounts of behavioral economics in *The Economist, New York Times* and *The Wall Street Journal* in the last 10 years; and the 2002 Nobel Prize being awarded to an experimental economist and a psychologist. The 1994 Nobel Prize was shared by another economist who is an active experimenter, Reinhard Selten.

Chapter 12

Personal Reflections on Theory and Psychology

When discussing psychological research, what surprises every economist or physicist is that psychology has no theory. It has many local ones but no overarching theory, not even a provisional one. Yet there is something even more surprising: a lack of awareness of the value of integration. Whereas the unification of theories, such as evolutionary theory and genetics, is a widely shared goal in physics and biology, it is barely visible in psychology. Few psychologists even consider theory integration as an objective. A textbook in economics starts with first principles that lead to an overarching theory and discusses how reality fits into this picture. A textbook in psychology lists dozens of theories in chapters on reasoning, intelligence, problem solving, and judgment and decision making—topics that appear closely related, but are populated by different researchers, published in different journals, and presented as independent enterprises. To the poor student, the relation between the various concepts in these different theories is never made clear. Why is present-day psychology such a patchwork of small territories, resembling, to use a political metaphor, Italy or Germany before unification around 1870? Why are psychologists so content working within the confines of their own small territories?

I have written a few pieces on this problem; these belong to my least cited papers. It appears to be a topic discussed less in journals than in letters. In response to my short essay entitled "Surrogates for Theory" in the APS *Observer* in 2009, which detailed some of psychologists' clever ways to avoid constructing theories, a professor from Indiana University wrote: "No field within psychology suffers more from the theoretical malaise than mine, educational psychology." Others working in social psychology, reasoning, and judgment and decision making claimed that the malaise is most virulent in their respective fields. A professor from a business school wrote: "I think

Originally published as Gigerenzer, G. (2010). Personal reflections on theory and psychology. *Theory and Psychology*, 20, 733–743.

the problem is that people in the social sciences have no idea how to build theory." Many letters came from distinguished professors emeritus. Young scholars appear to be busy with something else, building careers rather than theories.

When the editorial team for *Theory & Psychology* first met, we discussed what name to give the new journal. The common vision was to promote theory in a discipline where data are often the end rather than the means; some of us therefore proposed naming it "Journal of Theoretical Psychology." After all, there is a *Journal of Theoretical Biology* and an *International Journal of Theoretical Physics*; economic journals do not even need this qualifier because most economists would not consider publishing an article without formal theory. In many a discipline, those in the theoretical tower think more highly of themselves than of those in the empirical branch. That even holds for mathematics, where probability theorists tend to look down at statisticians and their "dirty hands" from dealing with the messy world. Einstein's assertion that he would reject the data before he would reject relativity theory illustrates the common trust in strong theory as opposed to noisy data (cited in Mahoney, 1979). Yet, as I recall, the title "Journal of Theoretical Psychology" encountered resistance, not only from the publisher. With such a precarious name, it would not sell. Few psychologists would open a journal with "theoretical" in the title, and even fewer would submit papers. The sad truth is that these objections were not off the mark. The final compromise was *Theory & Psychology*, which both distanced psychology from the problematic term and connected the two.[1] There appears to be something peculiar about psychology and its relation to theory. Here are some observations and proposals.

Teach Theory Construction?

The argument against teaching theory construction is that there is no single recipe for designing a theory. We had better teach experimental methodology and statistics instead, the argument continues, and hopefully theories will somehow emerge from significant results. The counterargument is that there is no single formula for methodology and statistics either. Generations of psychology students have been deceived about this fact: psychological textbooks on statistics are typically silent about different statistical theories and instead present an incoherent mishmash of Fisher, Neyman-Pearson, and Bayes as an apparent monolithic truth. The result is statistical rituals, not statistical thinking (Gigerenzer, 2004). We should not teach theory construction in the same mindless way as statistics, but as the art of scientific thinking and discovery.

1. There is of course the International Society for Theoretical Psychology, home to *Theory & Psychology*. Yet this society is only loosely connected with what one would expect to read in the *Psychological Review*, the theoretical flagship of psychology.

A curriculum on theory construction might start with good examples, first principles, and the ability to detect surrogates for theory (see below). First principles could be learned from a dip into the history of science, such as Kuhn's list of features of good theories: accuracy, consistency, broad scope, simplicity, and fruitfulness. But it should not begin and end with the history and philosophy of science, which for the most part does not deal with psychology. Rather, comparing research practices in the natural sciences with those in psychology is a first step toward getting a graduate student to think. For instance, many psychological theories—prospect theory is a prominent example—differ from those in the natural sciences in their abundant use of free parameters. Although the laws of classical physics do have parameters, the purpose is to measure these as precisely as possible rather than to fit them to each data set anew. In much of psychological research, however, the opposite is common practice: parameters are never fixed but are fitted to every new data set, with the goal to increase R^2, or some other measure of goodness of fit. The nature of parameters in a theory is only one dimension on which psychological theories differ from one another and from those of other disciplines. Others include: formal versus verbal theories, optimization versus heuristic processes, as-if models versus process models, and domain-specific versus domain-general theories.

To illustrate, the common approach in research on cognition and decision making builds formal models with many free parameters that assume some optimization processes (e.g., Bayesian or expected utility maximization); they predict behavior but do not model cognitive processes (as-if models; see Chapter 11) and are proposed as domain-general theories of cognition. In my own work on bounded rationality (Gigerenzer, 2008), I design formal models of heuristics without free parameters that reveal clearly what the model can predict and what it cannot; these models assume cognitive processes that are of a heuristic nature rather than concerned with optimizing some function (because optimization is rarely feasible in an uncertain world). That is, cognitive processes employ limited search and ignore information in order to be robust rather than "optimal." These models are process models that specify a search rule (the direction of search for information in the search space), a stopping rule (when search is stopped), and how a decision is finally made. Finally, there is not one general heuristic but an adaptive "toolbox" containing domain-specific heuristics. As a consequence, I study the "ecological rationality" of each heuristic, that is, the worlds in which they are more or less successful.

With these and other distinctions in mind, students can understand the structure of current theories and alternative ways of building theories.

From Tools to Theories

One of the most instructive ways of teaching theory construction is to investigate the origins of existing theories. The neglect of theory construction goes hand in hand with the inductive story that theories emerge from data, yet in spite of being repeatedly asserted, it is too simplistic an explanation (Holton, 1988). Discovery entails more than data and theories; it also involves scientific tools and practice. In fact, one major source of psychological theories is researchers' tools, such as statistics and computers. When psychologists grow accustomed to a new tool for data processing, the tool is repeatedly proposed as the way the mind works. I have called this principle of discovery the *tools-to-theories heuristic* (Gigerenzer, 1991). For instance, signal detection theory serves as a rich template for psychological theories, from sensory discrimination to recognition memory to eyewitness testimony in courts. It is a fruitful theory in the sense that it provides a highly useful conceptual language (hit rate, false alarm rate, decision criterion, sensitivity d′) and has been applied broadly. Originally proposed by Tanner and Swets (1954), signal detection theory did not emerge from data but from a statistical tool, the Neyman-Pearson hypotheses testing method. Tanner called his model mind a "Neyman-Pearson detector," and the origins of the theory are described in detail by Gigerenzer and Murray (1987, chap. 2). Instead of simply emerging from data, the new theory in fact changed the very nature of the data psychologists were generating. Just as in Neyman-Pearson theory, the novel data were hits and false alarms, unlike in the earlier work on sensory discrimination, from Fechner to Thurstone, which was based on measuring thresholds or psychological differences between stimuli. Here, we have a fascinating story about how a major new theory was discovered. After talking with some of today's major proponents of signal detection theory, I realized that few are aware of the origin of their own theory, with most believing that it stemmed from new experimental data. Yet the tool inspired the theory, and the theory inspired new kinds of data to be generated, which in turn were used to test the new theory.

The origin of a theory alone, however, does not tell us whether the theory is fruitful or accurate. But it does tell us something about its assumptions and possible alternatives. For instance, two quite different statistical theories of inference also turned into theories of the mind: Fisher's analysis of variance (ANOVA) and Bayes' probability updating rule. These two tools along with signal detection theory implied noticeably different pictures of the mind. Fisher's ANOVA provided the template for several cognitive theories, most prominently causal attribution theory (Kelley & Michaela, 1980). Unlike in signal detection theory, Fisher's null hypothesis testing deals with only one kind of error, and thus causal attribution theory has no decision criteria that balance hits and false alarms (Gigerenzer, 1991). Nor is causal attribution based on prior probabilities, using evidence to revise these into posterior probabilities, as suggested by Bayesian models of mind. Theories that originate from Fisherian statistics tend to picture the mind as mostly data driven, very different from earlier theories of causal perception by Piaget or Michotte. The fingerprint of the tool is both a theory's strength and its limitation. For Bayesian theories of cognition, which are presently in vogue, every mental task appears to involve computing a posterior probability distribution. The tools-to-theories process of discovery is also evident in exemplar models of categorization, where multidimensional scaling turned into a theory of mind, as well as in the analogy between the computer and the mind (Gigerenzer & Goldstein, 1996a).

By studying the features of statistical and computer tools that turned into theories of mind, students can understand the structures of current theories and alternative ways of building theories.

Surrogates for Theory

Opening students' eyes to the origins of theories and alternatives for construction of theories should set the stage to develop psychology into a more theoretical discipline. Opening their eyes to *what is not a theory* can be of equal help. In some areas of psychology, the development of theories goes backward, from extant genuine theories to surrogates. A surrogate is something that pretends to be a psychological theory. I describe three techniques for surrogates.

Circular Restatements as Explanations

The most primitive means of avoiding theories is to simply restate the phenomenon in question in different words and pretend to have offered an explanation. This technique is also known as redescription or tautology, and has a long tradition. Recall Molière's parody of the Aristotelian doctrine of substantial forms: Why does opium make us sleepy? Because of its dormative properties. Such circular restatements are not limited to attributing behavior *X* to an essence or trait *X*. A closer look at the structure of psychological explanations shows abundant uses of restatements in published research (Gigerenzer, 1996; Katzko, 2006; L. Wallach & Wallach, 1994; M.A. Wallach & Wallach, 1998). Here are examples from prominent research in the respective fields. In research on the "belief-bias effect," participants were instructed to judge the logical validity of syllogisms, such as: "No addictive things are inexpensive. Some cigarettes are inexpensive. Therefore, some addictive things are not cigarettes" (a logically invalid but believable conclusion). The observation was that judgments depended on both the logical validity and the believability of the conclusion. The explanation offered was: "Dual-process accounts propose that although participants attempt to reason logically in accord with the instructions, the influence of prior beliefs is extremely difficult to suppress and effectively competes for control of the response made" (Evans, 2003, p. 455). The phenomenon that both logical structure and prior belief influence judgments is "explained" by restating the phenomenon in other words. Dijksterhuis and van Knippenberg (1998) reported the interesting finding that priming intelligent behavior by inducing a professor stereotype resulted in more intelligent behavior. This result was explained in this way: "In concrete terms, activation of the professor stereotype is expected to result in intelligent behavior because activation of the professor stereotype leads to activation of intelligence" (p. 872). Katzko (2006) has analyzed this and other surrogate explanations in detail. There is nothing wrong about not having an explanation for how priming (writing down everything that comes to mind about a "professor") can lead to higher performance in a general knowledge task. Yet circular restatements pretend to have an explanation and thus distract from finding a model about how the specific priming task could have an effect on a general knowledge task. Restatements both create a theoretical void and cover it up.

One-Word Explanations

The second technique for avoiding theories is equally abundant. The observation that some factor *X* influences people's judgments more than factor *Y* is explained by saying that *X* is more "salient," "available," "relevant," "representative," or "vivid." One-word explanations are so perfectly flexible that they can, after the fact, account for almost every observed behavior. This technique is also known as the use of labels instead of models. To illustrate, one-word explanations can account for both phenomenon *A* and its opposite, *non-A* (see Ayton & Fischer, 2004). Consider the gambler's fallacy: after a series of *n* reds on the roulette table, the intuition is that the chance of another red *decreases*. This intuition was explained by people's reliance on "representativeness" by saying that "the occurrence of black will result in a more representative sequence than the occurrence of an additional red" (Tversky & Kahneman, 1974, p. 1125). Next consider the hot hand fallacy, which is the opposite belief: after a basketball player scores a series of *n* hits, the intuition is that the chance for another hit *increases* (see Chapter 9). This belief was also attributed to representativeness, because "even short random sequences are thought to be highly representative of their generating process" (Gilovich, Vallone, & Tversky, 1985, p. 295). No formal model of similarity ("representativeness") can predict a phenomenon and its contrary, but a label can do this by changing its meaning. To account for the gambler's fallacy, the term alludes to a higher similarity between the series of $n + 1$ outcomes and the underlying chance process, whereas to account for the hot hand fallacy, it alludes to a similarity between a series of *n* and a new observation $n + 1$ (Gigerenzer & Brighton, 2009). One-word explanations can be neither proved nor disproved, and hence do not enhance our understanding of how the mind works.

Lists of Dichotomies

A third way to avoid building theories are yin–yang lists of general dichotomies: *associative, unconscious, effortless, heuristic,* and *suboptimal* processes (assumed to foster "intuitive" judgments) versus *rule-based, conscious, effortful, analytic,* and *rational* processes (assumed to characterize "deliberate" judgments). The first list is called System 1, the second System 2, and both lists together are called dual-process theories of reasoning (e.g., Sloman, 1996; Kahneman, 2011). Today, we witness an avalanche of dual-systems or dual-process theories. In response to Sloman's original paper, Terry Regier and I (Gigerenzer & Regier, 1996) showed in some detail that much clarity is lost when one subsumes existing conceptual distinctions—such as Smolensky's intuitive processor versus rule interpreter, Hinton's intuitive versus rational processing, Schneider and Shiffrin's automatic versus controlled processes, and Freud's primary versus secondary processes—into one associative versus rule-based dichotomy, and the same holds for the other dichotomies. The problems we pointed out in our article have never been addressed or answered, nor is this article cited in current reviews on dual-process theories (e.g., Evans, 2008; Evans & Frankish, 2009). Dual-systems theories of reasoning are a surprising inversion of scientific progress: the backward development from precise theories to vague dichotomies. Consider the work on the adaptive decision maker (Payne, Bettman, & Johnson, 1993) and on the adaptive toolbox (Gigerenzer, Todd, & the ABC Research Group, 1999), which specifies a theory of heuristics, their building blocks, their ecological rationality, and the core cognitive capacities that heuristics exploit. I have seen none of this theoretical and experimental work incorporated into dual-process theories—this wealth of knowledge is now bundled together and renamed a "System 1" that is said to operate in a heuristic mode. The resulting problem with two-system theories "is the lack of any predictive power and the tendency to employ them as an after-the-fact explanation" (Keren & Schul, 2009, p. 544).

All three surrogates are obstacles to building rich and precise theories and, in fact, represent a step away from already existing theories. Moving "backward" from existing models to labels is an odd occurrence in science, which typically proceeds in the opposite direction.

Pursue Theory Integration?

In physics, the integration of extant theories is a primary goal. An example is the struggle to combine general relativity theory, which describes cosmic behavior on a large scale, with quantum theory, which describes behavior on a subatomic scale. In psychology, in contrast, the goal of bridging different theories is rarely pursued. Many years ago, Michael Watkins (1984) wrote that a cognitive theory "is a bit like someone else's toothbrush—it is fine for that individual's use, but for the rest of us . . . well, we would just rather not, thank you" (p. 86). That attitude has changed little over the last decades: 25 years later Walter Mischel (2009) repeated that many psychologists still tend to treat theories like toothbrushes—no self-respecting person wants to use anyone else's. This aversion of entering others' territory is associated with the widespread but flawed methodological practice of testing only one theory—one's own toothbrush—against data, as opposed to testing two or more theories comparatively. The toothbrush analogy may be amusing, but it is embarrassing that many researchers show little interest even in the neighboring subcommunities that work on the same topics, not to speak of other disciplines. For instance, psychologists who study intelligence using tests rarely talk to those who study thinking using experiments, and both communities avoid philosophers and economists who study rational decision making.

There are two paradigms for conducting psychological research. One is territorial, the second topical. In the first paradigm, researchers identify with a subdiscipline, such as developmental psychology or social psychology, including its methodology, publication outlets, and conferences, and ignore everything outside this professional frame. In the second paradigm, researchers identify with a topic, such as decision-making or moral behavior, and ignore disciplinary borders, using all methodologies and knowledge available to understand the topic. The first professional pathway is, in my observation, the one that most psychologists have set foot on. It is good enough for making a career but prevents psychology from ever becoming a cumulative science. The second professional pathway, less trod upon, opens the possibility for integrating different theories on the same topic.

Let me insert a personal note. Before I became director at the Max Planck Institute for Human Development, I was professor at the University of Chicago, my favorite university in the world. The reason why I left for Max Planck was the opportunity to set up an interdisciplinary, international research group that is funded long term and enables a few dozen smart researchers from different disciplines to work together to develop a theory of heuristic decision making under uncertainty. Currently, I have psychologists, mathematicians, computer scientists, behavioral economists, evolutionary biologists, philosophers, engineers, and others working together. This group has existed for 15 years to date and has been able to make major contributions, including to the study of the adaptive toolbox and ecological rationality, or what can be called the science of heuristics (Gigerenzer et al., 1999; Todd et al., 2012). Bringing researchers from different disciplines and subdisciplines together has enabled progress in both theory construction and theory integration. Examples for theory integration from our research group are the integration of ACT-R theory with the recognition heuristic model, resulting in new insights on when systematic forgetting helps to make better inferences (Schooler & Hertwig, 2005), and the integration of signal detection theory with fast-and-frugal trees, another class of heuristics, which allows the connections between the concepts in both frameworks to be understood in a way not possible before (Luan, Schooler, & Gigerenzer, 2011). None of this would have been possible if I had worked within the territorial paradigm. Last but not least, I learn something new every day from my colleagues with different backgrounds and am never bored.

Final Thought

Many researchers believe that in order to have a successful career, they need to publish as much as possible, the proliferation of so-called "least-publishable units." Yet excellent departments want to hire researchers who have made a theoretical contribution and usually discourage mere quantity by asking applicants to submit their six best papers only. This procedure generates incentives to write something better than one has written before, not just more of the same. Similarly, the editors of major journals might consider writing editorials that discourage toothbrush culture and surrogates, and explicitly inviting articles that make advances in theory integration. This will enable young researchers to combine making a career with developing psychology into a theoretical enterprise.

REFERENCES

Achter, J. A., Lubinski, D., & Benbow, C. P. (1996). Multipotentiality among intellectually gifted: It was never there and already it's vanishing. *Journal of Counseling Psychology, 43*, 65–76.

Adams, R. M. (1995). Momentum in the performance of professional tournament pocket billiards players. *International Journal of Sport Psychology, 26*, 580–587.

Aki, E. A., Oxman, A. D., Herrin, J., Vist, G. E., Terrenato, I., Sperati. F., et al. (2011). Using alternative statistical formats for presenting risks and risk reductions. *Cochrane Database of Systematic Reviews, 3*, CD006776.

Alter, A. L., & Oppenheimer, D. M. (2006). Predicting short-term stock fluctuations by using processing fluency. *Proceedings of the National Academy of Sciences of the United States of America, 103*, 9369–9372.

Altman, D. G., & Bland, J. M. (1991). Improving doctors' understanding of statistics. *Journal of the Royal Statistical Society, Series A, 154*, 223–267.

Amir, O., Ariely, D., Cooke, A., Dunning, D., Epley, N., Gneezy, U., et al. (2005). Psychology, behavioral economics, and public policy. *Marketing Letters, 16*, 443–454.

Anderson, J. R., & Lebiere, C. (1998). *The atomic components of thought.* Mahwah, NJ: Erlbaum.

Andreassen, H. K., Bujnowska-Fedak, M. M., Chronaki, C. E., Dumitru, R. C., Pudule, I., Santana, S., et al. (2007). European citizens' use of E-health services: A study of seven countries. *BMC Public Health, 7*, 53.

Andriole, G. L., Grubb, R. L., III, Buys, S. S., Chia, D., Church, T. R., Fouad, M. N., et al. (2009). Mortality results from a randomized prostate-cancer screening trial. *The New England Journal of Medicine, 360*, 1310–1319.

Anonymous. (1937). Mathematics and medicine. *The Lancet, 229*, 31.

Apotheken Umschau (2006). *Mammografie für alle—Ist das sinnvoll?* [*Mammograms for everyone: A good idea?*]. Retrieved April 8, 2008, from http://www.gesundheitpro.de/Onkologie-2-Mammografie-fuer-alle-Brustkrebs-A060425COCHP023606.html. (This article is no longer online, but an archived copy can been found at http://web.archive.org/web/20060806090802/http://www.gesundheitpro.de/Onkologie-2-Mammografie-fuer-alle-Brustkrebs-A060425COCHP023606.html.)

Appleton, D. R. (1990). What statistics should we teach medical undergraduates and graduates? *Statistics in Medicine, 9,* 1013–1021.

Ärztekammer Berlin. (2002, March 21). *Politischer Aktionismus führt zu Über- und Fehlversorgung* [*Political overreaction leads to overtreatment and mistreatment*]. Berlin: Author. (Press release).

Ashraf, N., Camerer, C. F., & Loewenstein, G. (2005). Adam Smith, behavioral economist. *Journal of Economic Perspectives, 19,* 131–145.

Astebro, T., & Elhedhli, S. (2006). The effectiveness of simple decision heuristics: Forecasting commercial success for early-stage ventures, *Management Science, 52,* 395–409.

Attneave, E. (1953). Psychological probability as a function of experienced frequency. *Journal of Experimental Psychology, 46,* 81–86.

Avon Health Foundation for Women (n.d.). *Breast health resource guide* (8th ed.). Retrieved November 4, 2014, from http://www.avonfoundation.org/assets/bccguide.pdf

Ayton, P., & Fischer, I. (2004). The hot hand fallacy and the gambler's fallacy: Two faces of subjective randomness? *Memory & Cognition, 32,* 1369–1378.

Bachmann, L. M., Gutzwiller, F. S., Puhan, M. A., Steurer, J., Steurer-Stey, C., & Gigerenzer, G. (2007). Do citizens have minimum medical knowledge? A survey. *BMC Medicine, 5,* 14.

Backlund, L. G., Bring, J., Skaner, Y., Strender, L.- E., & Montgomery, H. (2009). Improving fast and frugal in relation to regression analysis: Test of 3 models for medical decision making. *Medical Decision Making, 29,* 140–148.

Baines, C. J. (1992). Women and breast cancer: Is it really possible for the public to be well informed? *The Canadian Medical Association Journal, 142,* 2147–2148.

Bar-Eli, M., Avugos, S., & Raab, M. (2006). Twenty years of "hot hand" research. The hot hand phenomenon: Review and critique. *Psychology, Sport and Exercise, 7,* 525–553.

Baron, J., & Ritov, I. (2009). Protected values and omission bias as deontological judgments. In D. M. Bartels, C. W. Bauman, L. J. Skitka, & D. L. Medin (Eds.), *The psychology of learning and motivation (Vol. 50): Moral judgment and decision making* (pp. 133–167). San Diego, CA: Academic Press.

Barry, M. J., Fowler, F. J., Mulley, A. G., Henderson, J. V., & Wennberg, J. E. (1995). Patient reactions to a program designed to facilitate patient participation in treatment decisions for benign prostatic hyperplasia. *Medical Care, 33,* 771–782.

Bass, F. M., & Talarzyk, W. W. (1972). An attitude model for the study of brand preference. *Journal of Marketing Research, 9,* 93–96.

Baucells. M., Carrasco, J. A., & Hogarth, R. M. (2008). Cumulative dominance and heuristic performance in binary multiattribute choice. *Operations Research, 56,* 1289–1304.

Beckwith, N. E., & Lehmann, D. R. (1975). The importance of halo effects in multi-attribute attitude models. *Journal of Marketing Research, 12,* 265–275.

Beilock, S. L., Carr, T. H., MacMahon, C., & Starkes, J. L. (2002). When paying attention becomes counterproductive: Impact of divided versus skill-focused attention on novice and experienced performance of sensorimotor skills. *Journal of Experimental Psychology: Applied, 8,* 6–16.

Beisecker, A. E., & Beisecker, T. D. (1990). Patient information-seeking behaviors when communicating with physicians. *Medical Care, 28*, 19–28.

Benartzi, S., & Thaler, R. H. (2001). Naïve diversification strategies in defined contribution saving plans. *American Economic Review*, 91, 79–98.

Berg, N. (2003). Normative behavioral economics, *Journal of Socio-Economics, 32*, 411–427.

Berg, N., Biele, G., & Gigerenzer, G. (2013). *Does consistency predict accuracy of beliefs? Economists surveyed about PSA*. Working paper 1308. Dunedin, New Zealand: University of Otago Economics Department.

Berg, N., Eckel, C., & Johnson, K. (2009). *Inconsistency pays?: Time-inconsistent subjects and EU violators earn more*. Working paper. Dallas: University of Texas.

Berg, N., & Gigerenzer, G. (2007). Psychology implies paternalism?: Bounded rationality may reduce the rationale to regulate risk-taking. *Social Choice and Welfare, 28*, 337–359.

Berg, N., & Hoffrage, U. (2008). Rational ignoring with unbounded cognitive capacity. *Journal of Economic Psychology, 29*, 792–809.

Berg, N., & Lien, D. (2005). Does society benefit from investor overconfidence in the ability of financial market experts?. *Journal of Economic Behavior and Organization, 58*, 95–116.

Bergert, F. B., & Nosofsky, R., M. (2007). A response-time approach to comparing generalized rational and take-the-best models of decision making. *Journal of Experimental Psychology: Learning, Memory, and Cognition, 331*, 107–129.

Bernard, C. (1957). *An introduction to the study of experimental medicine* (H. C. Greene, Trans.). New York: Dover. (Original work published 1865).

Bernheim, B., & Rangel, A. (2005). *Behavioral public economics: Welfare and policy analysis with non-standard decision-makers*. NBER working paper no. 11518. Available at SSRN: http://ssrn.com/abstract=776006

Berwick, D. M., Fineberg, H. V., & Weinstein, M. C. (1981). When doctors meet numbers. *American Journal of Medicine, 71*, 991–998.

Beshears, J., Choi, J. J., Laibson, D., & Madrian, B. C. (2008). How are preferences revealed? *Journal of Public Economics, 92*, 1787–1794.

Bettman, J. R., & Park, C. W. (1980). Effects of prior knowledge and experience and phase of the choice process on consumer decision analysis: A protocol analysis. *Journal of Consumer Behavior, 7*, 234–248.

Biehler, R., Hofmann, T., Maxara, C., & Prömmel, A. (2006). *Fathom 2–Eine Einführung*. Heidelberg: Springer.

Binkowska, M., & Debski, R. (2005). Screening mammography in Polish female population aged 45 to 54. *Ginekologia Polska* (*Warszawa*), *76*, 871–878. (In Polish).

Binmore, K. (2009). *Rational decisions*. Princeton, NJ: Princeton University Press.

Binmore, K. G., & Samuelson, L. (1992). Evolutionary stability in repeated games played by finite automata. *Journal of Economic Theory, 57*, 278–305.

Binmore, K., & Shaked, A. (2010). *Experimental economics: Where next? Journal of Economic Behavior & Organization, 73*(1), 87–100.

Black, W. C., Nease, R. F., Jr., & Tosteson, A. N. A. (1995). Perceptions of breast cancer risk and screening effectiveness in women younger than 50 years of age. *Journal of the National Cancer Institute, 87*, 720–731.

Blume, L., Brandenburger, A., & Dekel, E. (1991). Lexicographic probability and choice under uncertainty, *Econometrica, 59*, 61–79.

Bond, M. (2009). Risk school. *Nature, 461*, 1189–1192.

Boyd, M. (2001). On ignorance, intuition and investing: A bear market test of the recognition heuristic. *Journal of Psychology and Financial Markets, 2*, 150–156.

Braddock, C. H., Edwards, K. A., Hasenberg, N. M., Laidley, T. L., & Levinson, W. (1999). Informed decision making in outpatient practice: Time to get back to basics. *Journal of the American Medical Association, 282*, 2313–2320.

Bramwell, R., West, H., & Salmon, P. (2006). Health professionals' and service users' interpretation of screening test results: Experimental study. *British Medical Journal, 333*, 284–286.

Brandstätter, E., Gigerenzer, G., & Hertwig, R. (2006). The priority heuristic: Making choices without trade-offs. *Psychological Review, 113*, 409–432.

Brandstätter, E., Gigerenzer, G., & Hertwig, R. (2008). Risky choice with heuristics: Reply to Birnbaum (2008), Johnson, Schulte-Mecklenbeck, and Willemsen (2008), and Rieger and Wang (2008). *Psychological Review, 115*, 281–290.

Brase, G. L. (2002). Ecological and evolutionary validity: Comments on Johnson-Laird, Legrenzi, Girotto, Legrenzi, & Caverni's (1999) mental model theory of extensional reasoning. *Psychological Review, 109*, 722–728.

Brase, G. L. (2008). Frequency interpretation of ambiguous statistical information facilitates Bayesian reasoning. *Psychonomic Bulletin & Review, 15*, 284–289.

BreastScreen Australia. (n.d.). *BreastScreen and you*. Retrieved November 4, 2014, from http://www.cancerscreening.gov.au/internet/screening/publishing.nsf/Content/A2DAE190D4675FC6CA257C07007E348C/$File/Breastscreen_Brochure_March_WEB.pdf

Bright, G. W., & Friel, S. N. (1998). Graphical representations: Helping students interpret data. In S. P. Lajoie (Ed.), *Reflections on statistics: Learning, teaching, and assessment in grades K–12* (pp. 63–88). Mahwah, NJ: Lawrence Erlbaum.

Brighton, H. (2006). Robust inference with simple cognitive models. In C. Lebiere & B. Wray (Eds.), *Between a rock and a hard place: Cognitive science principles meet AI-hard problems* (pp. 17–22). Papers from the AAAI Spring Symposium (AAAI technical report no. SS-06-03. Menlo Park, CA: AAAI Press.

Brighton, H., & Gigerenzer, G. (2008). Bayesian brains and cognitive mechanisms: Harmony or dissonance? In N. Chater & M. Oaksford (Eds.), *The probabilistic mind: Prospects for Bayesian cognitive science* (pp. 189–208). New York: Oxford University Press.

Brighton, H., & Gigerenzer, G. (2011). Towards competitive instead of biased testing of heuristics: A reply to Hilbig & Richter (2010). *Topics in Cognitive Science, 5*, 197–205.

Bröder, A. (2003). Decision making with the "adaptive toolbox": Influence of environmental structure, intelligence, and working memory load. *Journal of Experimental Psychology: Learning, Memory, and Cognition, 29*, 611–625.

Bröder, A., & Eichler, A. (2006). The use of recognition information and additional cues in inferences from memory. *Acta Psychologica. 121*, 275–284.

Bröder, A., & Gaissmaier, W. (2007). Sequential processing of cues in memory-based multi-attribute decisions. *Psychonomic Bulletin Review, 14*, 895–900.

Bröder, A., & Newell, B. R. (2008). Challenging some common beliefs: Empirical work within the adaptive toolbox metaphor. *Judgment and Decision Making, 3*, 205–214.

Bröder, A., & Schiffer, S. (2003). Take the best versus simultaneous feature matching: Probabilistic inferences from memory and effects of representation format. *Journal of Experimental Psychology: General, 132*, 277–293.

Bröder, A., & Schiffer, S. (2006). Stimulus format and working memory in fast and frugal strategy selection. *Journal of Behavioral Decision Making, 19*, 361–380.

Bröder, A., & Schütz, J. (2009). Recognition ROCs are curvilinear—or are they? On premature arguments against the two-high-threshold model of recognition. *Journal of Experimental Psychology: Learning, Memory, and Cognition, 35*, 587–606.

Broadbent, E., Petrie, K. J., Ellis, C. J., Anderson, J., Gamble, G., Anderson, D., et al. (2006). Patients with acute myocardial infarction have an inaccurate understanding of their risk of a future cardiac event. *Internal Medicine Journal, 36*, 643–647.

Bruni, L., & Porta, P. L. (2007). *Handbook on the economics of happiness.* Northhampton, MA: Elgar.

Bruni, L., & Sugden, R. (2007). The road not taken: How psychology was removed from economics, and how it might be brought back. *Economic Journal, 117*, 146–173.

Bundesministerium für Gesundheit (2002a, March 23). *Einführung von Mammographie Screening: Unberechtigte Kritik der Ärztekammer Berlin* [*Implementation of mammography screening: Unjustified criticism from the Berlin Chamber of Physicians*]. Press release. Berlin: Author.

Bundesministerium für Gesundheit (2002b, September 24). *Ulla Schmidt: Neue Schritte zur Qualitätssicherung bei Brustkrebs* [*New steps towards quality control with breast cancer*]. Press release. Berlin: Author.

Burns, B. D. (2001). The hot hand in basketball: Fallacy or adaptive thinking? In J. D. Moore & K. Stenning (Eds.), *Proceedings of the Cognitive Science Society* (pp. 152–157). Hillsdale, NJ: Lawrence Erlbaum.

Burns, B. D. (2004). Heuristics as beliefs and as behaviors: The adaptiveness of the "hot hand." *Cognitive Psychology, 48*, 295–331.

Burns, B. D., & Corpus, B. (2004). Randomness and inductions from streaks: "Gambler's fallacy" versus "hot hand." *Psychonomic Bulletin & Review, 11*, 179–184.

Butterworth, B. (1999). *What counts: How every brain is hardwired for math.* New York: Free Press.

Camerer, C. F. (1989). Does the basketball market believe in the "hot hand?" *American Economic Review, 79*, 1257–1261.

Camerer, C. (1999). Behavioral economics: Reunifying psychology and economics. *Proceedings of the National Academy of Sciences of the United States of America, 96*, 10575–10577.

Camerer, C. (2003). *Behavioral game theory: Experiments in strategic interaction.* Princeton, NJ: Princeton University Press.

Canadian Task Force on Preventive Health Care (n.d.). *Should I be screened with mammography for breast cancer?* Retrieved November 4, 2014, from http://canadiantaskforce.ca/perch/resources/breast-cancer-risks-benefits-40-49.pdf

Carlson, K. A., & Shu, S. B. (2007). The rule of three: How the third event signals the emergence of a streak. *Organizational Behavior and Human Decision Processes, 104*, 113–121.

Carpenter, J., & Seki, E. (2006). Competitive work environments and social preferences: Field experimental evidence from a Japanese fishing community. *Contributions to Economic Analysis & Policy Berkeley Electronic Press, 5*,(2), 1–25.

Caruso, E. M., & Epley, N. (2004). *Hot hands and cool machines: Perceived intentionality in the prediction of streaks.* Presented at the 5th annual meeting of the Society for Personality and Social Psychology (January 2004), Austin, Texas.

Casscells, W., Schoenberger, A., & Grayboys, T. (1978). Interpretation by physicians of clinical laboratory results. *The New England Journal of Medicine, 299*, 999–1000.

Cassels, A., Hughes, M. A., Cole, C., Mintzes, B., Lexchin, J., & McCormack, J. P. (2003). Drugs in the news: An analysis of Canadian newspaper coverage of new prescription drugs. *Canadian Medical Association Journal, 168*, 1133–1137.

Castaneda, J., & Rodrigo, M. J. (1998). Developmental effects of the content of visually presented base rates. *Current Psychology of Cognition, 3*, 555–576.

Center for the Evaluative Clinical Sciences Staff (Ed.). (1996). *The Dartmouth atlas of health care.* Chicago, IL: American Hospital Association.

Centers for Disease Control and Prevention. (2001). *HIV prevalence among selected populations: Low-risk populations: National serosurveillance 1993–1997.* Atlanta, GA: Atlanta Centers for Disease Control and Prevention.

Centers for Disease Control and Prevention. (2011, June 29). *Unintended pregnancy: Contraception.* Retrieved December 11, 2011, from http://www.cdc.gov/reproductivehealth/UnintendedPregnancy/Contraception.htm. (This article is no longer online, but an archived copy can be found at http://web.archive.org/web/20111216034315/http://www.cdc.gov/reproductivehealth/unintendedpregnancy/contraception.htm.)

Charles, C. A., Gafni, A., & Whelan, T. (1997). Shared decision-making in the medical encounter: What does it mean? (or, it takes at least two to tango). *Social Science and Medicine, 44*, 681–692.

Chater, N., Oaksford, M., Nakisa, R., & Redington, M. (2003). Fast, frugal and rational: How rational norms explain behavior. *Organizational Behavior and Human Decision Processes, 90*, 63–86.

Chemers, M. M. (2001). Leadership effectiveness: An integrative review. In M. A. Hogg & R. S. Tindale (Eds.), *Blackwell handbook of social psychology: Group processes* (pp. 376–399). Oxford, UK: Blackwell.

Clark, R. D. (2003). An analysis of streaky performance on the LPGA tour. *Perceptual and Motor Skills, 97*, 365–370.

Coates, S. L., Butler, L., T., & Berry, D. C. (2004). Implicit memory: A prime example for brand consideration and choice. *Applied Cognitive Psychology, 18*, 1195–1211.

Cohen, J. (1988). *Statistical power analysis for the behavioral* sciences (2nd ed.). Hillsdale, NJ: Lawrence Erlbaum.

Cohen, J., & Dickens, W. (2002). A foundation for behavioral economics. *American Economic Review, 92*, 335–338.

Coleman, W. (1987). Experimental physiology and statistical inference: The therapeutic trial in nineteenth-century Germany. In L. Krüger, G. Gigerenzer, & M. S. Morgan (Eds.), *The probabilistic revolution: Vol. II: Ideas in the Sciences* (pp. 201–226). Cambridge, MA: MIT Press.

Collins, E. D., Kerrigan, C. L., & Anglade, P. (1999). Surgical treatment of early breast cancer: What would surgeons choose for themselves? *Effective Clinical Practice, July/August*, 149–151.

Compte, O., & Postlewaite, A. (2004). Confidence-enhanced performance. *The American Economic Review, 94*, 1536–1557.

Concato, J., Wells, C. K., Horwitz, R. I., Penson, D., Fincke, G., Berlowitz, D. R., et al. (2006). The effectiveness of screening for prostate cancer: A nested case–control study. *Archives of Internal Medicine, 166*, 38–43.

Conlisk, J. (1996). Why bounded rationality? *Journal of Economic Literature, 34*, 669–700.

Cook, L. (2001). The world trade center attack: The paramedic response: An insider's view. *Critical Care*, 5, 301–303.

Cosmides, L., & Tooby, J. (1996). Are humans good intuitive statisticians after all? Rethinking some conclusions from the literature on judgment under uncertainty. *Cognition, 58*, 1–73.

Cosmides, L., & Tooby, J. (2006). Evolutionary psychology, moral heuristics, and the law. In G. Gigerenzer & C. Engel (Eds.), *Heuristics and the law (Dahlem workshop report 94)* (pp. 175–205). Cambridge, MA: MIT Press.

Council of Europe. (2010). *Parity democracy: A far cry from reality.* Strasbourg: Gender Equality Division, Directorate General of Human Rights and Legal Affairs. Retrieved October 30, 2014, from http://www.coe.int/t/DGHL/STANDARDSETTING/EQUALITY/03themes/women-decisionmaking/CDEG%20%282009%2917_en_corrected.pdf.

Covey, J. (2007). A meta-analysis of the effects of presenting treatment benefits in different formats. *Medical Decision Making, 27*, 638–654.

Croskerry, P. (2009). A universal model of diagnostic reasoning. *Academic Medicine, 84*, 1022–1028.

Cross, F. R., & Jackson, R. J. (2005). Spider heuristics. *Behavioural Processes, 69*, 125–127.

Crothers, T. (1998). Texas tornadoes. *Sports Illustrated, 88*, 98–101.

Cumming, G. (2008). Replication and p intervals: p values predict the future only vaguely, but confidence intervals do much better. *Perspectives on Psychological Science, 3*, 286–300.

Czerlinski, J., Gigerenzer, G., & Goldstein, D. G. (1999). How good are simple heuristics? In G. Gigerenzer, P. M. Todd, & the ABC Research Group, *Simple heuristics that make us smart* (pp. 97–118). New York: Oxford University Press

Czienskowski, U. (2002). *Hot-hand Program C++*. Berlin: Max Planck Institute for Human Development.

Daston, L. J. (1988). *Classical probability in the Enlightenment.* Princeton, NJ: Princeton University Press.

Dawes, R. M. (1979). The robust beauty of improper linear models in decision making. *American Psychologist, 34*, 571–582.

Dawes, R. M., & Corrigan, B. (1974). Linear models in decision making. *Psychological Bulletin, 81*, 95–106.

Dekel, E. (1999). On the evolution of attitudes toward risk in winner-take-all games. *Journal of Economic Theory, 87*, 125–143.

De Leeuw, E. D., & van der Zouwen, J. (1988). Data quality in telephone and face to face surveys: A comparative meta-analysis. In R. M. Groves, P. N. Biemer, L. E. Lyberg, J. T. Massey, W. L. Nichols, II, & J. Waksberg (Eds.), *Telephone survey methodology* (pp. 283–299). New York: Wiley.

DeMiguel, V., Garlappi, L., & Uppal, R. (2009). Optimal versus naive diversification: How inefficient is the 1/N portfolio strategy? *Review of Financial Studies, 22*, 1915–1953.

Dhami, M. K. (2003). Psychological models of professional decision making. *Psychological Science, 14*, 175–180.

Dhami, M. K., & Ayton, P. (2001). Bailing and jailing the fast and frugal way. *Journal of Behavioral Decision Making, 14*, 141–168.

Dhami, M. K., & Harries, C. (2001). Fast and frugal versus regression models of human judgment. *Thinking & Reasoning, 7*, 5–27.

Diamond, P. (2008). Behavioral economics. *Journal of Public Economics, 92*, 1858–1862.

Dieckmann, A., & Rieskamp, J. (2007). The influence of information redundancy on probabilistic inferences. *Memory & Cognition, 35*, 1801–1813.

Diekman, A. B., & Eagly, A. H. (2000). Stereotypes as dynamic constructs: Women and men of the past, present, and future. *Personality and Social Psychology Bulletin, 26*, 1171–1188.

Dijksterhuis, A., & van Knippenberg, A. (1998). The relation between perception and behavior or how to win a game of Trivial Pursuit. *Journal of Personality and Social Psychology, 74*, 865–877.

Dobbs, M. (2007, October 30). Rudy wrong on cancer survival chances. *Washington Post.* Retrieved October 30, 2014, from http://voices.washingtonpost.com/fact-checker/2007/10/rudy_miscalculates_cancer_surv.html.

Domenighetti, G., D'Avanzo, B., Egger, M., Berrino, F., Perneger, T., Mosconi, P., et al. (2003). Women's perception of the benefits of mammography screening: Population-based survey in four countries. *International Journal of Epidemiology, 32*, 816–821.

Dorsey-Palmateer, R., & Smith, G. (2004). Bowlers' hot hands. *The American Statistician, 58*, 38–45.

Dougherty, M. R., Franco-Watkins, A., & Thomas, R. P. (2008). Psychological plausibility of the theory of probabilistic mental models and the fast and frugal heuristics. *Psychological Review, 115*, 199–213.

Eagly, A. H. (1987). *Sex differences in social behavior: A social-role interpretation.* Hillsdale, NJ: Lawrence Erlbaum.

Eagly, A. H., Karau, S. J., & Makhijani, M. G. (1995). Gender and the effectiveness of leaders: A meta-analysis. *Psychological Bulletin, 117*, 125–145.

Eagly, A. H., Wood, W., & Diekman, A. B. (2000). Social role theory of sex differences and similarities: A current appraisal. In T. Eckes & H. M. Trautner (Eds.), *The developmental social psychology of gender* (pp. 123–174). Mahwah, NJ: Lawrence Erlbaum.

Eddy, D. M. (1982). Probabilistic reasoning in clinical medicine: Problems and opportunities. In D. Kahneman, P. Slovic, & A. Tversky (Eds.), *Judgment under uncertainty: Heuristics and biases* (pp. 249–267). Cambridge: Cambridge University Press.

Eddy, D. M. (1996). *Clinical decision making: From theory to practice: A collection of essays from the Journal of the American Medical Association.* Boston, MA: Jones and Bartlett.

Edwards, A., Elwyn, G. J., Covey, J., Mathews, E., & Pill, R. (2001). Presenting risk information: A review of the effects of "framing" and other manipulations on patient outcomes. *Journal of Health Communication, 6*, 61–82.

Egger, M., Bartlett, C., & Juni, P. (2001). Are randomised controlled trials in the BMJ different? *British Medical Journal, 323*, 1253.

Einhorn, H. J., & Hogarth, R. M. (1975). Unit weighting schemes for decision making. *Organizational Behavior and Human Performance, 13*, 171–192.

Elmore, J., & Gigerenzer, G. (2005). Benign breast disease—The risks of communicating risk. *The New England Journal of Medicine, 353*, 297–299.

Elmore, J. G., Barton, M. B., Moceri, V. M., Polk, S., Arena, P. J., & Fletcher, S. W. (1998). Ten-year risk of false positive screening mammograms and clinical breast examinations. *The New England Journal of Medicine, 338*, 1089–1096.

Elstein, A. S. (1999). Heuristics and biases: Selected errors in clinical reasoning. *Academic Medicine, 74*, 791–794.

Elwyn, G. J., Edwards, A., Eccles, M., & Rovner, D. (2001). Decision analysis in patient care. *The Lancet, 358*, 571–574.

Epstein, S., Pacini, R., Denes-Raj, V., & Heier, H. (1996). Individual differences in intuitive–experiential and analytical–rational thinking styles. *Journal of Personality and Social Psychology, 71*, 390–405.

Erev, I., Roth, A. E., Slonim, R. L., & Barron, G. (2002). *Combining a theoretical prediction with experimental evidence to yield a new prediction: An experimental design with a random sample of tasks.* Technion, Haifa, Israel: Columbia University and Faculty of Industrial Engineering and Management.

Ericsson, K. A., Prietula, M. J., & Cokely, E. T. (2007). The making of an expert. *Harvard Business Review, 85*, 114–121.

Estrada, C., Barnes, V., Collins, C., & Byrd, J. C. (1999). Health literacy and numeracy. *Journal of the American Medical Association, 282*, 527.

European Commission. (2009). *She figures 2009: Statistics and indicators on gender equality in science*. Brussels: Directorate-General for Research.

Eurostat. (2006). *A statistical view of the life of women and men in the EU25*. Luxembourg: Eurostat Press Office.

Eurostat. (2008a). *Childcare services in the EU: Memo 08/592*. Brussels: Commission of the European Communities.

Eurostat. (2008b). *The life of women and men in Europe: A statistical portrait*. Luxembourg: Office for the Official Publications of the European Communities.

Evans, J. S. (2003). In two minds: Dual process accounts of reasoning. *Trends in Cognitive Sciences, 7*, 454–459.

Evans, J. S. (2008). Dual-processing accounts of reasoning, judgment and social cognition. *Annual Review of Psychology, 59*, 255–278.

Evans, J. S., & Frankish, K. (2009). *In two minds: Dual processes and beyond*. New York: Oxford University Press.

Fagerlin, A., Ubel, P. A., Smith, D. M., & Zikmund-Fisher, B. J. (2007). Making numbers matter: Present and future research in risk communication. *American Journal of Health Behavior, 31*, 47–56.

Fahey, T., Griffiths, S., & Peters, T. J. (1995). Evidence based purchasing: Understanding results of clinical trials and systematic reviews. *British Medical Journal, 311*, 1056–1059.

Falk, R., & Konold, C. (1992). The psychology of learning probability. In F. S. Gordon & S. P. Gordon (Eds.), *Statistics for the twenty-first century* (pp. 151–164). Washington, DC: The Mathematical Association of America.

Federman, D. G., Goyal, S., Kamina, A., Peduzzi, P., & Concato, J. (1999). Informed consent for PSA screening: Does it happen? *Effective Clinical Practise, 2*, 152–157.

Fehr, E., & Schmidt, K. (1999). A theory of fairness, competition and cooperation. *Quarterly Journal of Economics, 114*, 817–868.

Felix Burda Stiftung. (2008, February 19). *Felix Burda Stiftung startet neue Medien-Kampagne* [*Felix Burda Foundation starts new media campaign*]. Press release. Retrieved March 2008 from http://www.felix-burda-stiftung.de/presseportal/felix-burda-stiftung/pressemeldungen/2008/index.php

Fiedler, K. (1996). Explaining and simulating judgment biases as an aggregation phenomenon in probabilistic, multiple-cue environments. *Psychological Review, 103*, 193–214.

Finzer, B., & Erickson, T. (2006). *Fathom*. Emeryville, CA: Key Curriculum Press.

Fischer, J. E., Steiner, F., Zucol, F., Berger, C., Martignon, L., Bossart, W., et al. (2002). Use of simple heuristics to target macrolide prescription in children with community-acquired pneumonia. *Archives of Pediatrics and Adolescent Medicine, 156*, 1005–1008.

Fischer, K., & Jungermann, H. (1996). Rarely occurring headaches and rarely occurring blindness: Is rarely = rarely? The meaning of verbal frequentistic labels in specific medical contexts. *Journal of Behavioral Decision Making, 9*, 153–172.

Fishbein, M. (1967). A consideration of beliefs and their role in attitude measurement. In M. Fishbein (Ed.), *Readings in attitude theory and measurement* (pp. 389–400). New York: Wiley.

Fishburn, P. C. (1974). Lexicographic orders, utilities and decision rules: A survey. *Management Science, 20*, 1442–1471.

Fiske, A. P., & Tetlock, P. E. (1997). Taboo trade-offs: Reactions to transactions that transgress the spheres of justice. *Political Psychology, 18*, 255–297.

Fletcher, S. W. (1997). Whither scientific deliberation in health policy recommendations? Alice in the Wonderland of breast-cancer screening. *The New England Journal of Medicine, 336*, 1180–1183.

Folkman, J., & Kalluri, R. (2004). Cancer without disease. *Nature, 427*, 787.

Ford, J. K., Schmitt, N., Schechtman, S. L., Hults, B. H., & Doherty, M. L. (1989). Process tracing methods: Contributions, problems, and neglected research questions. *Organizational Behavior and Decision Processes, 43*, 75–117.

Frank, R. H. (1991). *Microeconomics and behavior*. New York: McGraw-Hill.

Frank, R. H. (2008). Lessons from behavioral economics: Interview with Robert Frank. *Challenge, 51*, 80–92.

Franklin, B. (1987). *Writings*. New York: The Library of America. (Originally published in 1789).

Friedman, M. (1953). *Essays in positive economics*. Chicago, IL: University of Chicago Press.

Frome, P. M., & Eccles, J. S. (1998). Parents' influence on children's achievement-related perceptions. *Journal of Personality and Social Psychology, 74*, 435–452.

Frosch, C., Beaman, C. P., & McCloy, R. (2007). A little learning is a dangerous thing: An experimental demonstration of ignorance-driven inference. *Quarterly Journal of Experimental Psychology, 60*, 1329–1336.

Funder, D. C. (1987). Errors and mistakes: Evaluating the accuracy of social judgment. *Psychological Bulletin, 101*, 75–90.

Furedi, A. (1999). The public health implications of the 1995 "pill scare." *Human Reproduction Update, 5*, 621–626.

Gaissmaier W., & Marewski, J. (2011). Forecasting elections with mere recognition from lousy samples: A comparison of collective recognition, wisdom of crowds, and representative polls. *Judgment and Decision Making, 6*, 73–88.

Gaissmaier, W., & Schooler, L. J. (2008). The smart potential behind probability matching. *Cognition, 109*, 416–422.

Galbraith, J. K. (1938). Rational and irrational consumer preference. *The Economic Journal, 48*, 336–342.

Galef, B. G., Jr. (1987). Social influences on the identification of toxic food by Norway rats. *Animal Learning & Behavior, 15*, 327–332.

Galef, B. G., Jr., McQuoid, L. M., & Whiskin, E. E. (1990). Further evidence that Norway rats do not socially transmit learned aversions to toxic baits. *Animal Learning and Behavior, 18*, 199–205.

Galesic, M., Garcia-Retamero, R., & Gigerenzer, G. (2009). Using icon arrays to communicate medical risks: Overcoming low numeracy. *Health Psychology, 28*, 210–216.

Gallistel, C. R. (1993). Organization of learning. Cambridge, MA: MIT Press.

Galton, F. (1907). Vox populi. *Nature, 75*, 7.

García-Retamero, R., & Dhami, M. K. (2009). Take-the-best in expert–novice decision strategies for residential burglary. *Psychonomic Bulletin & Review, 16*, 163–169.

Garcia-Retamero, R., & López-Zafra, E. (2006a). Congruency between leadership and gender roles: Notes on causal attributions about success and failure. *Revista Latinoamericana de Psicología, 38*, 245–257.

Garcia-Retamero, R., & López-Zafra, E. (2006b). Prejudice against women in male-congenial environments: Perceptions of gender role congruity in leadership. *Sex Roles, 55*, 51–61.

Garcia-Retamero, R., & López-Zafra, E. (2008). Causal attributions about success and failure, and perceptions of leadership in women. *Estudios de Psicología, 15*, 273–287.

Garcia-Retamero, R., Müller, S. M., & López-Zafra, E. (2011). The malleability of gender stereotypes: Influence of population size on perceptions of men and women in the past, present, and future. *Journal of Social Psychology, 151*, 635–656.

García-Retamero, R., Takezawa, M., & Gigerenzer, G. (2009). Does imitation benefit cue order learning? *Experimental Psychology, 56*, 307–320.

Garfield, J., & Ben-Zvi, D. (2007). How students learn statistics revisited: A current review of research on teaching and learning statistics. *International Statistical Review, 75*, 372–396.

Gaskin, S., Evgeniou, T., Bailiff, D., & Hauser, J. (2007). Two-stage models: Identifying non-compensatory heuristics for the consideration set then adaptive polyhedral methods within the consideration set. *Proceedings of the Sawtooth Software Conference, 13*, 67–83.

Gelman, R., & Gallistel, C. R. (1978). *The child's understanding of number.* Cambridge, MA: Harvard University Press.

Geman, S. E., Bienenstock, E., & Doursat, R. (1992). Neural networks and the bias/variance dilemma. *Neural Computation, 4*, 1–58.

General Medical Council. (1998). *Seeking patients' consent: The ethical considerations.* London, UK: Author.

GfK-Nürnberg & Frank, R. (2007). *European consumer study 2007 Gesundheit in Europa.* Nuremberg, Germany: Author.

Ghosh, A. K., & Ghosh, K. (2005). Translating evidence-based information into effective risk communication: Current challenges and opportunities. *Journal of Laboratory and Clinical Medicine, 145*, 171–180.

Gigerenzer, G. (1991). From tools to theories: A heuristic of discovery in cognitive psychology. *Psychological Review, 98*, 254–267.

Gigerenzer, G. (1996). On narrow norms and vague heuristics: A reply to Kahneman and Tversky (1996). *Psychological Review, 103*, 592–596.

Gigerenzer, G. (2000). *Adaptive thinking: Rationality in the real world.* New York: Oxford University Press.

Gigerenzer, G. (2002). *Calculated risks: How to know when numbers deceive you.* New York: Simon & Schuster. (UK version: *Reckoning with risk: Learning to live with uncertainty.* London: Penguin.)

Gigerenzer, G. (2003). Why does framing influence judgment? *Journal of General Internal Medicine, 18*, 960–961.

Gigerenzer, G. (2004). Mindless statistics. *Journal of Socio-Economics, 33*, 587–606.

Gigerenzer, G. (2007). *Gut feelings: The intelligence of the unconscious.* New York: Viking. (UK version: London, UK: Allen Lane/Penguin).

Gigerenzer, G. (2008). *Rationality for mortals: How people cope with uncertainty.* New York: Oxford University Press.

Gigerenzer, G. (2009). Surrogates for theory. *Association for Psychological Science Observer, 22*, 21–23.

Gigerenzer, G. (2010). Moral satisficing: Rethinking morality as bounded rationality. *Topics in Cognitive Science, 2*, 528–554.

Gigerenzer, G. (2014). *Risk savvy: How to make good decisions.* New York: Viking.

Gigerenzer, G., & Brighton, H. (2009). *Homo heuristicus*: Why biased minds make better inferences. *Topics in Cognitive Science, 1*, 107–143.

Gigerenzer, G., & Edwards, A. G. K. (2003). Simple tools for understanding risks: From innumeracy to insight. *British Medical Journal, 327*, 741–744.

Gigerenzer, G., & Engel, C. (2006). *Heuristics and the law.* Cambridge, MA: MIT Press.

Gigerenzer, G., Fiedler, K., & Olsson, H. (2012). Rethinking cognitive biases as environmental consequences. In P. M. Todd, G. Gigerenzer, & the ABC Research Group, *Ecological rationality: Intelligence in the world* (pp. 80–110). New York: Oxford University Press.

Gigerenzer, G., Gaissmaier, W., Kurz-Milcke, E., Schwartz, L. M., & Woloshin, S. (2007). Helping doctors and patients to make sense of health statistics. *Psychological Science in the Public Interest, 8*, 53–96.

Gigerenzer, G., & Goldstein, D. G. (1996a). Mind as computer: Birth of a metaphor. *Creativity Research Journal, 9*, 131–144.

Gigerenzer, G., & Goldstein, D. G. (1996b). Reasoning the fast and frugal way: Models of bounded rationality. *Psychological Review, 103*, 650–669.

Gigerenzer, G., Hell, W., & Blank, H. (1988). Presentation and content: The use of base rates as a continuous variable. *Journal of Experimental Psychology: Human Perception and Performance, 14*, 513–525.

Gigerenzer, G., Hertwig, R., & Pachur, T. (Eds.). (2011). *Heuristics: The foundations of adaptive behavior.* New York: Oxford University Press.

Gigerenzer, G., Hertwig, R., van den Broek, E., Fasolo, B., & Katsikopoulos, K. V. (2005). "A 30% chance of rain tomorrow": How does the public understand probabilistic weather forecasts? *Risk Analysis, 25*, 623–629.

Gigerenzer, G., & Hoffrage, U. (1995). How to improve Bayesian reasoning without instruction: Frequency formats. *Psychological Review, 102*, 684–704.

Gigerenzer, G., & Hoffrage, U. (1999). Overcoming difficulties in Bayesian reasoning: A reply to Lewis & Keren and Mellers & McGraw. *Psychological Review, 106*, 425–430.

Gigerenzer, G., Hoffrage, U., & Ebert, A. (1998). AIDS counselling for low-risk clients. *AIDS Care, 10*, 197–211.

Gigerenzer, G., Hoffrage, U., & Goldstein, D. G. (2008). Fast and frugal heuristics are plausible models of cognition: Reply to Dougherty, Franco-Watkins, and Thomas. *Psychological Review, 115*, 230–239.

Gigerenzer, G., Hoffrage, U., & Kleinbölting, H. (1991). Probabilistic mental models: A Brunswikian theory of confidence. *Psychological Review, 98*, 506–528.

Gigerenzer, G., Mata, J., & Frank, R. (2009). Public knowledge of benefits of breast and prostate cancer screening in Europe. *Journal of the National Cancer Institute, 101*, 1216–1220.

Gigerenzer, G., & Muir Gray, J. A. (Eds.). (2011). *Better doctors, better patients, better decisions: Envisioning health care 2020.* Cambridge, MA: MIT Press.

Gigerenzer, G., & Murray, D. J. (1987). *Cognition as intuitive statistics.* Hillsdale, NJ: Erlbaum.

Gigerenzer, G., & Regier, T. (1996). How do we tell an association from a rule? Comment on Sloman (1996). *Psychological Bulletin, 119*, 23–26.

Gigerenzer, G., & Richter, H. R. (1990). Context effects and their interaction with development: Area judgments. *Cognitive Development, 5*, 235–264.

Gigerenzer, G., & Selten, R. (2001). *Bounded rationality: The adaptive toolbox.* Cambridge, MA: MIT Press.

Gigerenzer, G., Swijtink, Z., Porter, T., Daston, L., Beatty, J., & Krüger, L. (1989). *The empire of chance: How probability changed science and everyday life.* Cambridge: Cambridge University Press.

Gigerenzer, G., Todd, P. M., & the ABC Research Group. (1999). *Simple heuristics that make us smart.* Oxford: Oxford University Press.

Gilboa, I., Postlewaite, A., & Schmeidler, D. (2004). *Rationality of belief, or: Why Bayesianism is neither necessary nor sufficient for rationality.* Cowles Foundation discussion papers 1484. New Haven, CT: Cowles Foundation, Yale University.

Gilbride, T. J., & Allenby, G. M. (2004). A choice model with conjunctive, disjunctive, and compensatory screening rules. *Marketing Science, 23*, 391–406.

Gilden, D. L., & Wilson, S. A. (1995a). On the nature of streaks in signal detection. *Cognitive Psychology, 28*, 17–64.

Gilden, D. L., & Wilson, S. A. (1995b). Streaks in skilled performance. *Psychonomic Bulletin & Review, 2*, 260–265.

Gilovich, T. (1993). *How we know what isn't so: The fallibility of human reason in everyday life.* New York: Free Press.

Gilovich, T., Vallone, R., & Tversky, A. (1985). The hot hand in basketball: On the misperception of random sequences. *Cognitive Psychology, 17*, 295–314.

Glöckner, A., Betsch, T., & Schindler, N. (2010). Coherence shifts in probabilistic inference tasks. *Journal of Behavioral Decision Making, 23*, 439–462.

Glöckner, A., & Bröder, A. (2011). Processing of recognition information and additional cues: A model-based analysis of choice, confidence, and response time. *Judgment and Decision Making, 6*, 23–42.

Gnanadesikan, M., Scheaffer, R. L., & Swift, J. (1987). *The arts and techniques of simulation.* Palo Alto, CA: Dale Seymour Publications.

Goldsmith, E., & Goldsmith, R. E. (1997). Gender differences in perceived and real knowledge of financial investments. *Psychological Reports, 80*, 236–238.

Goldsmith, R. E., Goldsmith, E., & Heaney, J. G. (1997). Sex differences in financial knowledge: A replication and extension. *Psychological Reports, 81*, 1169–1170.

Goldstein, D. G. (1997). *Models of bounded rationality for inference.* (Doctoral dissertation, The University of Chicago, 1997). *Dissertation Abstracts International, 58*, 435B.

Goldstein, D. G., & Gigerenzer, G. (1999). The recognition heuristic: How ignorance makes us smart. In G. Gigerenzer, P. M. Todd, & the ABC Research Group, *Simple heuristics that make us smart* (pp. 37–58). New York: Oxford University Press.

Goldstein, D. G., & Gigerenzer, G. (2002). Models of ecological rationality: The recognition heuristic. *Psychological Review, 109*, 75–90.

Goldstein, D. G., & Gigerenzer, G. (2009). Fast and frugal forecasting. *International Journal of Forecasting, 25*, 760–772.

Goldstein, D. G., Gigerenzer, G., Hogarth, R. M., Kacelnik, A., Kareev, Y., Klein, G., et al. (2001). Group report: Why and when do simple heuristics work? In G. Gigerenzer & R. Selten (Eds.), *Bounded rationality: The adaptive toolbox* (pp. 173–190). Cambridge, MA: MIT Press.

Good, I. J. (1995). When batterer turns murderer. *Nature, 375*, 541.

Gøtzsche, P. C., & Jørgensen, K. J. (2013). Screening for breast cancer with mammography. *Cochrane Database of Systematic Reviews, 6*, CD001877.

Gould, D., Tammen, V., Murphy, S., & May, J. (1991). An evaluation of U.S. Olympic sport psychology consultant effectiveness. *The Sport Psychologist, 5*, 111–127.

Gould, S. J. (1992). *Bully for brontosaurus: Further reflections in natural history*. New York: Penguin Books.

Graefe, A., & Armstrong, J. S. (2013). Forecasting elections from voters' perceptions of candidates' ability to handle issues. *Journal of Behavioral Decision Making, 3*, 295–303.

Gray, J. A. M., Patnick, J., & Blanks, R. G. (2008). Maximising benefit and minimising harm of screening. *British Medical Journal, 336*, 480–483.

Green, L. A., & Mehr, D. R. (1997). What alters physicians' decisions to admit to the coronary care unit? *The Journal of Family Practice, 45*, 219–226.

Groner, M., Groner, R., & Bischof, W. F. (1983). Approaches to heuristics: A historical review. In R. Groner, M. Groner, & W. F. Bischof (Eds.), *Methods of heuristics* (pp. 1–18). Hillsdale, NJ: Erlbaum.

Gula, B., & Köppen, J. (2009). Einfluss von Länge und Perfektion einer "Hot-Hand" Sequenz auf Zuspielentscheidungen im Volleyball. [Influence of length and perfectionism of hot-hand sequences to playmakers' allocations in volleyball]. *Zeitschrift für Sportpsychologie, 16*, 1–6.

Gula, B., & Raab, M. (2004). Hot hand belief and hot hand behavior: A comment on Koehler and Conley. *Journal of Sport and Exercise Psychology, 26*, 167–170.

Güth, W. (1995). On ultimatum bargaining: A personal review. *Journal of Economic Behavior and Organization, 27*, 329–344.

Güth, W. (2008). (Non-) behavioral economics: A programmatic assessment. *Journal of Psychology, 216*, 244–253.

Haag, L., & Stern, E. (2003). In search of the benefits of learning Latin. *Journal of Educational Psychology, 95*, 174–178.

Hacking, I. (1975). *The emergence of probability*. Cambridge: Cambridge University Press.

Haggstrom, D. A., & Schapira, M. M. (2006). Black–white differences in risk perceptions of breast cancer survival and screening mammography benefit. *Journal of General Internal Medicine, 21*, 371–377.

Hahn, U., & Warren, P. A. (2009). Perceptions of randomness: Why three heads are better than four. *Psychological Review, 116*, 454–461.

Haidt, J., Baer, S., Cosmides, L., Epstein, R. A., Fikentscher, W., Johnson, E. J., et al. (2006). What is the role of heuristics in making law? In G. Gigerenzer & C. Engel (Eds.), *Heuristics and the law* (*Dahlem workshop report 94*) (pp. 239–257). Cambridge, MA: MIT Press.

Haidt, J., & Bjorklund, F. (2008). Social intuitionists answer six questions about moral psychology. In W. Sinnott-Armstrong (Ed.), *Moral psychology: Vol. 2. The cognitive science of morality: Intuition and diversity* (pp. 181–217). Cambridge, MA: MIT Press.

Hales, S. (1999). An epistemologist looks at the hot hand in sports. *Journal of the Philosophy of Sport, 25*, 79–87.

Hall, J. A. (1978). Gender effects in decoding nonverbal cues. *Psychological Bulletin, 85*, 845–857.

Hall, G. S. (1904). *Adolescence: Its psychology and its relations to physiology, anthropology, sociology, sex, crime, religion and education* (Vol. 2). New York: D. Appleton.

Hallowell, N., Statham, H., Murton, F., Green, J., & Richards, M. (1997). "Talking about chance:" The presentation of risk information during genetic counseling for breast and ovarian cancer. *Journal of Genetic Counseling, 6*, 269–286.

Hamm, R. M., & Smith, S. L. (1998). The accuracy of patients' judgments of disease probability and test sensitivity and specificity. *The Journal of Family Practice, 47*, 44–52.

Hammond, K. R. (1996). *Human judgment and social policy: Irreducible uncertainty, inevitable error, unavoidable injustice.* New York: Oxford University Press.

Hammond, K. R. (2007). *Beyond rationality: The search for wisdom in a troubled time.* Oxford: Oxford University Press.

Hanoch, Y., Miron-Shatz, T., & Himmelstein, M. (2010). Genetic testing and risk interpretation: How do women understand lifetime risk results. *Judgment and Decision Making, 5*, 116–123.

Hanselmann, M., & Tanner, C. (2008). Taboos and conflicts in decision making: Sacred values, decision difficulty, and emotions. *Judgment and Decision Making, 3*, 51–63.

Harrington, J., Noble, L. M., & Newman, S. P. (2004). Improving patients' communication with doctors: A systematic review of intervention studies. *Patient Education & Counseling, 52*, 7–16.

Hartmann, L. C., Schaid, D. J., Woods, J. E., Crotty, T. P., Myers, J. L., Arnold, P. G., et al. (1999). Efficacy of bilateral prophylactic mastectomy in women with a family history of breast cancer. *The New England Journal of Medicine, 340*, 77–84.

Hartz, J., & Chappell, R. (1997). *Worlds apart: How the distance between science and journalism threatens America's future.* Nashville, TN: First Amendment Center.

Hastie, R., & Kameda, T. (2005). The robust beauty of majority rules in group decisions. *Psychological Review, 112*, 494–508.

Hastie, R., & Wittenbrink, B. (2006). Heuristics for applying laws to facts. In G. Gigerenzer & C. Engel (Eds.), *Heuristics and the law* (*Dahlem workshop report 94*) (pp. 259–280). Cambridge, MA: MIT Press.

Hauser, J. R. (1978). Testing the accuracy, usefulness, and significance of probabilistic models: An information–theoretic approach. *Operations Research*, *26*, 406–421.

Hauser, J. R., Ding, M., & Gaskin, S. P. (2009.) Non-compensatory (and compensatory) models of consideration-set decisions. *Proceedings of the Sawtooth Software Conference*, *14*, 207–232.

Hayek, F. A. (1945). The use of knowledge in society. *American Economic Review*, *35*, 519–530.

Healthwise Staff. (2010). *Effectiveness rate of birth control methods* Retrieved October 30, 2014 from http://www.emedicinehealth.com/effectiveness_rate_of_birth_control_methods-health/article_em.htm.

Healy, M. J. R. (1979). Does medical statistics exist? *Bulletin of Applied Statistics*, *6*, 137–182.

Heilbronner, S. R., Rosati, A. G., Stevens, J. R., Hare, B., & Hauser, M. D. (2008). A fruit in the hand or two in the bush? Ecological pressures select for divergent risk preferences in chimpanzees and bonobos. *Biology Letters*, *4*, 246–249.

Hembroff, L. A., Holmes-Rovner, M., & Wills, C. E. (2004). Treatment decision-making and the form of risk communication: Results of a factorial survey. *BMC Medical Informatics and Decision Making*, *4*, 20.

Henneman, L., Timmermans, D. R., & van der Wal, G. (2004). Public experiences, knowledge and expectations about genetics and the use of genetic information. *Community Genetics*, *7*, 33–43.

Hepler, T. J. (2008). *Decision-making in sport: An examination of the take the first heuristic and self-efficacy theory*. Unpublished doctoral dissertation, Michigan State University.

Hertwig, R., Davis, J. N., & Sulloway, F. (2002). Parental investment: How an equity motive can produce inequality. *Psychological Bulletin*, *128*, 728–745.

Hertwig, R., Fischbacher, U., & Bruhin, A. (2012). Simple heuristics in a social game. In R. Hertwig, U. Hoffrage, & the ABC Research Group, *Simple heuristics in a social world* (pp. 39–65). New York: Oxford University Press.

Hertwig, R., Gigerenzer, G., & Hoffrage, U. (1997). The reiteration effect in hindsight bias. *Psychological Review*, *104*, 194–202.

Hertwig, R., & Herzog, S. M. (2009). Fast and frugal heuristics: tools of social rationality. *Social Cognition*, *27*, 661–698.

Hertwig, R., Herzog, S. M., Schooler, L. J., & Reimer, T. (2008). Fluency heuristic: A model of how the mind exploits a by-product of information retrieval. *Journal of Experimental Psychology: Learning, Memory, and Cognition*, *34*, 1191–1206.

Hertwig, R., & Ortmann, A. (2001). Experimental practices in economics: A methodological challenge for psychologists? *Behavioral and Brain Sciences*, *24*, 383–451.

Hertwig, R., & Todd, P. M. (2003). More is not always better: The benefits of cognitive limits. In D. Hardman & L. Macchi (Eds.), *Thinking: Psychological perspectives on reasoning, judgment and decision making* (pp. 213–231). Chichester, UK: Wiley.

Heukelom, F. (2007). *Who are the behavioral economists and what do they say?* Tinbergen Institute discussion papers 07-020/1. Tinbergen Institute, Amsterdam, Netherlands.

Hilbig, B. E., & Pohl, R. F. (2008). Recognizing users of the recognition heuristic. *Experimental Psychology, 55*, 394–401.

Hilbig, B. E., & Pohl, R. F. (2009). Ignorance- vs. evidence-based decision making: A decision time analysis of the recognition heuristic. *Journal of Experimental Psychology: Learning, Memory, and Cognition, 35*, 1296–1305.

Hilbig, B. E., Pohl, R. F., & Bröder, A. (2009). Criterion knowledge: A moderator of using the recognition heuristic? *Journal of Behavioral Decision Making, 22*, 510–522.

Hodgkinson, G. P., & Healey, M. P. (2008). Cognition in organizations. *Annual Review of Psychology, 59*, 387–417.

Hoffrage, U. (1995). *Zur Angemessenheit subjektiver Sicherheits-Urteile: Eine Exploration der Theorie der probabilistischen mentalen Modelle* [*The adequacy of subjective confidence judgments: Studies concerning the theory of probabilistic mental models*]. Unpublished doctoral dissertation, University of Salzburg, Austria.

Hoffrage, U. (2003). Risikokommunikation bei Brustkrebsfrüherkennung und Hormonersatztherapie. [Risk communication in the early identification of breast cancer and hormone-replacement therapy]. *Zeitschrift für Gesundheitspsychologie, 11*, 76–86.

Hoffrage, U. (2011). Recognition judgments and the performance of the recognition heuristic depend on the size of the reference class. *Judgment and Decision Making, 6*, 43–57.

Hoffrage, U., & Gigerenzer, G. (1998). Using natural frequencies to improve diagnostic inferences. *Academic Medicine, 73*, 538–540.

Hoffrage, U., Gigerenzer, G., Krauss, S., & Martignon, L. (2002). Representation facilitates reasoning: what natural frequencies are and what they are not. *Cognition, 84*, 343–352.

Hoffrage, U., Hertwig, R., & Gigerenzer, G. (2000). Hindsight bias: A by-product of knowledge updating? *Journal of Experimental Psychology: Learning, Memory, and Cognition, 26*, 566–581.

Hoffrage, U., Kurzenhäuser, S., & Gigerenzer, G. (2000). Wie kann man die Bedeutung medizinischer Testbefunde besser verstehen und kommunizieren? [How to better understand and communicate the meaning of test results]. *Zeitschrift für ärztliche Fortbilding und Qualitätssicherung, 94*, 713–719.

Hoffrage, U., Lindsey, S., Hertwig, R., & Gigerenzer, G. (2000). Communicating statistical information. *Science, 290*, 2261–2262.

Hogarth, R. M. (1980). *Judgement and choice: The psychology of decision.* Chichester, UK: Wiley.

Hogarth, R. M. (2012). When simple is hard to accept. In P. M. Todd, G. Gigerenzer, & the ABC Reseach Group, *Ecological rationality: Intelligence in the world* (pp. 61–79). Oxford: Oxford University Press.

Hogarth, R. M., & Karelaia, N. (2005). Simple models for multi-attribute choice with many alternatives: When it does and does not pay to face tradeoffs with binary attributes. *Management Science, 51*, 1860–1872.

Hogarth, R. M., & Karelaia, N. (2006). "Take-the-best" and other simple strategies: Why and when they work "well" with binary cues. *Theory and Decision, 61*, 205–249.

Hogarth, R. M., & Karelaia, N. (2007). Heuristic and linear models of judgment: Matching rules and environments. *Psychological Review, 114*, 733–758.

Holton, G. (1988). *Thematic origins of scientific thought* (2nd ed.). Cambridge, MA: Harvard University Press.

Horton, R. (2004 11). The dawn of McScience. *The New York Review of Books, 51*, 7–9.

House, R. J., Hanges, P. J., Javidan, M., Dorfman, P. W., & Gupta, V. (2004). *Culture, leadership, and organizations: The GLOBE study of 62 societies*. London: Sage.

Høye, S. P. H. (2002). "New wonder pill!" What do Norwegian newspapers write about new medications. *Tidsskr Nor Lægeforen, 122*, 1671–1676.

Hoyer, W. D., & Brown, S. P. (1990). Effects of brand awareness on choice for a common, repeat purchase product. *Journal of Consumer Research, 17*, 141–148.

Humphrey, N. K. (1988). The social function of intellect. In R. Byrne & A. Whiten (Eds.), *Machiavellian intelligence* (pp. 13–26). Oxford, UK: Clarendon Press. Reprinted from P. P. G. Bateson & R. A. Hinde (Eds.). (1976). *Growing points in ethology*, pp. 303–321. Cambridge, Cambridge University Press.

Humphrey, L. L., Helfand, M., Chan, B. K. S., & Woolf, S. H. (2002). Breast cancer screening: A summary of the evidence for the US Preventive Services Task Force. *Annals of Internal Medicine, 137*, 347–360.

Hutchinson, J. M. C., & Gigerenzer, G. (2005). Simple heuristics and rules of thumb: Where psychologists and behavioural biologists might meet. *Behavioural Processes, 69*, 97–124.

Impiccatore, P., Pandolfini, C., Casella, N., & Bonati, M. (1997). Reliability of health information for the public on the World Wide Web: Systemic survey of advice on managing fever in children at home. *British Medical Journal, 314*, 1875–1879.

Inglehart, R., Basanez, M., Diez-Medrano, J., Halman, L, & Luijkx, R. (2004). *Human values and beliefs: A cross-cultural sourcebook based on the 1999–2002 values surveys*. Bilbao, Spain: Fundación BBVA.

Inter-Parliamentary Union. (2010). *Women in national parliaments.* Retrieved April 2011 from http://www.ipu.org/wmn-e/classif.htm.

Jacoby, L. L., & Dallas, M. (1981). On the relationship between autobiographical memory and perceptual learning. *Journal of Experimental Psychology: General, 110*, 306–340.

Jacoby, L. J., Woloshyn, V., & Kelley, C. (1989). Becoming famous without being recognized: Unconscious influences of memory produced by dividing attention. *Journal of Experimental Psychology, 118*, 115–125.

Jahnke, T., & Wuttke, H. (Eds.). (2005). *Mathematik: Stochastik.* Berlin: Cornelsen Verlag.

Johnson, E. J., & Goldstein, D. G. (2003). Do defaults save lives? *Science, 302*, 1338–1339.

Johnson, J. G., & Raab, M. (2003). Take the first: Option generation and resulting choices. *Organizational Behavior and Human Decision Processes, 91*, 215–229.

Jolls, C., Sunstein, C. R., & Thaler, R. H. (1998). A behavioral approach to law and economics. *Stanford Law Review, 50*, 1471–1541.

Jørgensen, K. J., & Gøtzsche, P. C. (2004). Presentation on websites of possible benefits and harms from screening for breast cancer: Cross sectional study. *British Medical Journal, 328*, 148.

Jørgensen, K. J., & Gøtzsche, P. C. (2006). Content of invitations for publicly funded screening mammography. *British Medical Journal, 332*, 538–541.

Jorland, G. (1987). The Saint Petersburg Paradox 1713–1937. In L. Krüger, L. Daston, & M. Heidelberger (Eds.), *The probabilistic revolution: Vol. I. Ideas in history* (pp. 157–190). Cambridge, MA: MIT Press.

Judd, C. M., & Park, B. (1993). Definition and assessment of accuracy in social stereotypes. *Psychological Review, 100*, 109–128.

Juslin, P., & Persson, M. (2002). PROBabilities from EXemplars (PROBEX): A "lazy" algorithm for probabilistic inference from generic knowledge. *Cognitive Science, 26*, 563–607.

Kahneman D. (2003). Maps of bounded rationality: A perspective on intuitive judgment and choice. In T. Frangsmyr (Ed.), *Les Prix Nobel: The Nobel Prizes 2002* (pp. 449–489). Stockholm: Nobel Foundation.

Kahneman, D. (2011). *Thinking fast and slow.* London: Allen Lane.

Kahneman, D., & Frederick, S. (2002). Representativeness revisited: Attribute substitution in intuitive judgment. In T. Gilovich, D. Griffin, & D. Kahneman (Eds.), *Heuristics and biases: The psychology of intuitive judgment* (pp. 49–81). New York: Cambridge University Press.

Kahneman, D., Slovic, P., & Tversky, A. (Eds.). (1982). *Judgment under uncertainty: Heuristics and biases.* Cambridge: Cambridge University Press.

Kahneman, D., & Tversky, A. (1972). Subjective probability: A judgment of representativeness. *Cognitive Psychology, 3*, 430–454.

Kahneman, D., & Tversky, A. (1979). Prospect theory: An analysis of decision under risk. *Econometrica, 47*, 263–291.

Kahneman, D., & Tversky, A. (1982). Judgments of and by representativeness. In D. Kahneman, P. Slovic, & A. Tversky (Eds.), *Judgment under uncertainty: Heuristics and biases* (pp. 84–98). New York: Cambridge University Press.

Kaiser, T., Ewers, H., Waltering, A., Beckwermert, D., Jennen, C., & Sawicki, P. T. (2004). Sind die Aussagen medizinischer Werbeprospekte korrekt? *Arznei-Telegramm, 35*, 21–23.

Kalet, A., Roberts, J. C., & Fletcher, R. (1994). How do physicians talk with their patients about risks? *Journal of General Internal Medicine, 9*, 402–404.

Kant, E. (1784). Beantwortung der Frage: Was ist Aufklärung? [Answering the question: What is Enlightenment?]. *Berlinische Monatsschrift, December*, 481–494.

Kaphingst, K. A., DeJong, W., Rudd, R. E., & Daltroy, L. H. (2004). A content analysis of direct-to-consumer television prescription drug advertisements. *Journal of Health Communication, 9*, 515–528.

Kaphingst, K. A., Rudd, R. E., DeJong, W., & Daltroy, L. H. (2005). Comprehension of information in three direct-to-consumer television prescription drug advertisements among adults with limited literacy. *Journal of Health Communication, 10*, 609–619.

Kareev, Y. (2000). Seven (indeed, plus or minus two) and the detection of correlations. *Psychological Review, 107*, 397–402.

Karelaia, N. (2006). Thirst for confirmation in multi-attribute choice: Does search for consistency impair decision performance? *Organizational Behavior and Human Decision Processes, 100*, 128–143.

Kassenärztliche Bundesvereinigung. (2004). Einführung eines bundesweiten Mammographie-Screening-Programms [Introduction of a national mammography screening program]. *Beilage zum Deutschen Ärzteblatt, 4*, 1–44.

Katsikopoulos, K. V. (2010). The less-is-more effect: Predictions and tests, *Judgment and Decision Making, 5*(4), 244–257.

Katsikopoulos, K. V., & Gigerenzer, G. (2008). One-reason decision-making: Modeling violations of expected utility theory. *Journal of Risk and Uncertainty, 37*, 35–56.

Katsikopoulos, K. V., & Martignon, L. (2006). Naive heuristics for paired comparisons: Some results on their relative accuracy. *Journal of Mathematical Psychology, 50*, 488–494.

Katsikopoulos, K., Schooler, L. J., & Hertwig, R. (2010). The robust beauty of ordinary information. *Psychological Review, 117*, 1259–1266.

Kattah, J. C., Talkad, A. V., Wang, D. Z., Hsieh, Y.- H., & Newman-Toker, D. E. (2009). HINTS to diagnose stroke in the acute vestibular syndrome: Three-step bedside oculomotor examination more sensitive than early MRI diffusion-weighted imaging. *Stroke, 40*, 3504–3510.

Katzko, M. W. (2006). A study of the logic of empirical arguments in psychological research: "The automaticity of social behavior" as a case study. *Review of General Psychology, 10*, 210–228.

Kees, B. (2002, October 18). *Newsroom training: Where's the investment?* Retrieved November 4, 2014 from http://www.knightfoundation.org/publications/newsroom-training-wheres-investment.

Kelley, H. H., & Michaela, I. L. (1980). Attribution theory and research. *Annual Review of Psychology, 31*, 457–501.

Keren, G., & Schul, Y. (2009). Two is not always better than one: A critical evaluation of two-system theories. *Perspectives on Psychological Science, 4*, 533–550.

Kerlikowske, K., Grady, D., Barclay, J., Sickles, E. A., & Ernster, V. (1996a). Effect of age, breast density, and family history on the sensitivity of first screening mammography. *Journal of the American Medical Association, 276*, 33–38.

Kerlikowske, K., Grady, D., Barclay, J., Sickles, E. A., & Ernster, V. (1996b). Likelihood ratios for modern screening mammography: Risk of breast cancer based on age and mammographic interpretation. *Journal of the American Medical Association, 276*, 39–43.

Keynes, J. M. (1974). *The general theory of employment, interest and money.* London: Macmillan. (Original work published 1936).

Klaasen, F. J. G. M., & Magnus, J. R. (2001). Are points in tennis independent and identically distributed? Evidence from a dynamic binary panel data model. *Journal of the American Statistical Association, 96*, 500–509.

Kleffner, D. A., & Ramachandran, V. S. (1992). On the perception of shape from shading. *Perception & Psychophysics, 52*, 18–36.

Klein, G. (2004). *The power of intuition: How to use your gut feelings to make better decisions at work*. New York: Currency/Doubleday.

Kleinman, S., Busch, M. P., Hall, L., Thomson, R., Glynn, S., Gallahan, D., et al. (1998). False-positive HIV-1 test results in a low-risk screening setting of voluntary blood donation: Retrovirus Epidemiology Donor Study. *Journal of the American Medical Association, 280*, 1080–1085.

Kleiter, G. D. (1994). Natural sampling: Rationality without base rates. In G. H. Fischer & D. Laming (Eds.), *Contributions to mathematical psychology, psychometrics, and methodology* (pp. 375–388). New York: Springer.

Knight, F. H. (1921). *Risk, uncertainty and profit*. Boston, MA: Houghton Mifflin.

Koehler, J. J. (1996a). On conveying the probative value of DNA evidence: Frequencies, likelihood ratios, and error rates. *University of Colorado Law Review, 67*, 859–886.

Koehler, J. J. (1996b). The base rate fallacy reconsidered: Descriptive, normative, and methodological challenges. *Behavioral and Brain Sciences, 19*, 1–53.

Koehler, J. J. (1997). One in millions, billions, and trillions: Lessons from *People vs. Collins* (1968) for *People vs. Simpson* (1995). *Journal of Legal Education, 47*, 214–223.

Koehler, J. J., & Conley, C. A. (2003). The "hot hand" myth in professional basketball. *Journal of Sport and Exercise Psychology, 25*, 253–259.

Kohli, R., & Jedidi, K. (2007). Representation and inference of lexicographic preference models and their variants. *Marketing Science, 26*, 380–399.

Konold, C., & Miller, C. (2005). *TinkerPlots*. Emeryville, CA: Key Curriculum Press. (Computer software)

Köppen, J., & Raab, M. (2012). The hot and cold hand in volleyball: Individual expertise differences in a video-based playmaker decision test. *The Sport Psychologist, 26*, 167–185.

Krishnan, H. A., & Park, D. (2005). A few good women: On top management teams. *Journal of Business Research, 58*, 1712–1720.

Krueger, J., & Mueller, R.A. (2002). Unskilled, unaware, or both? The better-than-average heuristic and statistical regression predict errors in estimates of own performance. *Journal of Personality and Social Psychology, 82*, 180–188.

Kruger, J., & Dunning, D. (1999). Unskilled and unaware of it: How difficulties in recognizing one's own incompetence lead to inflated self-assessments. *Journal of Personality and Social Psychology, 77*, 1121–1134.

Krüger, L., Gigerenzer, G., & Morgan, M. S. (Eds.). (1987). *The probabilistic revolution: Vol. II: Ideas in the sciences*. Cambridge, MA: MIT Press.

Kruglanski, A. W., & Gigerenzer, G. (2011). Intuitive and deliberative judgments are based on common principles. *Psychological Review, 118*, 97–109.

Kurzenhäuser, S. (2003). Welche Informationen vermitteln deutsche Gesundheitsbroschüren über die Screening-Mammographie? [What information do German health brochures provide on mammography screening?]. *Zeitschrift für ärztliche Fortbildung und Qualitätssicherung, 97*, 53–57.

Kurzenhäuser, S., & Hoffrage, U. (2002). Teaching Bayesian reasoning: An evaluation of a classroom tutorial for medical students. *Medical Teacher, 24*, 516–521.

Kurz-Milcke, E., Gigerenzer, G., & Martignon, L. (2008). Transparency in risk communication: Graphical and analog tools. In W. T. Tucker, S. Ferson, A. Finkel, T. F. Long, D. Slavin, & P. Wright (Eds.), *Annals of the New York Academy of Sciences* (pp. 18–28). New York: Blackwell.

Kurz-Milcke, E., & Martignon, L. (2007). Stochastische Urnen und Modelle in der Grundschule [Stochastic urns and models in elementary school]. In G. Kaiser (Ed.), *Tagungsband der Jahrestagung der Gesellschaft für Didaktik der Mathematik, Berlin* (pp. 480–487). Hildesheim, Germany: Verlag Franzbecker.

Labarge, A. S., McCaffrey, R. J., & Brown, T. A. (2003). Neuropsychologists' ability to determine the predictive value of diagnostic tests. *Clinical Neuropsychology, 18*, 165–175.

Laibson, D. (1997). Golden eggs and hyperbolic discounting. *The Quarterly Journal of Economics, 112*, 443–477.

Laibson, D. (2002). Bounded rationality in economics. [PowerPoint slides], Retrieved from UC Berkeley 2002 Summer Institute on Behavioral Economics website, organized by the Behavioral Economics Roundtable, Russell Sage Foundation. Retrieved February 22, 2009, from http://elsa.berkeley.edu/symposia/sage02/slides/laibson3.pdf.

Lanzieri, G. (2006). Population in Europe 2005: First results. In: *Statistics in Focus: Population and Social Conditions, Vol. 16*. Luxembourg: Eurostat.

Laplace, P.- S. (1951). *A philosophical essay on probabilities* (F. W. Truscott & F. L. Emery, Trans.). New York: Dover. (Original work published 1814).

Larkey, P. D., Smith, R. A., & Kadane, J. B. (1989). It's okay to believe in the hot hand. *Chance: New Directions for Statistics and Computing, 4*, 22–30.

Laroche, M., Kim, C., & Matsui, T. (2003). Which decision heuristics are used in consideration set formation? *Journal of Consumer Marketing, 3*, 192–209.

Larson, R. J., Woloshin, S., Schwartz, B., & Welch, H. G. (2005). Celebrity endorsements of cancer screening. *Journal of the National Cancer Institute, 97*, 693–695.

Lauterbach, K. W. (2002). 100 000 überflüssige Operationen [100 000 Needless operations]. *Die Zeit*, August *28*, 16. (Letter to the editor).

Lee, M. D., & Cummins, T. D. R. (2004). Evidence accumulation in decision making: Unifying the "take the best" and the "rational" models. *Psychonomic Bulletin and Review, 11*, 343–352.

Lee, M. D., Loughlin, N., & Lundberg, I. B. (2002). Applying one reason decision-making: The prioritisation of literature searches. *Australian Journal of Psychology, 54*, 137–143.

Leland, J. W. (1994). Generalized similarity judgments: An alternative explanation for choice anomalies. *Journal of Risk and Uncertainty, 9*, 151–172.

Leland, J. W. (2002). Similarity judgments and anomalies in intertemporal choice. *Economic Inquiry, 40*, 574–581.

Lerman, C., Trock, B., Rimer, B. K., Jepson, C., Brody, D., & Boyce, A. (1991). Psychological side effects of breast cancer screening. *Health Psychology, 10*, 259–267.

Lerner, J. S., & Tetlock, P. E. (1999). Accounting for the effects of accountability. *Psychological Bulletin, 125*, 255–275.

Lichtenstein, S., Fischhoff, B., & Phillips, L. D. (1982). Calibration of probabilities: The state of the art to 1980. In D. Kahneman, P. Slovic, & A. Tversky (Eds.), *Judgment under uncertainty: Heuristics and biases* (pp. 306–334). Cambridge: Cambridge University Press.

Lichtman, A. J. (2008). The keys to the White House: An index forecast for 2008. *International Journal of Forecasting, 24*, 301–309.

Lindsey, S., Hertwig, R., & Gigerenzer, G. (2003). Communicating statistical DNA evidence. *Jurimetrics, 43*, 147–163.

Lipkus, I. M. (2007). Numeric, verbal, and visual formats of conveying health risks: Suggested best practices and future recommendations. *Medical Decision Making, 27*, 696–713.

Lipkus, I. M., Samsa, G., & Rimer, B. K. (2001). General performance on a numeracy scale among highly educated samples. *Medical Decision Making, 21*, 37–44.

Lipman, B. (1999). Decision theory without logical omniscience: Toward an axiomatic framework for bounded rationality. *Review of Economic Studies, 66*, 339–361.

Luan, S., Schooler, L., & Gigerenzer, G. (2011). A signal detection analysis of fast-and-frugal trees. *Psychological Review, 118*, 316–338.

Luce, R. D. (1956). Semiorders and a theory of utility discrimination. *Econometrica, 24*, 178–191.

Lyons, K. (2003). Performance analysis for coaches: game analysis. *Sports Coach, 26*, 30–31.

Macdonald, E., & Sharp, B. (2000). Brand awareness effects on consumer decision making for a common, repeat purchase product: A replication. *Journal of Business Research, 48*, 5–15.

Mahoney, M. J. (1979). Psychology of the scientist: An evaluative review. *Social Studies of Science, 9*, 349–375.

Malenka, D. J., Baron, J. A., Johansen, S., Wahrenberger, J. W., & Ross, J. M. (1993). The framing effect of relative and absolute risk. *Journal of General Internal Medicine, 8*, 543–548.

Malmberg, K. J. (2002). On the form of ROCs constructed from confidence ratings. *Journal of Experimental Psychology: Learning, Memory, and Cognition, 28*, 380–387.

March, J. G. (1978). Bounded rationality, ambiguity, and the engineering of choice. *Bell Journal of Economics, 9*, 587–608.

Marewski, J. N. (2010). On the theoretical precision, and strategy selection problem of a single-strategy approach: A comment on Glöckner, Betsch, and Schindler. *Journal of Behavioral Decision Making, 23*, 463–467.

Marewski, J. N., Gaissmaier, W., Schooler, L. J., Goldstein, D. G., & Gigerenzer, G. (2009). Do voters use episodic knowledge to rely on recognition? In N. A. Taatgen & H. van Rijn (Eds.), *Proceedings of the 31st annual conference of the Cognitive Science Society* (pp. 2232–2237). Austin, TX: Cognitive Science Society.

Marewski, J. N., Gaissmaier, W., Schooler, L. J., Goldstein, D. G., & Gigerenzer, G. (2010). From recognition to decisions: Extending and testing recognition-based models for multi-alternative inference. *Psychonomic Bulletin and Review, 17*, 287–309.

Marewski, J. N., & Schooler, L. J. (2011). Cognitive niches: An ecological model of strategy selection. *Psychological Review, 118*, 393–437.

Martignon, L., & Hoffrage, U. (1999). Why does one-reason decision making work? A case study in ecological rationality. In G. Gigerenzer, P. M. Todd, & the ABC Research Group, *Simple heuristics that make us smart* (pp. 119–140). New York: Oxford University Press.

Martignon, L., & Hoffrage, U. (2002). Fast, frugal, and fit: Lexicographic heuristics for paired comparison. *Theory and Decision, 52*, 29–71.

Martignon, L., Katsikopoulos, K. V., & Woike, J. (2008). Categorization with limited resources: A family of simple heuristics. *Journal of Mathematical Psychology, 52*, 352–361.

Martignon, L., Vitouch, O., Takezawa, M., & Forster, M. R. (2003). Naive and yet enlightened: From natural frequencies to fast and frugal trees. In D. Hardman & L. Macchi (Eds.), *Thinking: Psychological perspectives on reasoning, judgment, and decision making* (pp. 189–211). Chichester, UK: Wiley.

Martignon, L., & Wassner, C. (2005). Schulung frühen stochastischen Denkens von Kindern. *Zeitschrift für Erziehungswissenschaften, 8*, 202–222.

Mata, R., Schooler, L. J., & Rieskamp, J. (2007). The aging decision maker: Cognitive aging and the adaptive selection of decision strategies. *Psychology and Aging, 22*, 796–810.

Mayo Clinic. (2013). *Antidepressants for children and teens*. Retrieved November 4, 2014 from http://www.mayoclinic.org/diseases-conditions/teen-depression/in-depth/antidepressants/art-20047502.

Mazur, D. J., Hickam, D. H., & Mazur, M. D. (1999). How patients' preferences for risk information influence treatment choice in a case of high risk and high therapeutic uncertainty: Asymptomatic localized prostate cancer. *Medical Decision Making, 194*, 394–398.

McCallum, J. (1993). Hot hand, hot head. *Sports Illustrated, 78*, 22–24.

McCammon, I., & Hägeli, P. (2007). An evaluation of rule-based decision tools for travel in avalanche terrain. *Cold Regions Science and Technology, 47*, 193–206.

McCloy, R., Beaman, C. P., & Smith, P. T. (2008). The relative success of recognition-based inference in multichoice decisions. *Cognitive Science, 32*, 1037–1048.

McClure, E. B. (2000). A meta-analytic review of sex differences in facial expression processing and their development in infants, children, and adolescents. *Psychological Bulletin, 126*, 424–453.

McDonald, C. (1996). Medical heuristics: The silent adjudicators of clinical practice. *Annals of Internal Medicine, 124*, 56–62.

McGettigan, P., Sly, K., O'Connell, D., Hill, S., & Henry, D. (1999). The effects of information framing on the practices of physicians. *Journal of General Internal Medicine, 14*, 633–642.

McGrath, R. E. (2008). Predictor combination in binary decision-making situations. *Psychological Assessment, 20*, 195–205.

McLeod, P., & Dienes, Z. (1996). Do fielders know where to go to catch the ball or only how to get there? *Journal of Experimental Psychology: Human Perception and Performance, 22*, 531–543.

McNamara, J. M., & Houston, A. I. (2009). Integrating function and mechanism. *Trends in Ecology and Evolution, 24*, 670–675.

Messick, D. M. (1993). Equality as decision heuristic. In B. A. Mellers & J. Baron (Eds.), *Psychological perspectives on justice: Theory and application* (pp. 11–31). Cambridge: Cambridge University Press.

Miller, A. M., & Champion, V. L. (1997). Attitudes about breast cancer and mammography: Racial, income, and educational differences. *Women & Health, 26*, 41–63.

Miller, A. B., Wall, C., Baines, C. J., Sun, P., To, T., & Narod, S. (2014). Twenty-five year follow-up for breast cancer incidence and mortality of the Canadian National Breast Screening Study: randomised screening trial. *British Medical Journal, 348*, g366.

Mischel, W. (2006). Bridges toward a cumulative psychological science. In P. A. M. Van Lange (Ed.), *Bridging social psychology: Benefits of transdisciplinary approaches* (pp. 437–446). Mahwah, NJ: Lawrence Erlbaum.

Mischel, W. (2009). The toothbrush problem. *Association for Psychological Science Observer, 21*, 11.

Mooi, W. J., & Peeper, D. S. (2006). Oncogene-induced cell senescence: Halting on the road to cancer. *The New England Journal of Medicine, 355*, 1037–1046.

Moore, D. S. (1997). New pedagogy and new content: The case of statistics. *International Statistical Review, 65*, 123–165.

Morant, I. (2006). *Historia de las mujeres en España y América Latina* [The history of women in Spain and Latin America]. Madrid: Cátedra.

Morris, R. W. (2002). Does EBM offer the best opportunity yet for teaching medical statistics? *Statistics in Medicine, 21*, 969–977.

Moumjid, N., Gafni, A., Bremond, A., & Carrere, M.- O. (2007). Shared decision making in the medical encounter: Are we all talking about the same thing? *Medical Decision Making, 27*, 539–546.

Moxey, A., O'Connell, D., McGettigan, P., & Henry, D. (2003). Describing treatment effects to patients: How they are expressed makes a difference. *Journal of General Internal Medicine, 18*, 948–959.

Moyer, V. A. (2013). Screening for HIV: US Preventive Services Task Force recommendation. *Annals of Internal Medicine, 159*, 51–60.

Moynihan, R., Bero, L., Ross-Degnan, D., Henry, D., Lee, K., Watkins, J., et al. (2000). Coverage by the news media of the benefits and risks of medications. *The New England Journal of Medicine, 342*, 1645–1650.

Mueser, P. R., Cowan, N., & Mueser, K. T. (1999). A generalized signal detection model to predict rational variation in base rate use. *Cognition, 69*, 267–312.

Mühlhauser, I., & Höldke, B. (2002). Information zum Mammographiescreening: vom Trugschluss zur Ent-Täuschung [Information on mammography screening: From deception to dis-illusionment]. *Radiologe, 42*, 299–304.

Mühlhauser, I., Kasper, J., & Meyer, G. (2006). FEND: Understanding of diabetes prevention studies: questionnaire survey of professionals in diabetes care. *Diabetologia, 49*, 1742–1746.

Murphy, M. (1993). The contraceptive pill and women's employment as factors in fertility change in Britain 1963–1980: A challenge to the conventional view. *Population Studies. 47*, 221–243.

National Cancer Institute. (1998). Breast cancer risk tool: An interactive patient education tool. Bethesda, MD: Author. (Computer software).

National Cancer Institute. (2005). Fact sheet: Breast cancer prevention studies. Bethesda, MD: Author.

National Cancer Institute (2014). *Breast cancer: Early detection*. Bethseda, MD: Author. Retrieved November 4, 2014 from http://www.cancer.org/acs/groups/cid/documents/webcontent/003165-pdf.pdf

NCTM (National Council of Teachers of Mathematics). (2000). *Principles and standards for school mathematics*. Reston, VA: Author.

National Institutes of Health Consensus Conference. (1991). Treatment of early-stage breast cancer. *Journal of the American Medical Association, 265*, 391–395.

National Institutes of Health Consensus Development Panel. (1997). National Institutes of Health Consensus Development Conference statement: Breast cancer screening for women ages 40–49, January 21–23, 1997. *Journal of the National Cancer Institute, 89*, 1015–1020.

Naylor, C. D., Chen, E., & Strauss, B. (1992). Measured enthusiasm: Does the method of reporting trial results alter perceptions of therapeutic effectiveness? *Annals of Internal Medicine, 117*, 916–921.

Neurath, O. (1946). *From hieroglyphics to isotypes*. London: Future Books.

Newell, B. R. (2005). Re-visions of rationality? *TRENDS in Cognitive Sciences, 9*, 11–15.

Newell, B. R., & Fernandez, D. (2006). On the binary quality of recognition and the inconsequentiality of further knowledge: Two critical tests of the recognition heuristic. *Journal of Behavioral Decision Making, 19*, 333–346.

Newell, B. R., Rakow, T., Weston, N. J., & Shanks, D. R. (2004). Search strategies in decision-making: The success of "success." *Journal of Behavioral Decision Making, 17*, 117–137.

Newell, B. R., & Shanks, D. R. (2003). Take the best or look at the rest? Factors influencing "one-reason" decision making. *Journal of Experimental Psychology: Learning, Memory, and Cognition, 29*, 53–65.

Newell, B. R., & Shanks, D. R. (2004). On the role of recognition in decision making. *Journal of Experimental Psychology: Learning, Memory, and Cognition, 30*, 923–935.

New Zealand Breast Cancer Foundation (2011, February) *Mammogram–The facts*. Retrieved November 4, 2014 from http://nzbcf.org.nz/Portals/0/Documents/bhe_kit_2011_-_mammograms_-_the_facts_final.pdf

NHS Cancer Screening Programmes. (2013, June). *NHS breast screening: Helping you decide*. Retrieved November 4, 2014 from http://www.cancerscreening.nhs.uk/breastscreen/publications/nhsbsp.pdf

NHS Wales (n.d.). *Breast cancer explained: Family history*. Retrieved November 4, 2014 from http://www.breasttestwales.wales.nhs.uk/sitesplus/documents/1025/B705.v1.01-10%20BS%20Explained%20ENG%20ppp_PROOF_3.pdf

Noormofidi, D. (2006, May 3). Zank um Brustkrebs [Strife over breast cancer]. *Die Standard*. Retrieved November 3, 2014 from http://diestandard.at/?url=/?id=2431198.

Nosofsky, R. M., & Bergert, F. B. (2007). Limitations of exemplar models of multi-attribute probabilistic inference. *Journal of Experimental Psychology: Learning, Memory, and Cognition*, *33*, 999–1019.

Nuovo, J., Melnikow, J., & Chang, D. (2002). Reporting number need to treat and absolute risk reduction in randomized controlled trials. *Journal of the American Medical Association*, *287*, 2813–2814.

Nyström, L. (2002). Long-term effects of mammography screening: Updated overview of the Swedish randomised trials. *The Lancet*, *359*, 909–919.

Oppenheimer, D. (2003). Not so fast! (and not so frugal!): Rethinking the recognition heuristic. *Cognition*, *90*, B1–B9.

Ortmann, A., Gigerenzer, G., Borges, B., & Goldstein, D. G. (2008). The recognition heuristic: A fast and frugal way to investment choice? In C. R. Plott & V. L. Smith (Eds.), *Handbook of experimental economics results: Vol. 1* (*Handbooks in Economics no. 28*) (pp. 993–1003). Amsterdam: North-Holland.

Oskarsson, T., Van Boven, L., McClelland, G. H., & Hastie, R. (2009). What's next? Judging sequences of binary events. *Psychological Bulletin*, *135*, 262–285.

Österreichische Ärztekammer. (2005). *Österreichische Ärztekammer: Homöopathie kein Placebo. Viele internationale Studien ergeben positive Wirkungsweise* [*Austrian Chamber of Physicians: Homeopathy is not a placebo: Many international studies show positive results*]. Retrieved November 3, 2014 from http://www.ots.at/presseaussendung/OTS_20050916_OTS0118/oesterreichische-aerztekammer-homoeopathie-kein-placebo.

Pachur, T., & Biele, G. (2007). Forecasting from ignorance: The use and usefulness of recognition in lay predictions of sports events. *Acta Psychologica*, *125*, 99–116.

Pachur, T., Bröder, A., & Marewski, J. N. (2008). The recognition heuristic in memory-based inference: Is recognition a non-compensatory cue? *Journal of Behavioral Decision Making*, *21*, 183–210.

Pachur, T., & Hertwig, R. (2006). On the psychology of the recognition heuristic: Retrieval primacy as a key determinant of its use. *Journal of Experimental Psychology: Learning, Memory, and Cognition*, *32*, 983–1002.

Pachur, T., Mata, R., & Schooler, L. J. (2009). Cognitive aging and the adaptive use of recognition in decision making. *Psychology and Aging*, *24*, 901–915.

Pachur, T., Todd, P. M., Gigerenzer, G., Schooler, L. J., & Goldstein, D. G. (2012). When is the recognition heuristic an adaptive tool? In P. M. Todd, G. Gigerenzer, & the ABC Research Group, *Ecological rationality: Intelligence in the world* (pp. 113–143). New York: Oxford University Press.

Paiement, M., Baudin, P., & Boucher, J. (1993). Scouting and match preparation at the national and international level. *International Volleyball Technical Bulletin*, *2*, 4–14.

Paling, J. (2003). Strategies to help patients understand risks. *British Medical Journal, 327*, 745–748.

Pantaleoni, M. (1898). *Pure economics.* (T. B. Bruce, Trans.). London: Macmillan. (Original work published 1889).

Parikh, J. (1994). *Intuition: The new frontier of management.* Oxford, UK: Blackwell Business.

Payne, J. W. (1976). Task complexity and contingent processing in decision making: An information search and protocol analysis. *Organizational Behavior and Human Performance, 16*, 366–387.

Payne, J. W., Bettman, J. R., & Johnson, E. J. (1993). *The adaptive decision maker.* Cambridge: Cambridge University Press.

Payne, J. W., & Braunstein, M. L. (1978). Risky choice: An examination of information acquisition behavior. *Memory and Cognition, 6*, 554–561.

Peters, E., Hibbard, J., Slovic, P., & Dieckmann, N. (2007). Numeracy skill and the communication, comprehension, and use of risk and benefit information. *Health Affairs, 26*, 741–748.

Petrie, M., & Halliday, T. (1994). Experimental and natural changes in the peacocks (*Pavo cristatus*) train can affect mating success. *Behavioral and Ecological Sociobiology, 35*, 213–217.

Philips, E., Lappan, G., Winter, M. J., & Fitzgerald, G. (1986). *Middle grades mathematics project: Probability.* Menlo Park, CA: Addison-Wesley.

Piattelli-Palmarini, M. (1991). Probability blindness: Neither rational nor capricious. *Bostonia, March/April*, 28–35.

Pitt, M. A., Myung, I. J., & Zhang, S. (2002). Toward a method for selecting among computational models for cognition. *Psychological Review, 109*, 472–491.

Pleskac, T. J. (2007). A signal detection analysis of the recognition heuristic. *Psychonomic Bulletin & Review, 14*, 379–391.

Pohl, R. F. (2006) Empirical tests of the recognition heuristic. *Journal of Behavioral Decision Making, 19*, 251–271.

Politi, M. C., Han, P. K. J., & Col, N. F. (2007). Communicating the uncertainty of harms and benefits of medical interventions. *Medical Decision Making, 27*, 681–695.

Porter, T. M. (1995). *Trust in numbers: The pursuit of objectivity in science and public life.* Princeton, NJ: Princeton University Press.

Price, J. H., Desmond, S. H., Slenker, S., Smith, D. E., & Stewart, P. W. (1992). Urban black women's perceptions of breast cancer and mammography. *Journal of Community Health, 17*, 191–204.

Prostate.net. (2010). *Prostate cancer treatment side effects.* Retrieved November 3, 2014 from http://www.prostate.net/prostate-cancer/treatment-of-prostate-cancer/prostate-cancer-treatment-side-effects/.

Rabin, M. (1998). Psychology and economics. *Journal of Economic Literature, 36*, 11–46.

Rabin, M. (2002). A perspective on psychology and economics. *European Economic Review, 46*, 657–685.

Ransohoff, D. F., & Harris, R. P. (1997). Lessons from the mammography screening controversy: Can we improve the debate? *Annals of Internal Medicine, 127*, 1029–1034.

Rao, G. (2008). Physician numeracy: Essential skills for practicing evidence-based medicine. *Family Medicine, 40*, 354–358.

Rásky, É., & Groth, S. (2004). Informationsmaterialien zum Mammographiescreening in Österreich: Unterstützen sie die informierte Entscheidung von Frauen? [Mammography screening information in Austria: Does it facilitate women's informed decisions?]. *Sozial und Präventivmedizin, 49*, 301–397.

Ratcliff, R., & McKoon, G. (1989). Similarity information versus relational information: Differences in the time course of retrieval. *Cognitive Psychology, 21*, 139–155.

Read, D., & Grushka-Cockayne, Y. (2011). The similarity heuristic. *Journal of Behavioral Decision Making, 24*, 23–46.

Reimer, T., & Katsikopoulos, K. (2004). The use of recognition in group decision-making. *Cognitive Science, 28*, 1009–1029.

Reimer, L., Mottice, S., Schable, C., Sullivan, P., Nakashima, A., Rayfield, M., et al. (1997). Absence of detectable antibody in a patient infected with human immunodeficiency virus. *Clinical Infectious Diseases, 25*, 98–100.

Reisen, N., Hoffrage, U., & Mast, F. W. (2008). Identifying decision strategies in a consumer choice situation. *Judgment and Decision Making, 3*, 641–658.

Reyna, V. F., & Brainerd, C. J. (2007). The importance of mathematics in health and human judgment: Numeracy, risk communication, and medical decision making. *Learning and Individual Differences, 17*, 147–159.

Reyna, V. F., & Lloyd, F. J. (2006). Physician decision making and cardiac risk: Effects of knowledge, risk perception, risk tolerance, and fuzzy processing. *Journal of Experimental Psychology: Applied, 12*, 179–195.

Richter, T., & Späth, P. (2006). Recognition is used as one cue among others in judgment and decision making. *Journal of Experimental Psychology, Learning Memory and Cognition, 32*, 1501–1562.

Rieskamp, J., & Hoffrage, U. (1999). When do people use simple heuristics, and how can we tell? In G. Gigerenzer, P. M. Todd, & the ABC Research Group, *Simple heuristics that make us smart* (pp. 141–167). New York: Oxford University Press.

Rieskamp, J., & Hoffrage, U. (2008). Inferences under time pressure: How opportunity costs affect strategy selection. *Acta Psychologica, 127*, 258–276.

Rieskamp, J., & Otto, P. E. (2006). SSL: A theory of how people learn to select strategies. *Journal of Experimental Psychology: General, 135*, 207–236.

Rigby, M., Forsström, J., Roberts, R., Wyatt, J., & for the TEAC-Health Partners. (2001). Verifying quality and safety in health informatics services. *British Medical Journal, 323*, 552–556.

Rimm, A. A., & Bortin, M. (1978). Clinical trials as a religion. *Biomedicine Special Issue, 28*, 60–63.

Risueño d'Amador, B. J. I. (1836). Mémoire sur le calcul des probabilités appliqué à la médicine [Report on the calculation of probabilities applied to medicine]. *Bulletin de l'Academie Royale de Médicine, 1*, 622–680.

Roberts, J. H., & Lattin, J. M. (1991). Development and testing of a model of consideration set composition, *Journal of Marketing Research, 28*, 429–440.

Roberts, S., & Pashler, H. (2000). How pervasive is a good fit? A comment on theory testing. *Psychological Review, 107*, 358–367.

Rosati, A. G., Stevens, J. R., Hare, B., & Hauser, M. D. (2007). The evolutionary origins of human patience: Temporal preferences in chimpanzees, bonobos, and human adults. *Current Biology, 17*, 1663–1668.

Rosenfeld, A. A., Trappe, H., & Gornick, J. C. (2004). Gender and work in Germany: Before and after reunification. *Annual Review of Sociology, 30*, 103–124.

Rossmo, D. K. (2005). Geographic heuristics or shortcuts to failure? A response to Snook et al. 2004. *Applied Cognitive Psychology, 19*, 651–654

Roter, D. L., & Hall, J. A. (1993). *Doctors talking with patients/patients talking with doctors: Improving communication in medical visits.* London: Auburn House.

Rowe, G., Frewer, L., & Sjoberg, L. (2000). Newspaper reporting of hazards in the UK and Sweden. *Public Understanding of Science, 9*, 59–78.

Rozhkova, N. I., & Kochetova, G. P. (2005). Analysis of equipment of the Russian X-ray mammological service in 2003–2004. *Biomedical Engineering, 39*, 242–244.

Rubinstein, A. (1988). Similarity and decision-making under risk: Is there a utility theory resolution to the Allais-paradox? *Journal of Economic Theory, 46*, 145–153.

Rubinstein, A. (2003). Economics and psychology? The case of hyperbolic discounting. *International Economic Review, 44*, 1207–1216.

Ruscio, J. (2003). Comparing Bayes's Theorem to frequency-based approaches to teaching Bayesian reasoning. *Teaching of Psychology, 30*, 325–328.

Russo, J. E., & Dosher, B. A. (1983). Strategies for multiattribute binary choice. *Journal of Experimental Psychology, Learning, Memory, and Cognition, 9*, 676–696.

Sarfati, D., Howden-Chapman, P., Woodward, A., & Salmond, C. (1998). Does the frame affect the picture? A study into how attitudes to screening for cancer are affected by the way benefits are expressed. *Journal of Medical Screening, 5*, 137–140.

Sargent, T. J. (1993). *Bounded rationality in macroeconomics.* Oxford: Oxford University Press.

Särndal, C.- E., Swensson, B., & Wretman, J. (1992). *Model assisted survey sampling* (2nd ed.). New York: Springer-Verlag.

Savage, L. J. (1972). *The foundations of statistics* (2nd ed.). New York: Dover.

Schapira, M., Nattinger, A., & McHorney, C. (2001). Frequency or probability? A qualitative study of risk communication formats used in health care. *Medical Decision Making, 21*, 459–467.

Scheibehenne, B., & Bröder, A. (2007). Predicting Wimbledon 2005 tennis results by mere player name recognition. *International Journal of Forecasting, 23*, 415–426.

Schein, V. E., Mueller, R., Lituchy, T., & Liu, J. (1996). Think manager, think male: A global phenomenon? *Journal of Organizational Behavior, 17*, 33–41.

Schiebinger, L. L. (1989). *The mind has no sex? Women in the origins of modern science.* Cambridge, MA: Harvard University Press.

Schmitt, M., & Martignon, L. (2006). On the complexity of learning lexicographic strategies. *Journal of Machine Learning Research, 7*, 55–83.

Schmittlein, D. C., & Peterson, R. A. (1994). Customer base analysis: An industrial purchase process application. *Marketing Science, 13*, 41–67.

Schönemann, P. (1969). Review of "Faktorenanalyse" by K. Uberla and "Dimensionen des Verhaltens" by Kurt Pawlik. *Biometrics, 25*, 604–607.

Schooler, L. J., & Hertwig, R. (2005). How forgetting aids heuristic inference. *Psychological Review, 112*, 610–628.

Schroder, F. H., Hugosson, J., Roobol, M. J., Tammela, T. L., Zappa, M., Nelen, V., et al. (2014). Screening and prostate cancer mortality: results of the European Randomised Study of Screening for Prostate Cancer (ERSPC) at 13 years of follow-up. *Lancet*. doi: 10.1016/s0140-6736(14)60525-0

Schüssler, B. (2005). Im Dialog: Ist Risiko überhaupt kommunizierbar, Herr Prof. Gigerenzer?[Interview: Can risk be communicated, Prof. Ggigerenzer?] *Frauenheilkunde Aktuell, 14*, 25–31.

Schwartz, L. M., & Woloshin, S. (2000). *Physician grand round survey.* (Unpublished data).

Schwartz, L. M., & Woloshin, S. (2007). Participation in mammography screening. *British Medical Journal, 335*, 731–732.

Schwartz, L. M., & Woloshin, S. (2009). Using a drug facts box to communicate drug benefits and harms: Two randomized trials. *Annals of Internal Medicine, 5*, 516–527.

Schwartz, L. M., Woloshin, S., Black, W. C., & Welch, H. G. (1997). The role of numeracy in understanding the benefit of screening mammography. *Annals of Internal Medicine, 127*, 966–972.

Schwartz, L. M., Woloshin, S., Dvorin, E. L., & Welch, H. G. (2006). Ratio measures in leading medical journals: Structured review of accessibility of underlying absolute risks. *British Medical Journal, 333*, 1248–1252.

Schwartz, L. M., Woloshin, S., Fowler, F. J., Jr., & Welch, H. G. (2004). Enthusiasm for cancer screening in the United States. *Journal of the American Medical Association, 291*, 71–78.

Schwartz, L. M., Woloshin, S., Sox, H. C., Fischhoff, B., & Welch, H. G. (2000). U.S. women's attitudes to false positive mammography results and detection of ductal carcinoma in situ: Cross sectional survey. *British Medical Journal, 320*, 1635–1640.

Schwartz, L. M., Woloshin, S., & Welch, H. G. (1999a). Misunderstandings about the effect of race and sex on physicians' referrals for cardiac catheterization. *The New England Journal of Medicine, 341*, 279–283.

Schwartz, L. M., Woloshin, S., & Welch, H. G. (1999b). Risk communication in clinical practice: Putting cancer in context. *Monograph of the National Cancer Institute, 25*, 124–133.

Schwartz, L. M., Woloshin, S., & Welch, H. G. (2005). Can patients interpret health information? An assessment of the medical data interpretation test. *Medical Decision Making, 25*, 290–300.

Schwartz, L. M., Woloshin, S., & Welch, H. G. (2007). The drug facts box: Providing consumers with simple tabular data on drug benefit and harm. *Medical Decision Making, 27*, 655–662.

Sczesny, S., Bosak, J., Neff, D., & Schyns, B. (2004). Gender stereotypes and leadership: A cross-cultural comparison. *Sex Roles, 51*, 631–645.

Sedlmeier, P., & Gigerenzer, G. (2001). Teaching Bayesian reasoning in less than two hours. *Journal of Experimental Psychology: General, 130,* 380–400.

Sedlmeier, P., Hertwig, R., & Gigerenzer, G. (1998). Are judgments of the positional frequencies of letters systematically biased due to availability? *Journal of Experimental Psychology: Learning, Memory, and Cognition, 24,* 754–770.

Sedrakyan, A., & Shih, C. (2007). Improving depiction of benefits and harms: Analyses of studies of well-known therapeutics and review of high-impact medical journals. *Medical Care, 45,* 523–528.

Seeley, T. D. (2001). Decision making in superorganisms: How collective wisdom arises from the poorly informed masses. In G. Gigerenzer & R. Selten (Eds.), *Bounded rationality: The adaptive toolbox* (pp. 249–261). Cambridge, MA: MIT Press.

Serrano, M. (2007). Cancer regression by senescence. *The New England Journal of Medicine, 356,* 1996–1997.

Serwe, S., & Frings, C. (2006). Who will win Wimbledon? The recognition heuristic in predicting sports events. *Journal of Behavioral Decision Making, 19,* 321–322.

Shaffer, D. M., Krauchunas, S. M., Eddy, M., & McBeath, M. K. (2004). How dogs navigate to catch Frisbees. *Psychological Science, 15,* 437–441.

Shah, A. K., & Oppenheimer, D. M. (2008). Heuristics made easy: An effort-reduction framework. *Psychological Bulletin, 137,* 207–222.

Shang, A., Huwiler-Müntener, K., Nartey, L., Jüni, P., Dörig, S., Sterne, J., et al. (2005). Are the clinical effects of homoeopathy placebo effects? Comparative study of placebo-controlled trials of homoeopathy and allopathy. *The Lancet, 366,* 726–732.

Shanteau, J. (1992). How much information does an expert use? Is it relevant? *Acta Psychologica, 81,* 75–86.

Shaughnessy, J. M. (1992). Research on probability and statistics: Reflections and directions. In D. A. Grouws (Ed.), *Handbook of research on mathematical teaching and learning* (pp. 465–499). New York: Macmillan.

Shepard, R. N. (2001). Perceptual-cognitive universals as reflections of the world. *Behavioral and Brain Sciences, 24*(4), 581–601.

Sheridan, S., Pignone, M. P., & Lewis, C. L. (2003). A randomized comparison of patients' understanding of number needed to treat and other common risk reduction formats. *Journal of General Internal Medicine, 18,* 884–892.

Shettleworth, S. J. (2005). Taking the best for learning. *Behavioural Processes, 69,* 147–149.

Shibata, A., & Whittemore, A. S. (2001). RE: Prostate cancer incidence and mortality in the United States and the United Kingdom. *Journal of the National Cancer Institute, 93,* 1109–1110.

Shocker, A. D., Ben-Akiva, M., Boccara, B., & Nedungadi, P. (1991). Consideration set influences on consumer decision making and choice: Issues, models, and suggestions. *Marketing Letters, 2,* 181–197.

Shrager, J., & Siegler, R. S. (1998). SCADS: A model of strategy choice and strategy discovery. *Psychological Science, 9,* 405–410.

Simon, H. A. (1955). A behavioral model of rational choice. *Quarterly Journal of Economics, 69,* 99–118.

Simon, H. A. (1979). Rational decision making in business organizations. *American Economic Review, 69*, 493–513.

Simon, H. A (1989). The scientist as problem solver. In D. Klahr & K. Kotovsky (Eds.), *Complex information processing: The impact of Herbert A. Simon* (pp. 373–398). Hillsdale, NJ: Lawrence Erlbaum.

Simon, H. A. (1990). Invariants of human behavior. *Annual Review of Psychology, 41*, 1–19.

Simon, H. A. (1992). What is an "explanation" of behavior? *Psychological Science, 3*, 150–161.

Simon, H. A. (1999). Appraisal. Back cover of G. Gigerenzer, P. M. Todd, & the ABC Research Group, *Simple heuristics that make us smart*. New York: Oxford University Press.

Şimşek, Ö. (2013). Linear decision rule as aspiration for simple decision heuristics. *Advances in Neural Information Processing Systems, 26*, 2904–2912.

Singh, V., & Vinnicombe, S. (2005). *The female FTSE report 2005*. Bedfordshire, UK: Cranfield School of Management, Cranfield University.

Slaytor, E. K., & Ward, J. E. (1998). How risks of breast cancer and benefits of screening are communicated to women: Analysis of 58 pamphlets. *British Medical Journal, 317*, 263–264.

Sleath, B., Roter, D. L., Chewning, B., & Svarstad, B. (1999). Question-asking about medications: Physician experiences and perceptions. *Medical Care, 37*, 1169–1173.

Sloman, S. A. (1996). The empirical case for two systems of reasoning. *Psychological Bulletin, 119*, 3–22.

Slovic, P., Finucane, M. L., Peters, E., & MacGregor, D. G. (2002). The affect heuristic. In T. Gilovich, D. Griffin, & D. Kahneman (Eds.), *Heuristics and biases: The psychology of intuitive judgment* (Vol. 2, pp. 397–420). New York: Cambridge University Press.

Slovic, P., Monahan, J., & MacGregor, D. G. (2000). Violence risk assessment and risk communication: The effects of using actual cases, providing instruction, and employing probability versus frequency formats. *Law and Human Behavior, 24*, 271–296.

Smith, A. (1997). *The theory of moral sentiments*. Washington, DC: Regnery Publishing. (Original work published 1759).

Smith, G. (2003). Horse shoe pitchers' hot hands. *Psychonomic Bulletin & Review, 10*, 753–758.

Smith, D. E., Wilson, A. J., & Henry, D. A. (2005). Monitoring the quality of medical news reporting: Early experience with media doctor. *The Medical Journal of Australia, 183*, 190–193.

Smith, L., & Gilhooly, K. (2006). Regression versus fast and frugal models of decision-making: The case of prescribing for depression. *Applied Cognitive Psychology, 20*, 265–274.

Smith, R. (2005). Medical journals are an extension of the marketing arm of pharmaceutical companies. *PLoS Medicine, 2*, e138.

Smith, V. L. (2003). Constructivist and ecological rationality in economics. *The American Economic Review, 93*, 465–508.

Snijders, R. J., Noble, P., Sebire, N., Souka, A., & Nicolaides, K. H. (1998). UK multicentre project assessment of risk of trisomy 21 by maternal age and fetal nuchal-translucency thickness at 10–14 weeks of gestation. *The Lancet, 352*, 343–346.

Snook, B., & Cullen, R. M. (2006). Recognizing national hockey league greatness with an ignorance-based heuristic. *Canadian Journal of Experimental Psychology, 60*, 33–43.

Snook, B., Taylor, P. J., & Bennell, C. (2004). Geographic profiling: The fast, frugal and accurate way. *Applied Cognitive Psychology, 18*, 105–121.

Snook, B., Zito, M., Bennell, C., & Taylor, P. J. (2005). On the complexity and accuracy of geographic profiling strategies. *Journal of Quantitative Criminology, 21*, 1–26.

Sone, S., Li, F., Yang, Z., Honda, T., Maruyama, Y., & Takashima, S. (2001). Results of three-year mass screening programme for lung cancer using mobile lowdose spiral computed tomography scanner. *British Journal of Cancer, 84*, 25–32.

Soros, G. (2003). *The alchemy of finance* (2nd ed.). Hoboken, NJ: Wiley.

Soros, G. (2009). *The crash of 2008 and what it means: The new paradigm for financial markets*. New York: Public Affairs.

Starmer, C. (2004). Friedman's risky methodology. Working paper. Nottingham, United Kingdom: University of Nottingham

Starmer, C. (2005). Normative notions in descriptive dialogues. *Journal of Economic Methodology, 12*, 277–289.

Steckelberg, A., Berger, B., Köpke, S., Heesen, C., & Mühlhauser, I. (2005). Kriterien für evidenzbasierte Patienteninformationen [Criteria for evidence-based information for patients]. *Zeitschrift für ärztliche Fortbildung und Qualität im Gesundheitswesen, 99*, 343–351.

Steckelberg, A., Hülfenhaus, C., Kasper, J., Rost, J., & Mühlhauser, I. (2007). How to measure critical health competences: Development and validation of the Critical Health Competence Test (CHC Test). *Advances in Health Sciences Education, 14*, 11–22.

Steckelberg, A., Hülfenhaus, C., Kasper, J., Rost, J., & Mühlhauser, I. (2009). Ebm@school: A curriculum of critical health literacy for secondary school students: Results of a pilot study. *International Journal of Public Health, 54*, 158–165.

Steimle, S. (1999). U.K.'s Tony Blair announces crusade to fight cancer. *Journal of the National Cancer Institute, 91*, 1184–1185.

Steurer, J., Held, U., Schmidt, M., Gigerenzer, G., Tag, B., & Bachmann, L. M. (2009). Legal concerns trigger PSA testing. *Journal of Evaluation in Clinical Practice, 15*, 390–392.

Stevens, J., & King, A. (2012). The life of others: Social rationality in animals. In R. Hertwig, U. Hoffrage, & the ABC Research Group, *Simple heuristics in a social world* (pp. 409–431). New York: Oxford University Press.

Stiftung Warentest. (2004). Urologen im Test: Welchen Nutzen hat der PSA-Test? [Testing urologists: What are the benefits of a PSA test?]. *Stiftung Warentest, February*, 86–89.

Stiglitz, J. E. (2010). *Freefall: America, free markets, and the sinking of the world economy*. New York: Norton.

Stine, G. J. (1999). *AIDS update 1999: An annual overview of acquired immune deficiency syndrome*. Upper Saddle River, NJ: Prentice-Hall.

Stolberg, S. G. (2002, February 6). Study says clinical guides often hide ties of doctors, *The New York Times*. Retrieved November 3, 2014 from http://www.nytimes.com/2002/02/06/health/06DRUG.html?module=Search&mabReward=relbias%3As

Street, R. L. J. (2001). Active patients as powerful communicators. In W. P. Robinson & H. Giles (Eds.), *The new handbook of language and social psychology* (pp. 541–560). New York: Wiley.

Sundali, J., & Croson, R. (2006). Biases in casino betting: The hot hand and the gambler's fallacy. *Judgment and Decision Making, 1*, 1–12.

Sunstein, C. R. (2005). Moral heuristics. *Behavioral and Brain Sciences, 28*, 531–542.

Swim, J. K. (1994). Perceived vs. meta-analytic effect sizes: An assessment of the accuracy of gender stereotypes. *Journal of Personality and Social Psychology, 66*, 21–36.

Takezawa, M., Gummerum, M., & Keller, M. (2006). A stage for the rational tail of the emotional dog: Roles of moral reasoning in group decision making. *Journal of Economic Psychology, 27*, 117–139.

Tan, S. B., Goh, C., Thumboo, J., Che, W., Chowbay, B., & Cheung, Y. B. (2005). Risk perception is affected by modes of risk presentation among Singaporeans. *Annals of the Academy of Medicine, 34*, 184–187.

Tanner, W. P., Jr., & Swets, J. A. (1954). A decision-making theory of visual detection. *Psychological Review, 61*, 401–409.

Tenenbaum, H. R., & Leaper, C. (2003). Parent–child conversations about science: The socialization of gender inequities? *Developmental Psychology, 39*, 34–47.

Thaler, R. H. (1991). *Quasi rational economics*. New York: Russell Sage Foundation.

Thaler, R. H., & Sunstein, C. R. (2008). *Nudge: Improving decisions about health, wealth, and happiness*. New Haven, CT: Yale University Press.

Todd, P. M., Billari, F. C., & Simao, J. (2005). Aggregate age-at-marriage patterns from individual mate-search heuristics. *Demography, 42*, 559–574.

Todd, P. M., & Dieckmann, A. (2005). Heuristics for ordering cue search in decision making. In L. K. Saul, Y. Weiss, & L. Bottou (Eds.), *Advances in neural information processing systems* (Vol. 17, pp. 1393–1400). Cambridge, MA: MIT Press.

Todd, P. M., Gigerenzer, G., & the ABC Reseach Group. (2012). *Ecological rationality: Intelligence in the world*. New York: Oxford University Press.

Towle, A., & Godolphin, W. (1999). Framework for teaching and learning informed shared decision making. *British Medical Journal, 319*, 766–771.

Trevena, L. J., Davey, H. M., Barratt, A., Butow, P., & Caldwell, P. (2006). A systematic review on communicating with patients about evidence. *Journal of Evaluation in Clinical Practice, 12*, 13–23.

Tsyska, T., Zielonka, P., Decey, R., & Sawicki, P. (2007). Perception of randomness and predicting uncertain events. *Thinking & Reasoning, 14*, 83–110.

Tversky, A. (1972). Elimination by aspects: A theory of choice. *Psychological Review, 79*, 281–299.

Tversky A. (1977). Features of similarity. *Psychological Review, 84*, 327–352.

Tversky, A., & Gilovich, T. (1989). The cold facts about the "hot hand" in basketball. *Chance*, *2*, 16–21.

Tversky, A., & Kahneman, D. (1973). Availability: A heuristic for judging frequency and probability. *Cognitive Psychology*, *5*, 207–232.

Tversky, A, & Kahneman, D. (1974). Judgment under uncertainty: Heuristics and biases. *Science*, *185*, 1124–1131.

Tversky, A., & Kahneman, D. (1986). Rational choice and the framing of decisions. *The Journal of Business*, *59*, S251–S278.

Tversky, A., & Kahneman, D. (1992). Advances in prospect theory: Cumulative representation of uncertainty. *Journal of Risk and Uncertainty*, *5*, 297–323.

U.S. Equal Employment Opportunity Commission. (2003). *Investment banking*. Retrieved November 3, 2014 from http://www.eeoc.gov/eeoc/statistics/reports/investmentbanking/.

U.S. Preventive Services Task Force. (1996). *Guide to clinical preventive services: Report of the U.S. Preventive Services Task Force* (2nd ed.). Alexandria, VA: International Medical Publishing.

U.S. Preventive Services Task Force. (2002). *Guide to clinical preventive services: Report of the U.S. Preventive Services Task Force* (3rd ed.). Baltimore, MD: Williams & Wilkins.

U.S. Preventive Services Task Force. (2012). Screening for prostate cancer: U.S. Preventive Services Task Force recommendation station. *Annals of Internal Medicine*, *157*, 120–134.

Veblen, T. (1994). *The theory of the leisure class*. New York: Dover Publications (Original work published 1899).

Villanueva, P., Peiró, S., Librero, J., & Pereiró, I. (2003). Accuracy of pharmaceutical advertisements in medical journals. *The Lancet*, *361*, 27–32.

Volz, K. G., Schubotz, R. I., Raab, M., Schooler, L. J., Gigerenzer, G., & von Cramon, D. Y. (2006). Why you think Milan is larger than Modena: Neural correlates of the recognition heuristic. *Journal of Cognitive Neuroscience*, *18*, 1924–1936.

von Helversen, B., & Rieskamp, J. (2008). The mapping model: A cognitive theory of quantitative estimation. *Journal of Experimental Psychology: General*, *137*, 73–96.

von Helversen, B., & Rieskamp, J. (2009). Predicting sentencing for low level crimes: Comparing models of human judgment. *Journal of Experimental Psychology: Applied*, *15*, 375–395.

Voss, M. (2002). Checking the pulse: Midwestern reporters' opinions on their ability to report health care news. *American Journal of Public Health*, *92*, 1158–1160.

Wallach, L., & Wallach, M. A. (1994). Gergen versus the mainstream: Are hypotheses in social psycholgy subject to empirical test? *Journal of Personality and Social Psychology*, *67*, 233–242.

Wallach, M. A., & Wallach, L. (1998). When experiments serve little purpose: Misguided research in mainstream psychology. *Theory & Psychology*, *8*, 183–194.

Wallsten, T. S. (1983). The theoretical status of judgmental heuristics. In R. W. Scholz (Ed.), *Decision making under uncertainty* (pp. 21–39). Amsterdam: Elsevier.

Wallsten, T. S., Budescu, D. V., Zwick, R., & Kemp, S. M. (1993). Preference and reasons for communicating probabilistic information in numerical or verbal terms. *Bulletin of the Psychonomic Society, 31*, 135–138.

Wang, X. T. (2008). Decision heuristics as predictors of public choice. *Journal of Behavioral Decision Making, 21*, 77–89.

Wansink, B. (2006) *Mindless eating: Why we eat more than we think.* New York: Bantam Books.

Warner, J. H. (1986). *The therapeutic perspective: Medical practice, knowledge, and identity in America, 1820–1885.* Cambridge, MA: Harvard University Press.

Wason, P. C. (1971). Natural and contrived experience in a reasoning problem. *Quarterly Journal of Experimental Psychology, 23*, 63–71.

Watkins, M. J. (1984). Models as toothbrushes. *Behavioral & Brain Sciences, 7*, 86.

Watson, J. B. (1913). Psychology as the behaviorist views it. *Psychological Review, 20*, 158–177.

Weber, E. U., & Johnson, E. J. (2009). Mindful judgment and decision making. *Annual Review of Psychology, 60*, 53–86.

Wegwarth, O., Gaissmaier, W., & Gigerenzer, G. (2011). Deceiving numbers: Survival rates and their impact on doctors' risk communication. *Medical Decision Making, 31*, 386–394.

Wegwarth, O., Schwartz, L. M., Woloshin, S., Gaissmaier, W., & Gigerenzer, G. (2012). Do physicians understand cancer screening statistics? A national survey of primary care physicians in the United States. *Annals of Internal Medicine*, 156, 340–349.

Welch, H. G. (2004). *Should I be tested for cancer?* Berkeley, CA: University of California Press.

Welch, H. G., Schwartz, L. M., & Woloshin, S. (2000). Are increasing 5-year survival rates evidence of success against cancer? *Journal of the American Medical Association, 283*, 2975–2978.

Welch, H. G., Schwartz, L. M., & Woloshin, S. (2007, January 2). What's making us sick is an epidemic of diagnoses. *New York Times.* Retrieved November 3, 2014 from http://www.nytimes.com/2007/01/02/health/02essa.html?_r=0.

Welch, H. G., Woloshin, S., Schwartz, L. M., Gordis, L., Gøtzsche, P. C., Harris, R., et al. (2007). Overstating the evidence for lung cancer screening: The International Early Lung Cancer Action Program (I-ELCAP) Study. *Archives of Internal Medicine, 167*, 2289–2295.

Wells, H. G. (1994). *World brain.* London: Cambridge University Press. (Original work published 1938).

Whiten, A., & Byrne, R. W. (1997). *Machiavellian intelligence II: Evaluations and extensions.* Cambridge: Cambridge University Press

Wilke, A., & Barrett, H. C. (2009). The hot hand phenomenon as a cognitive adaption to clumped resources. *Evolution and Human Behavior, 30*, 161–169.

Williams, J. E., & Best, D. L. (1990). *Measuring sex stereotypes: A multination study* (Rev. ed.). Beverly Hills, CA: Sage Publications.

Winter, S. G., Jr. (1964). Economic "natural selection" and the theory of the firm. *Yale Economic Essays, 4*, 225–272.

Wolff, A. (1998). No question. *Sports Illustrated, 88*, 70–72.

Woloshin, S., & Schwartz, L. M. (1999). How can we help people make sense of medical data? *Effective Clinical Practice, 2*, 176–183.

Woloshin, S., & Schwartz, L. M. (2002). Press releases: Translating research into news. *Journal of the American Medical Association, 287*, 2856–2858.

Woloshin, S., & Schwartz, L. M. (2006a). Giving legs to restless legs: A case study of how the media helps make people sick. *PLoS Medicine, 3*, e170.

Woloshin, S., & Schwartz, L. M. (2006b). Media reporting of research presented at scientific meetings: More caution needed. *The Medical Journal of Australia, 184*, 576–580.

Woloshin, S., Schwartz, L. M., Black, W. C., & Welch, H. G. (1999). Women's perceptions of breast cancer risk: How you ask matters. *Medical Decision Making, 19*, 221–229.

Woloshin, S., Schwartz, L. M., Tremmel, J., & Welch, H. (2001). Direct to consumer drug advertisements for prescription drugs: What are Americans being sold? *The Lancet, 358*, 1141–1146.

Woloshin, S., Schwartz, L. M., & Welch, H. G. (2004, April 28). The value of benefit data in direct-to-consumer drug ads. *Health Affairs*, W234–245.

Woloshin, S., Schwartz, L. M., & Welch, H. G. (2008). The risk of death by age, sex, and smoking status in the United States: Putting health risks in context. *Journal of the National Cancer Institute, 100*, 845–853.

Wong, N., & King, T. (2008). The cultural construction of risk understandings through illness narratives. *Journal of Consumer Research, 34*, 579–594.

Wood, W., & Eagly, A. H. (2002). A cross-cultural analysis of the behavior of women and men: Implications for the origins of sex differences. *Psychological Bulletin, 28*, 699–727.

World Health Organization. (2008). *World health statistics*. Geneva: WHO Press.

Wübben, M., & Wangenheim, F. v. (2008). Instant customer base analysis: Managerial heuristics often "get it right." *Journal of Marketing, 72*, 82–93.

Wundt, W. (1874). *Grundzüge der physiologischen Psychologie* [Principles of physiological psychology]. Leipzig: Engelmann.

Yee, M., Dahan, E., Hauser, J. R., & Orlin, J. (2007). Greedoid-based noncompensatory inference. *Marketing Science, 26*, 532–549.

Yonelinas, A. P., Otten, L. J., Shaw, K. N., & Rugg, M. D. (2005). Separating the brain regions involved in recollection and familiarity in recognition memory. *Journal of Neuroscience, 25*, 3002–3008.

Young, J. M., Glasziou, P., & Ward, J. E. (2002). General practitioners' self rating of skills in evidence based medicine: A validation study. *British Medical Journal, 324*, 950–951.

Young, S. D., & Oppenheimer, D. M. (2006). Different methods of presenting risk information and their influence on medication compliance intentions: Results of three studies. *Clinical Therapeutics, 28*, 129–139.

Zajonc, R. B. (1968). Attitudinal effects of mere exposures. *Journal of Personality and Social Psychology Monograph Supplement, 9*, 1–27.

Zhu, L., & Gigerenzer, G. (2006). Children can solve Bayesian problems: The role of representation in mental computation. *Cognition, 98*, 287–308.

INDEX

Page numbers followed by an f *or a* t *indicate figures and tables respectively.*

www.ingramcontent.com/pod-product-compliance
Lightning Source LLC
LaVergne TN
LVHW011914200926
848484LV00002B/616